Lecture Notes in Computer Science　5412

Commenced Publication in 1973
Founding and Former Series Editors:
Gerhard Goos, Juris Hartmanis, and Jan van Leeuwen

Ioannis Tomkos   Maria Spyropoulou
Karin Ennser   Martin Köhn
Branko Mikac (Eds.)

Towards Digital Optical Networks

COST Action 291 Final Report

 Springer

Volume Editors

Ioannis Tomkos
Maria Spyropoulou
Athens Information Technology Centre
19002 Peania-Attica, Greece
E-mail: {itom,mspi}@ait.edu.gr

Karin Ennser
Swansea University
Institute of Advanced Telecommunications
SA2 8PP, Swansea, UK
E-mail: k.ennser@swansea.ac.uk

Martin Köhn
University of Stuttgart
IKR
70569 Stuttgart, Germany
E-mail: martin.koehn@ikr.uni-stuttgart.de

Branko Mikac
University of Zagreb
Faculty of Electrical Engineering and Computing (FER)
Department of Telecommunications
10000 Zagreb, Croatia
E-mail: branko.mikac@fer.hr

Library of Congress Control Number: Applied for

CR Subject Classification (1998): C.2, B.4.3, H.3.4

LNCS Sublibrary: SL 5 – Computer Communication Networks and
Telecommunications

ISSN   0302-9743
ISBN-10   3-642-01523-9 Springer Berlin Heidelberg New York
ISBN-13   978-3-642-01523-6 Springer Berlin Heidelberg New York

Typesetting: Camera-ready by author, data conversion by Markus Richter, Heidelberg
Printed on acid-free paper   SPIN: 12627301   06/3180   5 4 3 2 1 0

Foreword

COST – the acronym for European COoperation in Science and Technology – is the oldest and widest European intergovernmental network for cooperation in research. Established by the Ministerial Conference in November 1971, COST is presently used by the scientific communities of 35 European countries to cooperate in common research projects supported by national funds.

The funds provided by COST – less than 1% of the total value of the projects – support the COST cooperation networks (COST Actions) through which, with € 30 million per year, more than 30,000 European scientists are involved in research having a total value which exceeds € 2 billion per year. This is the financial worth of the European added value which COST achieves.

A "bottom up approach" (the initiative of launching a COST Action comes from the European scientists themselves), "à la carte participation" (only countries interested in the Action participate), "equality of access" (participation is open also to the scientific communities of countries not belonging to the European Union) and "flexible structure" (easy implementation and light management of the research initiatives) are the main characteristics of COST.

As a precursor of advanced multidisciplinary research, COST has a very important role in the realization of the European Research Area (ERA) anticipating and complementing the activities of the Framework Programmes, constituting a "bridge" towards the scientific communities of emerging countries, increasing the mobility of researchers across Europe and fostering the establishment of "Networks of Excellence" in many key scientific domains such as: biomedicine and molecular biosciences; food and agriculture; forests, their products and services; materials, physical and nanosciences; chemistry and molecular sciences and technologies; earth system science and environmental management; information and communication technologies; transport and urban development; individuals, societies, cultures and health. It covers basic and more applied research and also addresses issues of pre-normative nature or of societal importance.

More information is available at: http://www.cost.esf.org/.

ESF provides the COST Office through an EC contract. COST is supported by the EU RTD Framework programme.

List of Contributors

Slavisa Aleksic
Vienna University of Technology
Institute of Broadband
Communications,
1040 Vienna, Austria

Carlos Almeida
Istituto de Telecomunicacoes
Universidade De Aveiro, Campus
Universitário
3810193 Aveiro, Portugal

Ramon Aparicio-Pardo
Universidad Politecnica de Cartagena
30202 Cartagena, Spain

Javier Aracil
Universidad Autónoma de Madrid
Madrid, Spain

Siamak Azodolmolky
Athens Information Technology
Peania-Attica 19002
Athens, Greece

G. Bosco
Dipartimento di Elettronica
Politecnico di Torino
10129 Turin, Italy

Davide Careglio
Universitat Politècnica de Catalunya
08034 Barcelona, Spain

Ching-Hung Chang
University of Hertfordshire
Science and Technology Research
Institute, Herts AL10 9AB, UK

Jiajia Chen
The Royal Institute of Technology
KTH, The School of Information and
Communication Technology
164 40 Kista, Sweden

Walter Colitti
Vrije Universiteit Brussel
1050 Elsene, Brussels, Belgium

Didier Colle
IBBT, Ghent University
9050 Ghent-Ledeberg, Belgium

Franco Curti
ISCOM - Italian Ministry of
Economic Development Sector
Communication
00144 Rome, Italy

Grzegorz Danilewicz
Poznan University of Technology
Chair of Communication and
Computer Networks
60965 Poznan, Poland

Piet Demeester
IBBT, Ghent University
9050 Ghent-Ledeberg, Belgium

Silvia Di Bartolo
Università di Roma "Tor Vergata"
Dipartimento di Ingegneria
Elettronica
Via del Politecnico 1
00133 Rome, Italy

Taisir E.H. El-Gorashi
School of Electronic and Electrical
Engineering, University of Leeds
Leeds, UK

Jaffar M.H. Elmirghani
School of Electronic and Electrical
Engineering, University of Leeds
Leeds, UK

Karin Ennser
Institute of Advanced
Telecommunications,
Swansea University, UK

P. Fasser
TU Graz,
Technical University of Graz,
8010 Graz, Austria

Daniel Fonseca
Nokia Siemens Networks
2720093 Amadora, Portugal and
Istituto de Telecomunicacoes,
Optical Communications, Instituto
Superior Técnico,
1049001 Lisbon, Portugal

D.M. Forin
ISCOM - Italian Ministry of
Economic Development Sector
Communication,
00144 Rome, Italy

Maurice Gagnaire
Telecom ParisTech (ENST),
Computer Science and Networks
75013 Paris, France

Michael Galili
DTU Fotonik, Technical University
of Denmark
800 Kgs. Lyngby, Denmark

Joan García-Haro
Universidad Politécnica de
Cartagena,
30202 Cartagena, Spain

Belen Garcia-Manrubia
Universidad Politecnica de Cartagena
30202 Cartagena, Spain

Christian Gaumier
Ecole Polytechnique Fédérale de
Lausanne
1015 Lausanne, Switzerland

Sebastian Gunreben
University of Stuttgart
Stuttgart, Germany

Pasquale Gurzì
Vrije Universiteit Brussel,
1050 Elsene, Brussels, Belgium

Guoqiang Hu
University of Stuttgart
Stuttgart, Germany

Mikel Izal
Public University of Navarra
31006 Pamplona, Spain

Marek Jaworski
National Institute of
Telecommunications,
Warsaw, Poland

Wojciech Kabacinski
Poznan University of Technology
Chair of Communication and
Computer Networks
60965 Poznan, Poland

M. Karasek
Institute of Photonics and
Electronics, Academy of Sciences of
the Czech Republic
18251 Prague 8, Czech Republic

Kostas Katrinis
Athens Information Technology
Peania-Attica 19002
Athens, Greece

Andreas Kimsas
Norwegian University of Science and
Technology,
Trondheim, Norway

Martin Köhn
University of Stuttgart
Stuttgart, Germany

Pandelis Kourtessis
Science & Technology Research
Institute (STRI),
University of Hertfordshire
Herts AL10 9AB, UK

Marko Lackovic
University of Zagreb
Faculty of Electrical Engineering and
Computing (FER)
10000 Zagreb, Croatia

Ozren Lapcevic
University of Zagreb
Faculty of Electrical Engineering and
Computing (FER)
10000 Zagreb, Croatia

Janusz Kleban
Poznan University of Technology
Chair of Communication and
Computer Networks
60965 Poznan, Poland

M. Klinkowski
Universitat Politècnica de Catalunya,
Barcelona, Spain and
National Institute of
Telecommunications,
Warsaw, Poland

Dimitrios Klonidis
Athens Information Technology
19002 Peania-Attica, Athens
Greece

Jochen Leibrich
University of Kiel
Kiel, Germany

Erich Leitgeb
TU Graz,
Technical University of Graz,
8010 Graz, Austria

Ilse Lievens
IBBT, Ghent University
9050 Ghent-Ledeberg, Belgium

Mário Lima
Istituto de Telecomunicacoes
Universidade De Aveiro, Campus
Universitário
3810193 Aveiro, Portugal

Markus Löschnigg
TU Graz,
Technical University of Graz,
8010 Graz, Austria

Eduardo Magaña
Public University of Navarra
31006 Pamplona, Spain

Marian Marciniak
National Institute of
Telecommunications, Transmission
and Optical Technologies
04894 Warsaw, Poland

George Markidis
Athens Information Technology
19002 Peania-Attica
Athens, Greece

Juan Antonio Martínez-León
Universidad Politecnica de Cartagena
30202 Cartagena, Spain

Branko Mikac
University of Zagreb
Faculty of Electrical Engineering and
Computing (FER)
10000 Zagreb, Croatia

Daniel Morató
Public University of Navarra
31006 Pamplona, Spain

Ann Nowé
Vrije Universiteit Brussel,
1050 Elsene, Brussels, Belgium

Leif K. Oxenløwe
DTU Fotonik, Technical University
of Denmark
2800 Kgs. Lyngby, Denmark

Ioannis Papagiannakis
Athens Information Technology
19002 Peania-Attica
Athens, Greece

Francesca Parmigiani
Optoelectronics Research Centre
University of Southampton
SO17 1BJ Southampton, UK

Natasa Pavlovic
Istituto de Telecomunicacoes
Universidade De Aveiro, Campus
Universitário
3810193 Aveiro, Portugal

Pablo Pavón-Mariño
Universidad Politécnica de
Cartagena,
30202 Cartagena, Spain

Jordi Perelló
Universitat Politècnica de Catalunya,
08034 Barcelona, Spain

Periklis Petropoulos
Optoelectronics Research Centre
University of Southampton
SO17 1BJ Southampton, UK

Mario Pickavet
IBBT, Ghent University
9050 Ghent-Ledeberg, Belgium

P. Poggiolini
Dipartimento di Elettronica,
Politecnico di Torino
10129 Turin, Italy

Bart Puype
IBBT, Ghent University
9050 Ghent-Ledeberg, Belgium

Werner Rosenkranz
Chair for Communications
University of Kiel
24143 Kiel, Germany

Sébastien Rumley
Ecole Polytechnique Fédérale de
Lausanne (EPFL)
1015 Lausanne, Switzerland

Joachim Scharf
Institute of Communication Networks
and Computer Engineering (IKR)
Stuttgart, Germany

Y. Shachaf
Science & Technology Research
Institute (STRI), University of
Hertfordshire
Herts AL10 9AB, UK

Nina Skorin-Kapov
University of Zagreb
Faculty of Electrical Engineering and
Computing (FER)
10000 Zagreb, Croatia

Radan Slavík
Institute of Photonics and Electronics
AS CR, v.v.i.,
Prague, Chaberska 57, 182 51,
Czech Republic

Salvatore Spadaro
Universitat Politècnica de Catalunya,
08034 Barcelona, Spain

Maria Spyropoulou
Athens Information Technology
19002 Peania-Attica
Athens, Greece

Dimitri Staessens
IBBT, Ghent University
9050 Ghent-Ledeberg, Belgium

Kris Steenhaut
Vrije Universiteit Brussel
1050 Elsene, Brussels, Belgium

Stefano Taccheo
Politecnico di Milano,
Milan, Italy

Antonio Luis Jesus Teixeira
Instituto de Telecomunicações (IT)
Universidade de Aveiro
Campus Universitário
3810-193 Aveiro, Portugal

Torger Tokle
Research Center COM
Technical University of Denmark
2800 Lyngby, Denmark

Ioannis Tomkos
Athens Information Technology
19002 Peania-Attica
Athens, Greece

Giorgio Maria Tosi Beleffi
ISCOM - Italian Ministry of
Economic Development Sector
Communication,
00144 Rome, Italy

Anna Tzanakaki
Athens Information Technology
19002 Peania-Attica
Athens, Greece

Emmanouel Varvarigos
Department of Computer
Engineering and Informatics,
University of Patras Research
Academic Computer Technology
Institute,
Rio 26500, Greece

Javier Veiga-Gontán
Universidad Politécnica de
Cartagena,
30202 Cartagena, Spain

Kyriakos Vlachos
Department of Computer
Engineering and Informatics,
University of Patras Research
Academic Computer Technology
Institute,
Rio 26500, Greece

Cedric Ware
Institut TELECOM, TELECOM Paris
Tech (formerly ENST), France

Slawomir Weclewski
Poznan University of Technology
Chair of Communication and Computer Networks
60-965 Poznan, Poland

Lena Wosinska
KTH-Royal Institute of Technology,
Information & Communication
Technology
SE-164 40 Kista, Sweden

Chunmin Xia
Chair for Communications
University of Kiel
24143 Kiel, Germany

Konstantinos Yiannopoulos
Department of Computer Engineer-
ing and Informatics, University of
Patras Research Academic Computer
Technology Institute,
Rio 26500, Greece

Table of Contents

Part III

Introduction

I. Tomkos and M. Spyropoulou

The explosive growth of data, particularly internet traffic has led to a dramatic increase in demand for transmission bandwidth imposing an immediate requirement for broadband networks. An additional driving force for higher capacity, enhanced functionality and flexibility networks is the increased trend for interactive exchange of data and multimedia communications. Due to the unpredictable and ever growing size of data files and messages exchanged over global distances the future communication network must be able to react rapidly to support end-to-end bandwidth requirements for transmission of messages and data files of any conceivable size encountered in real-life communications.

This publication is supported by COST. The primary objective of the COST 291 Action "Towards Digital Optical Networks" was to focus on novel network concepts, subsystems and architectures to enable future telecommunication networks, exploiting the features and properties of photonic technologies. These need to be very flexible and rapidly reactive to efficiently accommodate the abrupt and unpredictable changes in traffic statistics introduced by current and future applications with low end-to-end latency. They will enable advanced features such as efficient and simple multicasting and broadcasting of broadband signals. They need to support a future proof, flexible, efficient and bandwidth-abundant fiber-optic network infrastructure capable of supporting ubiquitous services in a resilient manner offering protection and restoration capabilities as well as secure services to the users.

Three Working Groups (WGs) were set up from the start of the Action to deal with the above research objectives:

- WG1: "Optical processing for digital network performance". This WG aimed to deal with the physical layer and implementation related issues of transparent optical networks such as optical signal per bit processing, optical switch architecture designs and implementations as well as transmission related issues.
- WG2: "Novel network architectures". This WG aimed to focus on the evolution of network scenarios including novel network architectures. Also different node architectures and technologies in terms of network performance and functionality were investigated. Three different architectures were studied and compared: circuit (wavelength, waveband etc), optical burst and optical packet switched networks.
- WG3: "Unified control plane, network resilience and service security". This WG focused on two directions. The former dealt with the impact of transparency on photonic network architectures and the associated control and protocol issues and the latter focused on network survivability and security issues, covering topics such as protection and restoration, its impact on routing and wavelength assignment algorithms, fault isolation, disaster recovery, etc.

I. Tomkos et al. (Eds.): COST 291 – Towards Digital Optical Networks, LNCS 5412, pp. 1–2, 2009.

The results obtained within the three working groups (WGs) are collected and reported in three Parts, each organized in individual Chapters.

Part I covers advanced modulation formats for transmission in future optical networks, all-optical channel equalization and regeneration techniques, novel switch buffer architectures, architectures/protocols for access optical networks including FTTX developments, 100Gbp/s Ethernet, radio over fibre and dynamic bandwidth allocation.

Part II covers transparent wavelength routed networks, optical burst switching (OBS) and optical packet switching (OPS) architectures, multi-layer network engineering, network resilience and storage area networks.

Part III covers software tools and methods for modelling physical layer and network layer issues.

Part I

Introduction (Part I)

K. Ennser (part editor)

The Part I of this book consists on the research activities from the partners of COST291 on the physical layer of the network. The contributions are distributed in five chapters addressing key issues of the network evolution. The chapter 1 presents the novel transponder interfaces to achieve ultra-high capacity network at reduced cost. Advanced modulation formats are considered as for example, multilevel formats and coherent approach. Another mechanism to obtain very high data rates beyond electronic limitations is the use of optical signal processing to convert low bit rate data streams into high bit rate. Then techniques for optical rate conversion are explored.

The chapter 2 focuses on the electronic equalization techniques to compensate transmission impairment distortions by electronic means. The main impairments addressed are chromatic dispersion and nonlinearities. In particular, FFE/DFE and MLSE based equalizers are explored with the assistance of different modulation formats.

The chapter 3 presents optical processing techniques and digital logic based on nonlinear phenomena of material to avoid expensive optical-electronic-optical conversion and achieve ultra-high speed processing. Optical regeneration and wavelength conversion based on high nonlinear fiber and semiconductor optical amplifier are reported. In addition optical time domain demultiplexing, optoelectronic clock recovery and ultra-high speed retiming up to 640Gb/s are achieved.

The chapter 4 reports an evolution of optical access networks. Starting from a review of current development in access network and protocols several standards and deployment issues are reviewed to enable interoperability of TDM and WDM PON by means of coarse routing. Hybrid wireless and free space optical technology are discussed to provide flexible connectivity. More specifically, dynamic bandwidth allocation algorithms are implemented to manage the bandwidth independently of the constituent sectors of the architecture.

The chapter 5 discusses novel switch architectures in particular optical buffering and optical packet switches. The potentiality of quantum-dot semiconductor technology is explored. Moreover, several photonic asynchronous packet switching architectures are proposed.

In conclusion, several key issues in the physical layer towards next generation network are presented and discussed as part of the research development within COST291 mobility action.

1 Novel Transponder Interfaces: Novel Modulation Formats

W. Rosenkranz (chapter editor), S. Aleksic, and T. Tokle

1.1 Introduction

For a long period of time, the modulation format in optical communications technology was based on intensity modulation with an on/off - keying signalling, using NRZ and sometimes RZ pulse-shapes. However with today's requirements for very high capacity networks with constraints on cost efficiency, more advanced solutions are required. Advanced modulation formats are considered as one of the key issue in the networks of the future.

With novel modulation formats we have the general options to modulate the amplitude or the phase or both of the optical carrier signal of a coherent laser. An additional degree of flexibility can be achieved by using polarization (division) multiplexing (PDM). The table gives an overview of the currently discussed options.

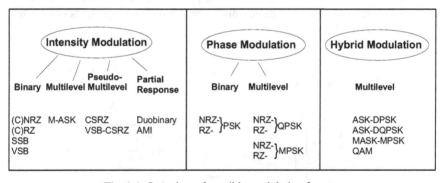

Fig. 1.1. Overview of possible modulation formats

Generally one has to distinguish between binary and multilevel modulation formats. Using binary formats, only two different levels are encoded onto the amplitude or phase of the optical carrier. Using multilevel modulation, $\log_2(M)$ data bits are encoded on M symbol levels. The data is therefore transmitted at a reduced symbol rate of $R/\log_2(M)$, R being the initial data rate.

At the receiver side the information can be retrieved from the carrier by either coherent or non-coherent (auto-correlating) approaches, where the coherent receiver generally is more complex as a local laser source is required at the receiver

I. Tomkos et al. (Eds.): COST 291 – Towards Digital Optical Networks, LNCS 5412, pp. 7–21, 2009.

as well as some means for carrier synchronization. With PDM additional effort for polarization control is required.

Another approach to generate, process and transmit very high data rates beyond limitation of electronics is to use optical signal processing. There are several techniques that can be used for conversion of low bit rate data streams into high bit rate (up-conversion) and vice versa (down-conversion) directly in the optical domain. For example, up-conversion can be accomplished easily in the optical domain by using a technique similar to the bit-interleaved optical time-division multiplexing (OTDM). Here, it is important to generate sufficiently short optical pulses and to use precisely adjusted optical delay elements in order to meet strong timing requirements associated with the short bit-period of the high-speed signal. If low bit rate packets are acquired from a transmitting buffer in parallel, the same structure such as those used in OTDM multiplexer can be used for up-conversion (see Fig. 1.2). The only difference to OTDM multiplexing, where N completely separated electrical signals at the basic bit rate that represent N OTDM channels, the structure shown in Fig. 1.2 uses parallel electrical signals read out from a single transmitting buffer that represent a single parallelized low-speed packet. The number of branches, N, has to be the same as the conversion rate $K = T_0/\tau_0$, where T_0 is the bit-period of the low-speed data and $\tau_0$is the bit-period of the high-speed signal (H-S Out). In principle, the up-conversion scheme is similar to the electronic serializers, which transform N-bit wide parallel data into a serial signal with an N-times higher bit rate. Therefore, this scheme can be referred to as optical serializing method.

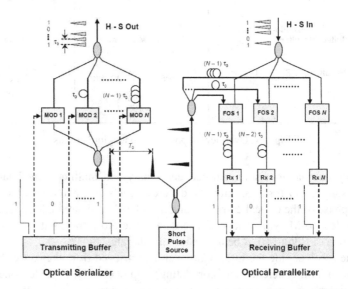

Fig. 1.2. Optical serializer/parallelizer (MOD: modulator, FOS: fast optical switch, Rx: receiver, H-S: high-speed).

Down conversion is more complex because single bits have to be separated from the incoming high-speed signal. For down-conversion of high bit-rate optical signal, a number of high-speed all-optical switches are required. These fast optical switches (FOSs) must be able to switch between two consecutive bits of the high-speed optical signal. Different technologies and structures that have already been proposed for OTDM demultiplexers can also directly be used for this purpose. In the optical parallelizer as illustrated in Fig. 1.1, the high-speed data pulses (H-S In) are split into N branches and selected by all-optical switches such that the first pulse is selected by the first FOS, the second pulse by the second FOS and so on. The selected pulses are then delayed by the delay lines in order to ensure that all pulses arrive concurrently at the receivers. Thus, the high-speed data packet dropped by the node is down-converted and can be processed electronically in a bit-parallel manner at an N-times lower bit rate.

In the following sections, we report on results that have been achieved with substantial support by the COST 291 action.

1.2   Transmission of 8-Level 240 Gb/s RZ-DQPSK-ASK

During an STSM in July 2005, researcher Murat Serbay from Lehrstuhl für Nachrichten- und Übertragungstechnik (LNT), University of Kiel, Germany was hosted by Technical University of Denmark (COM), Copenhagen.

The main goal of the STSM mission was to investigate a high spectral efficient multilevel modulation format which combines an amplitude modulation (ASK) together with a differential quaternary phase modulation (DQPSK). In order to increase the receiver sensitivity we combined the modulation format with additional return-to-zero (RZ) -pulse carving. We call this modulation format RZ-DQPSK-ASK. In order to investigate this, we had to combine equipment from COM and the University of Kiel.

Together we managed to set-up a 240Gb/s RZ-DQPSK-ASK transmission in combination with polarisation multiplexing. This was at that time the highest achieved data rate for a single channel transmission, if optical time division multiplex is not considered. Our results were accepted for the post deadline session of the European Conference on Optical Communication (ECOC 2005). The accepted paper is attached to this report and describes the main results obtained during the STSM. In addition to that we also determined the receiver sensitivity of RZ-DPSK, RZ-DQPSK and RZ-ASK-DPSK with the setup described in the post deadline paper.

The experimental setup uses a first Mach-Zehnder (MZ) modulator driven with the 40GHz clock signal as a pulse carver with a pulse width equal to 50% of the symbol period. This pulse train was then phase modulated using a second MZ modulator biased at a null point and driven with a differential 2_7-1 bit pseudo-

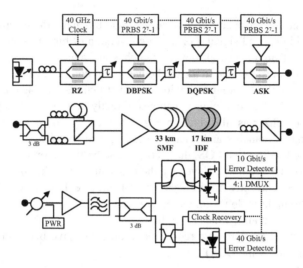

Fig. 1.3. Simplified experimental setup showing the transmitter (top), transmission span with polarisation multiplexing (middle) and the receiver (bottom).

random bit sequence (PRBS) data signal. Then, an 80 Gbit/s RZ-DQPSK signal was generated by a successive phase modulator driven by an inverted 2^7–1 bit PRBS signal. Finally, amplitude modulation was added in a third MZ modulator driven with a 2^7–1 bit PRBS data signal, where the modulator bias and drive signal amplitude were adjusted to obtain the desired extinction ratio on the ASK signal. The resulting signal was thus RZ-DQPSK- ASK with a bit rate of 120 Gbit/s, which was polarisation multiplexed to 240 Gbit/s.

At the receiver input, the signal was polarisation demultiplexed using a polariser. Then the signal was split into two branches, one for DQPSK detection and one for ASK detection. In the ASK branch, the signal was directly detected using a 50 GHz photodetector and errors detected using a 40 Gbit/s error detector. The DQPSK signal was demodulated using a one-bit delay demodulator and received by a 45 GHz balanced photodetector.

We measured the BER curves after transmission, and the results are presented below. Measurements for all ASK and DQPSK tributaries and both states of polarisation are shown, in addition to the average value. There was almost no degradation of the signal quality after polarisation multiplexing, transmission and polarisation demultiplexing, as the receiver sensitivity was found to be –16.5 dBm. The inset shows the optical power spectrum of 40 Gbit/s RZ-DBPSK and 120 Gbit/s RZ-DQPSK-ASK. As the symbol rate is the same for both formats, the spectral width is the same. Thus, even if the bit rate is three times higher (six with polarisation multiplexing), the spectral width and thus many of the transmission impairments such as chromatic dispersion are almost identical.

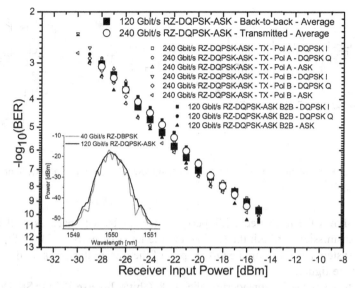

Fig. 1.4. BER measurements for the back-to-back case and after transmission. The inset shows the optical power spectrum of 40 Gbit/s RZ-DBPSK and 120 Gbit/s RZ-DQPSK-ASK.

1.3 Four Bits per Symbol 16-ary Transmission Experiments

Using multiple bits per symbol is a key technology for achieving high spectral efficiency modulation formats. After the first STSM there was a continuing contact established between LNT and COM. During this work, the inverse RZ format with both 4-ary ASKS and 4-ary PSK based on non-coherent detection was investigated by joint experiments. Thus we call this novel format IRZ-QASK-DQPSK.

The experimental setup with transmitter and receiver is shown below. Light from a distributed feedback (DFB) laser is first modulated by a Mach-Zehnder modulator (MZM), driven with a 10.7 Gbaud electrical quaternary RZ signal to generate an optical 21.4 Gbit/s Inverse-RZ-QASK signal with an Inverse-RZ pulse width equal to 40% of the symbol period. DQPSK information is added to the signal using two phase modulators, each driven with a 10.7 Gbit/s NRZ electrical drive signal, that added π and $\pi/2$ phase modulation, respectively. Decorrelation of all signals is ensured by different path lengths. Thus, at the transmitter output a 42.8 Gbit/s (10.7 Gbaud) an Inverse-RZ-QASK-DQPSK signal is generated.

Transmission over a 75 km fibre span with only minor degradations has been demonstrated.

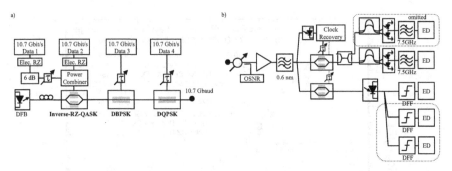

Fig. 1.5. The experimental setup for 42.8 Gb/s IRZ-QASK-QPSK modulation without polarization multiplexing.

Fig. 1.6 shows the measured BER performance for back-to-back as well as for 75 km transmission for both the QASK and the DQPSK parts. A 21.4 Gbit/s RZ-DQPSK system (by omitting the QASK part of the transmitter), was used as a reference in the figure.

With this approach, considering the 42.8 Gbit/s Inverse-RZ-QASK-DQPSK signal, the additional amplitude modulation degrades the optical signal-to-noise ratio (OSNR) performance of the DQPSK part by less than 2 dB, whereas the QASK tributary determines the overall system performance with a substantial OSNR degradation due to the reduced distance of the amplitude levels in the QASK modulator. We think that the performance of the QASK tributaries can be significantly increased by using further developments like high speed digital signal processing units and special QASK modulators.

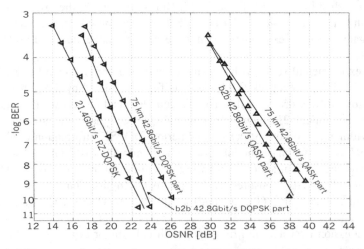

Fig. 1.6. BER curves for 16-ary Inverse RZ-QASK-DQPSK modulation in the back-to-back case and after transmission of 75 km standard fibre.

1.4 Optical Rate Conversion Units

The rate conversion scheme depicted in Fig. 1.2 allows up conversion and down conversion of optical packets not limited in length if the level of parallelism is high enough, i.e., if the number of branches equals the conversion rate ($N = K$). For $N < K$, only short packets with a maximum length of N bits can be handled. In practical systems, conversion rates of ten, hundred or even several hundreds are required. For example, a possible application could be conversion of parallel data clocked at several hundreds of MHz into a data stream at 100 Gbit/s or above, which is then transmitted through a high capacity optical network. Thus, applications requiring conversion rates of several tens to several hundreds are likely to be the most relevant. To reach such high conversion rates, a very large number of fast optical switches need to be used, which results in a quite complex arrangement and difficult stabilization. Furthermore, the incoming optical signal has to be split in a large number of branches, thereby causing large splitting losses. Therefore, the optical serializer/parallelizer shown Fig. 1.2 may be an impractical solution when large packets and high conversion rates are required, which is the case in most practical applications.

1.4.1 Optical Packet Compression and Expansion

Another possibility to convert the data rate directly in the optical domain is to use the so called optical packet conversion/expansion technique. Several methods for optical packet compression and expansion have been proposed [8,9,10]. Most of these methods are based on an optical buffer (optical recirculation loop) and a sampling technique. However, there are some restrictions concerning bit rate and packet size when optical recirculation loop is used. Those restrictions are caused by difficulties in buffering of optical packets for a longer period because of the impairments caused by dispersion in optical fibres and amplified spontaneous emission (ASE) noise accumulation during amplification. A further technique for optical packet compression is a feed-forward delay-line structure consisting of $q = \log_2(N)$ stages reported in [10]. This technique allows simultaneous compression and expansion of N-bit large optical packets using the same device. However, the maximal packet size, N_{max}, is limited by the compression rate, K, as follows

$$N_{max} = K - 2 \cdot \left\lceil \frac{t_{gate}}{\tau_0} \right\rceil \tag{1.1}$$

where $K = T_0/\tau_0$ is the compression rate and t_{gate} denotes the response time of the optical gate. Consequently, for $K = 100$, $\tau_0 = 10$ ps, and $t_{gate} = 40$ ps, N_{max} is limited to 91 bits. Such short packets are usually impractical in many applications. The rate conversion time increases linearly with increasing packet length (N) and conversion rate (K) accordingly to $T_{prc} = K \cdot N \cdot \tau_0$. In case of high conversion rates and large packet lengths, the resulting large processing latency can significantly impair the network performance.

In this section, several basic methods for optical packet compression/expansion are described. Some results of an analytical analysis concerning power budget and ASE noise accumulation are shown. Furthermore, the scheme using the feed-forward delay-line structure (optical delay line structure, ODLS) is modelled by means of numerical simulations. Main simulation results are shown and discussed.

1.4.2 Optical Compression/Expansion Loop

The optical packet rate conversion scheme based on a recirculation loop [8,9] is shown in Fig. 1.7. The loop length is chosen to be T_0 - τ_0 in order to allow a compression rate of $K = T_0/\tau_0$. The principal operation of an optical packet compressor based on the recirculating loop scheme is shown in Fig. 1.7(a). Low-speed input packets (1) enter the loop through the coupler. The switch (SW) is set to be in the "bar" state during the compression operation. After the first pulse has finished a round-trip and passed the coupler ($t = T_0$ seconds), the second pulse enters the loop. It follows the first pulse by the bit-period of the compressed packet τ_0. After $N \cdot T_0$ seconds (N is the packet length in bits), N bits spaced by τ_0 are circulating in the loop (2). The switch is then set into the "cross" state, whereby the whole compressed packet is coupled at the output (3). Note that the pulses, from which the packets are generated, must be sufficiently short to produce high-speed output packets with the bit-period τ_0 without inter-bit interferences, i.e., $\tau_p < \tau_0$.

Fig. 1.7(b) shows an optical packet expander using the recirculating loop scheme. High-speed optical packets (1) enter the recirculating loop through the switch (SW), which is initially set to be in the "cross" state. After the whole high-speed packet has entered the loop, the switch is set to the "bar" state and remains in this state up to the end of the expansion operation. The packet circulates in the loop N-times (2), while in each round-trip a fraction of the high-speed packet is

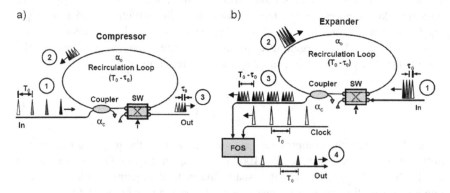

Fig. 1.7. Optical packet (a) compressor and (b) expander using recirculation loop; α_0 denotes the overall loss in traversing the loop once (FOS: fast optical switch, SW: switch).

coupled at the output of the loop. Consequently, N copies of the packet are produced (3). Because the round-trip propagation time in the loop is adjusted to be T_0 - τ_0, each copy of the packet is delayed with respect to the next copy by T_0 - τ_0. A fast optical switch (FOS) selects bits spaced at the bit-period T_0, thereby expanding the whole high-speed input packet (4). However, this method leads to some restrictions concerning bit rate and packet size. To prevent bit overlapping in the loop, the number of bits in a packet (N) must be smaller than the compression rate, i.e., $N < K$ - 1. Additionally, the switch must change its state arbitrarily quickly, namely in the time gap between the last bit and the first bit of the packet in the loop. Since the recirculation loop length is chosen to be T_0 - τ_0 and the length of the high-speed packet is $N \cdot \tau_0$, the switching time can be calculated from $t_{sw} = T_0$ - τ_0 - $N \cdot \tau_0 = \tau_0 \cdot (K - 1 - N)$. For example, assuming a compression rate $K = 100$, a switching time $t_{sw} = 200$ ps, and $\tau_0 = 10$ ps, the maximal packet length, N_{max}, is limited to 79 bits.

1.4.3 Optical Delay Line Structure

The optical packet compressor/expander comprising an optical delay line structure (ODLS), an optical gate, and a FOS for the expansion operation is shown in Fig. 1.8. It allows compression of packets to be transmitted simultaneously with expansion of the received packets using the same device. The number of stages increases logarithmically with the number of bits to be processed. Because of the fact that the complete compressed packet occurs in the gap between two bits of the low-speed signal, the number of bits in the packet is, similar to the recirculating loop scheme, limited by the compression rate as follows $N < K$ - 1. If the response time of the gate (t_{gate}) is taken into account, the limitation of the packet size is given by $N_{max} < K - 1 - 2 \cdot (t_{gate} / \tau_0)$. If larger packet sizes are needed, a larger compression rate or a parallel arrangement of optical delay line structures (ODLS) can be used.

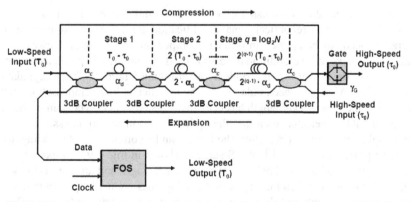

Fig. 1.8. Packet compressor/expander using an optical delay line structure (ODLS).

The principal operation of an optical packet rate conversion unit based on an *optical delay line structure* is as follows. For the compression operation, a low-speed input packet is split into the two arms by the 3 dB couplers in every stage. The signal in the upper arm of the delayed by $2^{(q-1)} \cdot (T_0 - \tau_0)$ and combined with the undelayed packet by the next coupler. At the output, each input pulse is copied N-times and spaced from its neighbour by $T_0 - \tau_0$. An optical gate at the output of the delay line structure selects the complete compressed packet, which occurs in the middle of the output sequence.

By adding an additional fast optical switch, the high-speed packets can be expanded using the same device in the inverse direction. The high-speed optical packet enters the device at the high-speed input port. In each stage, it is delayed by an appropriate delay line in the upper arm and combined by a 3 dB coupler with the undelayed signal, thereby making N copies of the high-speed input packet at the output of the ODLS. Each copy of the input packet is delayed with respect to the next copy by $T_0 - \tau_0$. A fast all-optical switch selects bits spaced at the bit-period $T0$ by a very narrow switching window. Thus, the expanded low-speed output signal has the same bit pattern sequence as the compressed packet.

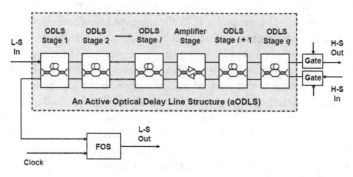

Fig. 1.9. Compressor/expander using an active optical delay line structure (aODLS).

Optical losses in the compressor/expander can limit the maximum achievable packet length. Therefore, optical signal amplification during compression/expansion operation needs to be applied in order to compensate these losses. However, the added amplifiers induce additional noise caused by amplified spontaneous emission (ASE). The impact of the optical signal-to-noise ratio (OSNR) degradation caused by the ASE noise accumulation and by insufficient loss compensation on the maximal achievable packet length has been investigated in [11]. The gain of the amplifiers should be selected such that the optical losses in the compressor/expander are completely compensated. Thus, the losses can be compensated by adding additional amplifiers after each ith ODLS stage as shown in Fig. 1.9. For example, the overall gain in a segment consisting of four ODLS stages and an amplifying stage is given by $\delta = \alpha_c^5 \cdot G$ when the losses in the delay lines are neglected because of $\alpha_d \ll \alpha_c$. The total number of these segments in a compressor-expander constellation is $2 \cdot |q/4|$, where the symbol ”| |” denotes the rounding down operation.

1.4.4 Scalable Packet Compression/Expansion Units

A scalable optical packet rate conversion scheme based on a parallel arrangement of optical delay line structures is shown in Fig. 1.10. This scheme allows high compression rates and large packet sizes, thereby reducing the impact of the "time out" phenomenon by processing packets in a parallel manner. The "time out" phenomenon [12] prevents receiving or transmitting of two successive packets by a node within a time period equal to the packet processing time ($T_{prc} = K \cdot N \cdot \tau_0$). For example, for a conversion rate of $K = 100$, the receiver can access only each 100-th packet. The remaining 99 packets arriving at the node in the meantime can not be processed because the rate conversion unit is occupied with processing a single packet. This hard restriction leads to inefficient bandwidth utilization. The scalable optical packet compression/expansion unit [13] consists of M bidirectional optical gates, M fast optical switches (FOSs), and M parallel active ODLSs, each of them responsible for the rate conversion of a part (an N-bit sequence) of the optical packet. Thus, the packet length is $N \cdot M$, where $M = 1, 2, \dots K$, while the packet processing time remains the same as for one N-bit sequence. An active ODLS that processes an N-bit sequence is composed of q delay stages and an amplifier stage placed after each i-th delay stage to compensate for the optical losses (see Fig. 1.9).

The operation of the device can be described as follows. In the compression direction, the parallel data acquired from the transmitting queue can be used to modulate M parallel short pulse trains, thereby generating low-speed optical sequences.

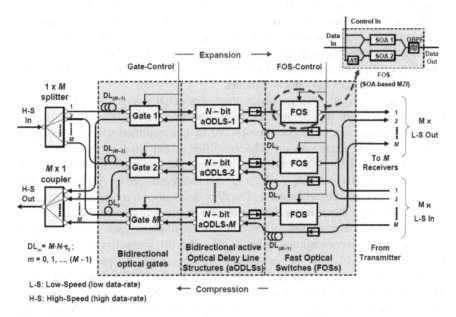

Fig. 1.10. Scalable optical packet compression/expansion unit (SOA: semiconductor optical amplifier).

In the compression/expansion unit, M low-speed sequences are first compressed in the ODLSs. The fully compressed sequences are then selected by the gates and de-layed by the delay lines to meet the timing requirements for combining the com-pressed sequences at the output. The compressed sequences are finally combined by an $M \times 1$ coupler. The resulting signal at H-S Out represents an $M \cdot N$-bit com-pressed packet.

High-speed packets can be expanded using the same device in the reverse di-rection. A high-speed input packet is divided into M separate N-bit sequences us-ing a splitter, M optical delay lines, and M bi-directional optical gates. Those se-quences are then copied N times within the ODLSs. Each copy is delayed by T_0 - τ_0 with respect to the next copy of the packet. A fast optical switch selects bits separated by the bit period T_0 within a very narrow switching window, thereby ex-panding the high-speed sequence. Thus, the whole high-speed input packet is down-converted and divided into M low-speed sequences, which can be received in a parallel manner by M receivers.

Numerical simulations were carried out in order to investigate the feasibility of the proposed scheme. In our simulation set-up, 2.5 ps full width at half maximum (FWHM) optical pulses were generated at 1.55 μm wavelength with a repetition frequency of $f_{L-S} = 1/T_0$, and split into M ways. The M pulse trains were then modulated in a parallel manner using pseudo random bit sequences (PRBS) to generate M N-bit optical sequences. The low-speed optical sequences were first compressed using the proposed scheme, then transmitted over a transmission line composed of 300 m standard single mode fibre (SSMF), an optical amplifier (EDFA) and an optical band-pass filter (OBPF), and finally expanded using the same compression/expansion scheme in reverse direction. LiNbO$_3$ modulators were used for implementing the gating functionality, while Mach-Zehnder inter-ferometers with semiconductor optical amplifiers in its arms (SOA-based symmet-ric MZI) were used as fast optical switches

First, we investigated the maximum number of aODLS stages. It is mainly lim-ited by the chromatic dispersion in the fibre delay lines and additionally by the op-tical signal-to-noise ratio degradation caused by the amplified spontaneous emis-sion (ASE) noise accumulation. The amplifier model used in the simulations represents an ideal amplifier with frequency-dependent gain and noise figure. The output signal is obtained as $E_{out} = E_{in}\sqrt{G(f)}$ and then corrupted by ASE noise that is modelled by a Gaussian white noise. The fibre delay lines were modelled employ-ing the split-step Fourier method and using the parameters of commercially avail-able standard single mode fibres (SSMFs). The small-signal gain and noise figure of the amplifiers were assumed to be 20 dB and 5 dB, respectively. As it can be seen from Fig. 1.11, nine stages (capable to process 512-bit packets) induce a power penalty of approximately 0.8 dB and 1.2 dB for 80 Gbit/s and 100 Gbit/s data rates at the H-S Out, respectively. That is, $M \cdot 512$-bit packets can be provided using the proposed scheme without any dispersion compensation. Because the number of employed amplifiers is q/i, the OSNR degradation has less influence on the scalability than the chromatic dispersion.

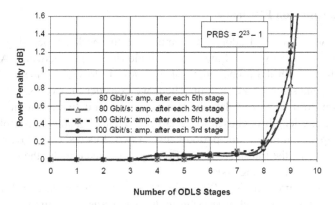

Fig. 1.11. Power penalty vs. number of ODLS stages (ODLS: optical delay line structure).

Second, impact of the gate-control signal synchronization has been investigated. It must be well-synchronized with the incoming/outgoing high-speed optical packets. The impact of the gate-control signal adjustment and broadening was investigated recently [13, 14]. The obtained power penalties (see Fig. 1.12(a)) have shown that a gate-control signal detuning between −7.5 ps and +1.9 ps (an interval of 9.4 ps) as well as a pulse broadening between −1.65 ps and +2.8 ps (in total 4.45 ps) results in a power penalty lower than 3 dB for a 100 Gbit/s signal. These results show that the proposed scheme is sensitive to the gate-control signal synchronization; hence very fast and well-synchronized gates should be used or a separation gap between the sequences has to be added.

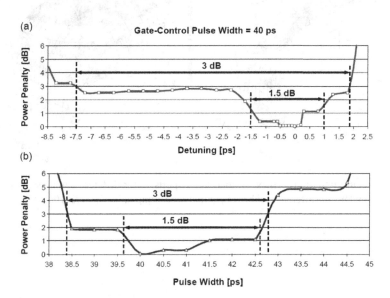

Fig. 1.12. Impact of the gate-control signal (a) adjustment and (b) broadening.

1.4.5 Transmission Efficiency

In the following, we will address the impact of the packet processing time on transmission efficiency in a time-slotted single-ring metro network. A simple TDMA scheme with the frame length of $N_s = K - 1$ slots is assumed, where the slot length equals the packet length. Then, we consider an overloaded network with uniform traffic and calculate the rate of successful deliveries in the network. Since the frame length is chosen to be large enough to guarantee completion of packet compression in the time period between two slots dedicated to a particular transmitting node, the transmission inefficiency is caused only by the rate conversion latency of the receiving node.

The calculated transmission efficiencies for different compression rates (K) and different number of parallel processing units (M) versus total number of ports in the network are shown in Fig. 1.13. It can be seen that the transmission efficiency can be greatly improved by using a larger number of parallel aODLSs; thus by simultaneous increasing the slot size and reducing the packet processing time [14]. Reduced transmission efficiency can be expected for very large compression rates and a high number of ports. Because the total number of ports in system is expected to be large enough (several tens to few hundreds), a transmission efficiency higher than 90 % can be achieved by using packet processing units with more than 10 parallel aODLSs.

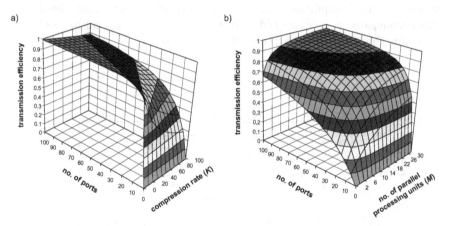

Fig. 1.13. Transmission efficiency vs. total no. of ports a) for different compression rates when M=8 and b) for different number of parallel processing units when K=100.

References

1. Jensen, J.B., Tokle, T., Geng, Y., Jeppesen, P., Serbay, M., Rosenkranz, W.: Dispersion Tolerance of 40 Gbaud Multilevel Modulation Formats with up to 3 bits per Symbol. In: Leos Annual 2006, Montreal, QC, Canada, 29 Oct.–2 Nov. 2006, paper WH 4 (2006)
2. Tokle, T., Serbay, M., Jensen, J.B., Geng, Y., Rosenkranz, W., Jeppesen, P.: Investigation of multilevel phase and amplitude modulation formats in combination with polarisation multiplexing up to 240 Gbit/s. IEEE Photonics Technology Letters 18(20), 2090–2092 (2006)
3. Tokle, T., Serbay, M., Rosenkranz, W., Jeppesen, P.: 32.1 Gbit/s InverseRZ-ASK-DQPSK Modulation with Low Implementation penalty. In: Leos Annual 2006, Montreal, QC, Canada, 29 Oct.–2 Nov. 2006, paper WH 2 (2006)
4. Serbay, M., Tokle, T., Jeppsen, P., Rosenkranz, W.: 42.8 Gbit/s, 4 Bits per Symbol 16-ary Inverse-RZ-QASK-DQPSK Transmission Experiment without Polmux. In: Optical Fibre Conference, OFC 2007, Anaheim, USA, 25-29 March, OThL2 (2007)
5. Tokle, T., Serbay, M., Geng, Y., Jensen, J.B., Rosenkranz, W., Jeppesen, P.: Penalty-free Transmission of Multilevel 240 Gbit/s RZ-DQPSK-ASK using only 40 Gbit/s Equipment. In: ECOC 2005, Glasgow, Scotland, Sep. 2005, post deadline paper Th4.1.6 (2005)
6. Xia, C., Rosenkranz, W.: Statistical Analysis of Electrical Equalization of Differential Mode Delay in MMF Links for 10-Gigabit Ethernet. In: OFC 2005, Anaheim, USA (2005)
7. Xia, C., Rosenkranz, W.: Electrical Equalization for Duobinary and Phase Shift Keyed Modulation Formats (oral presentation). In: Workshop on Design of Next Generation Optical Networks: from the Physical up to the Network Level Perspective, Ghent, Belgium, 6 Feb. (2006)
8. Toda, H., Nakada, F., Suzuki, M., Hasegawa, A.: An Optical Packet Compressor Using a Fibre Loop for a feasible all optical TDM Network. In: 25th European Conference on Optical Communication (ECOC 1999), Nice, France, 26-30 September 1999, vol. 2(I), pp. 256–257 (1999)
9. Patel, N.S., Hall, K.L., Rauschenbach, K.A.: Optical Rate Conversion for High-Speed TDM Networks. IEEE Photonics Technology Letters 9(9), 1277–1279 (1997)
10. Toliver, P., Deng, K.L., Glesk, I., Prucnal, P.R.: Simultaneous Optical Compression and Decompression of 100 Gb/s OTDM Packets Using a Single Bidirectional Optical Delay Line Lattice. IEEE Photonics Technology Letters 11(9), 1183–1185 (1999)
11. Aleksic, S., Krajinovic, V., Bengi, K.: A Novel Scalable Optical Packet Compression/ Decompression Scheme. In: 27th European Conference on Optical Communication (ECOC 2001), Amsterdam, Netherlands, 30 September – 04 October 2001, vol. 3(1), pp. 478–479 (2001)
12. Acampora, A.S., Shah, S.I.A.: A packetcompression/decompression approach for very high speed optical networks. In: Proceedings of SBT/IEEE ITS, Rio de Janeiro, Brasil, September 1990, pp. 38–48 (1990)
13. Aleksic, S.: Packet-Switched OTDM Networks Employing the Packet Compression/Expansion Technique. Photonic Network Communications 5(3), 273–288 (2003)
14. Aleksic, S.: Design Considerations for a High-Speed Metro Network using All-Optical Packet Processing. In: 8th International Conference on Transparent Optical Networks (ICTON2006), Nottingham, United Kingdom (invited), June 2006, pp. 82–86 (2006)

2 Electronic Channel Equalization Techniques

*I. Papagiannakis (chapter editor), G. Bosco, D. Fonseca, D. Klonidis,
P. Poggiolini, W. Rosenkranz, A. Teixeira, I. Tomkos, and C. Xia*

Abstract. This paper presents the key design approaches and results in the
field of optical impairment distortion compensation by electronic means, as
an outcome of the studies and research innovations developed within the
COST 291 action. The research topics addressed are related with chromatic
dispersion and nonlinearities, with particular reference on FFE/DFE and
MLSE-based equalizers as well as with the assistance of different modulation formats. Additionally, the use of electronic compensation in metro-
access applications is examined with reference on studies related with the
performance enhancement of DML transmitters.

2.1 Introduction

Driven by the rapidly increasing traffic demands, stemming from the broadband
applications in the access networks, optical transmission technologies are widely
used in today's telecommunications networks, providing link connections closer
to the end user at higher data rates and with larger number of multiplexed channels
per fibre. However, optical networks are analogue in nature and suffering from a
variety of linear and nonlinear transmission impairments. These effects have a direct impact on the bit-error-rate performance of the system and most importantly,
this impact increases in systems supporting higher data rates and larger number of
channels.

A variety of techniques have been proposed and implemented in order to reduce the effect of physical impairments present in optical transmission system.
These techniques can be classified as a) purely optical and b) optoelectronic based
solutions. Typical purely optical solutions are the dispersion compensating modules used to reduce signal broadening due to dispersion effects and the optical regenerators proposed to cope with signal distortion effects like additive noise and
nonlinear impairments. However, the complexity and the cost of these implementations (especially in the case of regenerators) reduce their usefulness in the network. On the other hand, electronic-based solutions can be applied in order to
mitigate transmission impairments and enhance either the transmission distance or
the span budget. These end node solutions relax the network operators from strict
network design rules and allow in many cases the use of already installed non-
optimized infrastructures to be operated at higher data rates, accommodating more
users per link.

I. Tomkos et al. (Eds.): COST 291 – Towards Digital Optical Networks, LNCS 5412, pp. 23–47, 2009.

For core network applications, where the filtering characteristics of the transmission medium (dispersion, polarization) impose significant constrains on the shape of the transmitted signal, electronic based solutions are expected to increase the transparency distance and simultaneously reduce the equipment count per signal path, resulting in significant cost advantage in applications like submarine links where the cost of undersea repeaters is higher. Additionally, ease of applicability of adaptive schemes is particularly important in order to compensate for time varying effects like polarization mode dispersion (PMD). Electronic equalizers may significantly improve the limited reach of these systems, allowing their applicability in core networks.

For Metro and Access network applications the role of electronic equalizers is to enhance the performance and reach of low cost transmitters according to modulation properties and the transmitted medium. In this case, when the electronic equalization process is combined with certain limited performance components and media, like direct modulated lasers (DMLs), enhanced transmission capabilities and extended reach can be achieved. This allows the implementation of extended access networks (i.e. more users and/or longer distances) and the transparency in metro networks in a cost effective way.

Advanced solutions for the research topics presented in the last two paragraphs have been studied and implemented within the COST 291. These solutions are addressed next in this paper which is organized as follows: First, section 2 presents an overview on the most relevant types of equalizers used in the applications that follow. Section 3 focuses on the performance of maximum likelihood sequence estimator equalizer (MLSE), whereas section 4 shows the performance of nonlinear electrical equalizer for different modulation formats. Section 5 presents the impact of equalization for optical single sideband systems. Finally, section 6 shows the enhancement of the system performance in metro, access and local area network applications with the use of equalizers, focusing on the performance improvement of DML type of transmitters.

2.2   Electronic Equalizers

Equalization techniques have been widely used in digital communications over microwave wireless links, as well as in modem designs. The idea behind the equalizer is to create the inverse transfer function of the optical communication channel, implementing the proper filter shape in order to compensate for the introduced impairments (chromatic dispersion (CD) and PMD), cancelling out the filtering effects of the channel.

There are three main architectures for electrical channel equalizer (ECE). The most basic scheme is the feed-forward equalizer (FFE) (Fig. 2.1(a)). In this set-up, a finite-impulse-response (FIR) filter is added to the transmission line after the optical-to-electrical conversion. The filter has several stages, each consisting of a delay element, a multiplier and an adder. After every delay element, an image of the

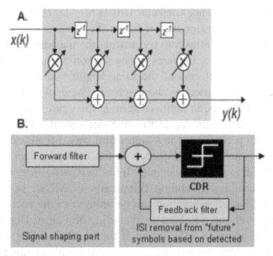

Fig. 2.1. The structure of: (a) FFE and (b) DFE. {CDR: Clock and data recovery module, ISI: Intersymbol Interference}.

non-delayed input is multiplied with a coefficient and added to the signal. The number of filter stages used and the coefficients chosen are crucial parameters for effective dispersion cancellation. A second approach is the decision-feedback equalizer (DFE) (Fig. 2.1(b)). This structure is an FFE with a second FIR filter added to form a feedback loop. The coefficients of both filters require active control, in order the filter to be adapted to time-varying changes. Today's integration levels permit either FFE or DFE to be built into a clock and data recovery or demultiplexer chip with an extra power requirement of about 500 mW.

The tap coefficients of the filter are calculated and are always adjusted in an adaptive operation according to an algorithm that runs in parallel with the purpose to minimize the error. The type of the algorithm and more significantly the way that this algorithm is optimized are particularly important in order to minimize the error and enhance the transmission properties of the system. The general operating modes of the adaptive equalizer include training and decision mode. During the training mode the algorithm adjusts to the channel characteristics and calculates the filter taps to compensate for the introduced impairments. In the decision mode, small variations in the taps allows for the compensation of the time varying effects of the channel. The most common algorithm that is used in order to calculate the taps is the least mean square (LMS). The goal of that algorithm is to minimize the mean squared error (MSE) between the desired equalizer output and the actual equalizer output. It is controlled by the error signal which is derived by the output of the equalizer with some other signal which is the replica of transmitted signal.

The DFE version of the equalizer is a non-linear process that uses the same algorithm but it subtracts the interference by the already detected data offering advanced performance characteristics [1], [2].The most sophisticated structure is the maximum-likelihood (sequence) detector (MLD/MLSD) (Fig. 2.2). This is effectively a digital signal processor that performs the necessary mathematical operations on the incoming data stream to reconstruct the transmitted signal. More specifically, an MLSE receiver compares a long section of the noisy received signal with all the possible waveforms of the same length that could be received and chooses the one that is "closer" to the received. The drawback of this method is that it needs the signal to be in digitized form at the input, which requires an analogue-to-digital converter running at full line rate and at a high resolution. ECE can improve system performance in a significant way. MLSE estimates the channel and decides the most likely sequence sent at the transmitter based on the received signal. MLSE is a Viterbi decoder. It has two parts: channel estimation (ISI estimation) and decoding. Decoding complexity of MLSE is the same as the Viterbi decoders. Channel estimation can be thought as the encoder that encodes the original signal with the weighted neighbouring bit values. Due to encoders' nature, the channel is assumed to be linear. If the channel is time varying, then the estimation should be able to track the changes. MLSE also requires soft decisions for decoding to maximize the gain. This is analogous to hard and soft decision Viterbi decoding performance difference. The effectiveness of MLSE in the compensation of the CD may extend into several hundred kilometres range, but at such long range, the complexity of the MLSE processor is very large. It is important to examine and optimize the effectiveness of MLSE equalization and provide techniques to reduce complexity without impacting performance [1], [2].

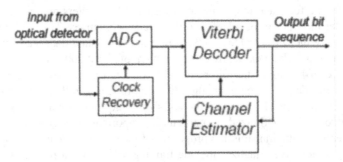

Fig. 2.2. Digital implementation of MLSE. (ADC: Analog–to-digital converter).

An alternative technique to mitigate optical impairments by means of an electrical process consists of the association of optical single sideband (OSSB) modulations to electrical dispersive components [3]. This technique has the advantage of using simple electrical components (e.g. microstrip lines or FFE) to mitigate the accumulated optical dispersion (in the form of group velocity dispersion). Fig. 2.3 presents an example of a transmission system using OSSB modulation and electrical dispersion compensation. Contrary to the other cases of equalizer, phase preservation is observed after direct detection due to the absence of spectrum back folding (valid in the case of carrier-unsuppressed formats such as the conventional on-off

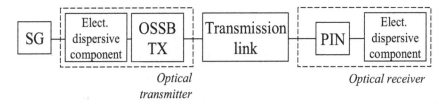

*Optical
transmitter*   *Optical receiver*

Fig. 2.3. Transmission system using OSSB modulation and electrical dispersion compensation (at the transmitter and receiver sides). SG – Signal Generator; OSSB – OSSB transmitter.

keying). As a consequence, placing an electrical dispersive component after the photodetector, almost complete optical dispersion mitigation is observed and large distances of standard single mode fibre (SSMF) can be achieved without optical dispersion compensation. Moreover, placing the same type of electrical dispersive component at the transmitter side, optical dispersion mitigation is also observed. As the compensation is done prior to the direct detection, optical dispersion can be mitigated with enhanced efficiency compared to the case using electrical dispersion compensation at the receiver side. Additionally, in the case of electrical dispersion compensation at the transmitter side, carrier unsuppressed modulation (such as alternate mark inversion (AMI)) can be used.

2.3   Fundamental Limits of MLSE Performance with Large Number of States

The theoretical possibility of long-haul uncompensated IMDD transmission at 10 Gb/s using an MLSE processor has been investigated up to 700 km of SSMF (11,700 ps/nm) in [4],[5]. Simulations predicted that, provided that enough complexity was allowed for the MLSE processor, the OSNR penalty with respect to back-to-back would saturate at about 2 to 3 dB, depending on system set-up, at a distance between 300 and 400 km, and would not further grow for longer link lengths. We show here the results of a thorough transmitter (TX) and receiver (RX) filter optimization study, carried out in order to identify the best operating set-up and the corresponding OSNR-penalty with respect to back-to-back, under severe CD distortion. A minimum asymptotic penalty of about 2 dB for large CD values was found (up to about 11,200 ps/nm, or 700 km of SSMF), in good agreement with recent results based on Information Rate estimates [6].

The analyzed system setup is shown in Fig. 2.4. The electrical filters (at TX and RX) were 5-pole Bessel with bandwidth $B_{TX,elt}$ and BPD, respectively, and the optical RX filter was second-order super Gaussian with bandwidth Bo. The MLSE processor used the (square root) SQRT metric [4]. For all simulations, the transmitted sequence was a full PRBS cycle of length $2^{20}-1$ bits (about 1 million).

We considered four different system set-up configurations:

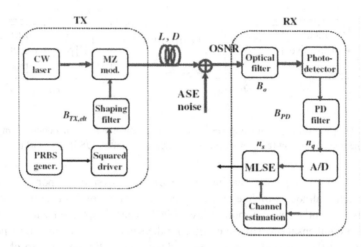

Fig. 2.4. System schematics.

- *No PD-filter, 8 samples/bit.* No electrical post-detection (PD) filter is present in the RX. The use of substantial over-sampling (8 samples/bit) enables the MLSE processor to perform a sort of 'digital' PD (non-linear) filtering. The optical filter bandwidth is variable.
- *With PD-filter, 2 samples/bit.* The sampling rate is reduced to the customary 2 samples/bit. An electrical PD filter is present in the RX, with variable bandwidth. The optical filter bandwidth is variable as well.
- *Wide optical filter, PD-filter, 2 samples/bit.* The PD filter is present in the RX, with variable bandwidth. The processor uses 2 samples/bit. The optical filter bandwidth is set at the relatively large value of 35 GHz.
- *Wide optical filter, PD-filter, 1 sample/bit.* This configuration is the same as the previous one, but the sampling rate is reduced to only 1 sample/bit.

2.3.1   No PD-Filter, 8 Samples/Bit

In this system configuration the role of the PD filter is taken over by the MLSE processor itself. Thanks to the substantial over-sampling (8 samples/bit), the MLSE processor can effectively perform a sort of 'digital' PD filtering on each branch.

The optimum filter bandwidth values at 400 km are shown in Table 1. We selected this length because it is a value for which the OSNR versus system length curve has already reached its 'saturated' flat region. We then evaluated the system performance from 0 to 700 km of SSMF using the optimum filter values found above.

Fig. 2.5 (solid line with triangles) shows the OSNR required to obtain a BER=10^{-3} versus link length. The number of trellis states was dynamically adapted as necessary for best performance and ranged from 2 (at 0 km) to 2048 (at 700 km). In Fig. 2.5, the star-shaped marker represents the performance of a non-MLSE

Table 2.1. Optimum filter bandwidth values at 400 km for the four analyzed system configurations.

System	$B_{TX,elt}$ (GHz)	B_o (GHz)	B_{PD} (GHz)	n_{samp}
A	18	12.5	-	8
B	18	12.5	14	2
C	8	35	8	2
D	4.5	35	3.5	1

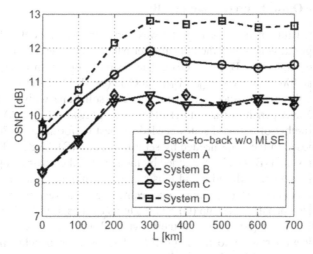

Fig. 2.5. OSNR values needed to obtain a BER equal to 10-3 vs. fibre length for the four analyzed system configurations (A) through (D), with the optimum filter bandwidths shown in Table 1.

benchmark system, obtained with standard filter bandwidths ($B_{TX,elt}$=B_{PD}=7.5 GHz, B_o=35 GHz) using optimized-threshold hard decision.

The plot clearly shows penalty saturation upward of 200 km. The saturated penalty with respect to back-to-back is 2 dB, in agreement with the results of [6], that were obtained with a different system set-up. A remarkable result is that, if we compare the saturated OSNR performance of the optimized set-up (A) with the back-to-back performance of the benchmark system w/o MLSE in Fig. 2.5, the penalty is 0.5 dB only, up to 700 km.

2.3.2 With PD-Filter, 2 Samples/Bit

In this configuration, the burden of PD filtering is shared between the electrical PD filter and the 2 samples/bit MLSE processor. As for System (A), an optimization of the filter bandwidths was performed at 400 km. This time, three bandwidths had to be varied, rather than just two: the TX and the PD electrical filter bandwidths BTX,elt, BPD, and the RX optical filter bandwidth Bo. The resulting optimum bandwidth values are shown in Table 1, set-up (B). Using these values, the performance shown in Fig. 2.5 (dashed line with diamonds) is obtained, which is virtually coincident with the one of set-up (A).

As we remarked for (A), we point out that the saturated OSNR performance of the optimized set-up (B) is only 0.5 dB away from the non-MLSE back-to-back benchmark. This result is even more significant for (B) because, as opposed to (A), (B) is a more 'realistic' and practically implementable configuration.

2.3.3 Large Optical Filter, 2 Samples/Bit

The optimum value of B_o for set-up (B) was very narrow, so we decided to consider a configuration in which the optical filter bandwidth is set to a more standard and easier to implement $B_o = 35$ GHz. The resulting optimum electrical filters bandwidths at 400 km were $B_{TX,elt} = B_{PD} = 8$ GHz (see Table 1, set-up (C)). Using these values, the performance shown in Fig. 2.5 (solid line with circles) is obtained. An OSNR penalty of about 1.1 dB with respect to set-ups (A) and (B) is found. The number of required states needed for best performance at 700 km was 8192, as opposed to the 2048 needed by set-ups (A) and (B). The smaller number of states for (A) and (B) can be explained by the fact that their narrow optical filter curtails the signal spectral width and thus somewhat mitigates the impact of dispersion. This reduces system memory in agreement to what was found in [7].

The saturated penalty of set-up (C) relative to its own back-to-back is still 2 dB, as it was for (A) and (B) relative to their own back-to-back. However, the saturated penalty with respect to the non-MLSE back-to-back benchmark (the star marker) increases, reaching about 1.6 dB.

2.3.4 Large Optical Filter, 1 Sample/Bit

Set-up (D) is the same as set-up (C), except the MLSE processor uses only 1 sample/bit. The resulting optimum electrical filters bandwidths at 400 km, shown in Table 1, set-up (D), are much smaller than the ones for 2 samples/bit. In particular, the optimum PD filter bandwidth is only 3.5 GHz. The PD filter must be so narrow because it has to make up for the lack of PD-filtering by the MLSE processor, which now operates with only 1 sample/bit.

The optimum performance is plotted in Fig. 2.5 (dashed line with squares), showing an OSNR penalty of 1.2 dB with respect to set-up (C), and of 2.3 dB with respect to (A) and (B). This result shows that the narrow electrical PD filter is not as effective as over-sampled MLSE in performing PD filtering. The number of required states at 700 km is 4096, vs. the 8192 needed when 2 samples/bit are used. The system memory is lower than in (C) because of the narrower TX electrical filter of (D), that curtails the TX signal spectral width and thus somewhat mitigates the impact of dispersion.

2.3.5 Compensation of SPM Using MLSE

We investigated experimentally the effectiveness of MLSE-EDC for application with signals distorted by self-phase modulation, and compare the results with those obtained using a variety of optical dispersion compensation maps.

We analyzed a system scenario in which, as in [8], high power is launched. In contrast with [8], neither optical dispersion compensation nor management is used. We carried out the experiment using IMDD at 10Gb/s over 800 km (10x80 km) of standard fibre (SMF). MLSE processing was done off-line on the samples of the received signal to assess the capability of MLSE to deal with the challenging distortion caused by significant SPM-induced chirp interacting with the overall-link uncompensated dispersion. The experimental results, confirmed by numerical simulations, show that a 1024-state MLSE processor allowed to keep the BER below the FEC threshold of 10^{-3} up to 10 dBm of launched power into each span. The results are also compared with simulations using optimized optical dispersion management, taken as an ideal benchmark.

The experiment setup is shown in Fig. 2.6. A pulse pattern generator (Anritsu MP1763B) was employed to generate a 10 Gb/s, 2^{20}-1 PRBS. The tunable laser (Agilent 8164A) was set at 1551.65nm. A conventional chirpless LiNbO$_3$ Mach-Zehnder modulator, biased at the -3dB point, was used to generate the NRZ-OOK format signal. A 4 GHz low-pass Bessel filter (LPF) was placed between the driver and the modulator. The bandwidth of this filter was optimized through computer simulations of the experiment. No optical filter was present at the transmitter (TX). The signal was then input to a re-circulating loop with 80-km span length. The fibre was SMF G.652 (17 ps/nm·km). The signal was re-circulated 10 times, for a total distance of 800 km (13,360 ps/nm). The average power

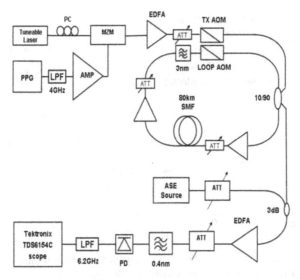

Fig. 2.6. System set-up.

launched into the loop was varied from -3 dBm to 12 dBm. No DCF was installed in the loop. The receiver (RX) consisted of an EDFA pre-amplifier, followed by a 50 GHz Gaussian optical filter, a photo-detector and a 6.2 GHz Bessel LPF. The optical power input to the detector was kept constant at -3dBm.

A noise-loading stage was placed immediately before the pre-amplifier to set the desired optical signal-to-noise ratio (OSNR). The electrical RX signal was sampled using a Tektronix TDS6154C real-time oscilloscope. The OSNR, over a 0.1 nm noise bandwidth, was set at 16 dB.

Two full 2^{20}-1 PRBS instances were sampled and stored to disk for each transmitted power value. An MLSE processor using 2 samples per bit was then run off-line. The algorithm employed the SQRT branch metric proposed in [4] and a reduced-state algorithm was used to minimize the number of MLSE states [9]. The trellis was first instructed on one PRBS instance of the sampled signal and then run on the other. The number of trellis states was at first set to 2^{10} (1024), yielding the results shown in Fig. 2.7 (solid line). We then simulated the same link. To get a good agreement at low power, the simulation OSNR was set to 14.5 dB (solid line with circles in Fig. 2.8). Such OSNR discrepancy was also found in back-to-back operation and therefore could seemingly be ascribed to TX and RX imperfections. By increasing the number of MLSE states, the BER could be significantly improved both in the experiment (Fig. 2.7) and the simulations, though this was much more effective in the simulations (not shown for clarity) where BER=10^{-3} could be reached even at 12 dBm launch power, with 16384 states.

To obtain a suitable benchmark, a 10 Gb/s IMDD optically dispersion-compensated (ODC) system was simulated, next. The analyzed system is similar

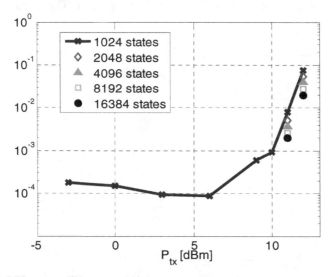

Fig. 2.7. BER versus TX power, 800 km uncompensated SMF transmission experiment, OSNR=16 dB (0.1nm), MLSE RX with variable number of states.

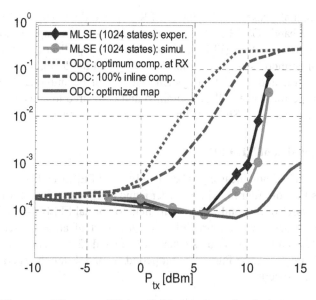

Fig. 2.8. BER versus TX power, 800 km SMF, optical vs. electrical compensation; OSNR (0.1 nm) at RX: 16 dB (MLSE experiment), 14.5 dB (MLSE simulations), 11.2 dB (ODC simulations).

to the experiment, except the TX and RX electrical filters were replaced by 7.5 GHz Bessel and the OSNR was reduced to obtain the same BER as the MLSE-based system at low launch powers. Three different configurations were analyzed: (i) dispersion compensation performed all at the RX, optimized vs. TX power; (ii) 100% compensation span-by-span; (iii) optimized dispersion map for each value of TX power. The obtained results are shown in Fig. 2.8. Despite the fact that the MLSE system operates under conditions that are similar to configuration (i), i.e., all compensation at the receiver, it greatly outperforms it and achieves perform-ance midway between (ii) and (iii). In fact, for higher number of states, MLSE simulation results improved to almost coincide with (iii). On the other hand, it must be mentioned that, as observed in other papers [10], there is an approxi-mately 3 dB OSNR penalty between ODC and MLSE even in the linear case (here 11.2 dB OSNR for ODC vs. 14.5 MLSE).

2.4 Nonlinear Electrical Equalization for Different Modulation Formats

Based on bit error ratio (BER) simulations, we investigate electrical dispersion compensation (EDC) performance by using a nonlinear electrical equalizer based on nonlinear Volterra theory for different modulation formats: On-off keying

(OOK), differential-phase-shift-keying (DPSK), optical duobinary (ODB) and optical single sideband (OSSB). This nonlinear equalizer is compared to conventional feedforward equalizer (FFE), decision feedback equalizer (DFE) as well as to maximum-likelihood sequence estimation (MLSE). First of all, nonlinear equalizer NL-FFE-DFE and simulation system setups will be introduced. Secondly, NL-FFE-DFE performance will be examined and compared to conventional FFE, FFE-DFE as well as to MLSE for different modulation formats.

2.4.1   Introduction of NL-FFE-DFE

We call this kind of nonlinear equalizer as NL[x,y]-FFE[m]-DFE[n]. This designation means FFE of order m, DFE of order n and nonlinear order x for FFE and y for DFE. The case for x=1 and y=1 is equivalent to normal FFE[m]-DFE[n]. As an example, the signal outputs from FFE and DFE part of an equalizer NL[3,2]-FFE[4]-DFE[2] can be expressed as equation (2.1) and (2.2), respectively. Where, $y_r(t)$ and d are defined in Fig. 2.1 T :bit period, e and b are the coefficients for FFE and DFE, respectively.

$$y_e(t) = \sum_{k=0}^{4} e_k y_r(t-kT) + \sum_{k=0}^{4}\sum_{l=k}^{4} e_{k,l} y_r(t-kT)y_r(t-lT)$$
$$+ \sum_{k=0}^{4}\sum_{l=k}^{4}\sum_{m=l}^{4} e_{k,l,m} y_r(t-kT)y_r(t-lT)y_r(t-mT) \tag{2.1}$$

$$y_b(q) = \sum_{k=1}^{2} b_k d_{q-k} + \sum_{k=1}^{2}\sum_{l=k+1}^{2} b_{k,l} d_{q-k} d_{q-l} \tag{2.2}$$

We can see that the first terms on the right in equation (2.1) and (2.2) correspond to the conventional FFE-DFE. Nonlinear systems including optical channels together with some distortions such as chromatic dispersion and narrowband optical filtering result in not only linear distortions but also nonlinear distortions, namely, the nonlinear interactions between neighbouring bits. Conventional FFE-DFE does not consider the interactions of the signals with different delay. However, NL-FFE-DFE takes into account these interactions of signals, as the second and third terms shown in (2.1) and the second term shown in (2.2). Therefore, the other terms can be considered as nonlinear terms, which take into account the nonlinear distortion mitigation. For illustration, we show this kind of equalizer setup with the simplest structure NL[2,2]-FFE[1]-DFE[2] in Fig. 2.9. It turns into normal FFE[1]-DFE[2] without the nonlinear parts in the dotted line boxes.

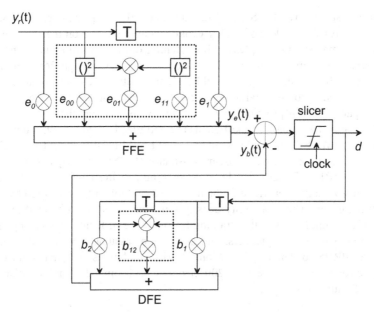

Fig. 2.9. NL[2,2]-FFE[1]-DFE[2] setup. Nonlinear parts with dot line boxes.

2.4.2   System Setups and Parameters

We assume a 10 Gb/s optically pre-amplified system with a NRZ pulse shape. The system setups (transmitter, receiver and channel) for different modulation formats are shown in Fig. 2.10. All modulation formats are generated based on Mach-Zehnder modulator (MZM). ODB is based on a fifth-order Bessel electrical low-pass filter (ELF) with a 3dB-bandwidth of 2.5GHz. A differential encoder is used for both ODB and DPSK to avoid the error propagation. The OSSB signal is generated in two steps. First, the conventional double sideband (DSB) signal is generated (similar to OOK). Then the DSB signal is phase modulated by the Hilbert transformed electrical data signal [24]. The Hilbert transformer (HT) is approximated by a FIR filter [24]. ODB and OSSB share the same receiver with OOK, while DPSK is based on balanced detection by using a Mach-Zehnder delay-interferometer (MZDI). A third-order Butterworth ELF (3dB cut-off frequency 7GHz) is applied after detection for all receivers. ASE noise from EDFA is assumed to be dominant and other noises are omitted. Two Gaussian optical band-pass filter (OBF1 and OBF2) with a same bandwidth of 50GHz are assumed at the transmitter and receiver sides, respectively. The fibre is the standard single mode fibre (SSMF) with only considering chromatic dispersion (dispersion coefficient D=17ps/nm/km).

For the equalizers, we assume the following structures: FFE[6] (6-delay tap FFE), FFE[4]-DFE[2] (4-delay tap FFE and 2-delay tap DFE), NL[3,2]-FFE[4]-DFE[2](4-delay tap FFE and 2-delay tap DFE, nonlinear order 3 and 2 for FFE

36   I. Papagiannakis et al.

and DFE, respectively), MLSE[2], MLSE[3] and MLSE[4] (memory of 2, 3, and 4, respectively). The orders of analog equalizers are assumed on the trade off between performance and complexity. Importantly, our additional simulations have demonstrated that further increased order does not result in significant improvement, especially for FFE and FFE-DFE. Memory of 2 is chosen for MLSE because it is already commercial available, on one hand, and it will be compared to MLSE[3] and MLSE[4] (which are expected to be available in the near future), on the other hand. The equalizer coefficients are optimized based on minimum-mean-square-error (MMSE) for analogy equalizers (FFE, FFE-DFE and NL-FFE-DFE). Adaptive algorithms such as least-mean-square (LMS) [19, 20] and recursive-least-square (RLS) [19,20] can be used to optimize the coefficients. Infinitive resolution is assumed for MLSE, which is based on lookup table method by using Viterbi algorithm. The probability density functions (PDF) are obtained by using Monte-Carlo simulations with a training bit sequence of length. Two samples per bit are assumed for all equalizers. Our simulations assume a pseudo-bit-random-sequence (PRBS) of length 2^{10}-1. Monte-Carlo simulations are carried out to obtain an error count of more than 200 errors at each optical signal-to-noise-ratio (OSNR) for all cases.

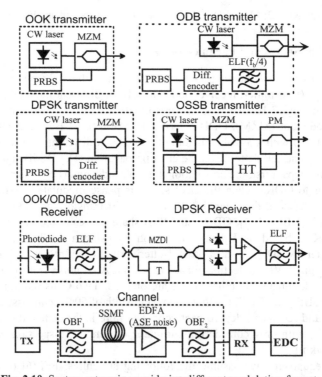

Fig. 2.10. System setups in considering different modulation formats.

2.4.3 EDC Performance for Different Modulation Formats

Based on bit error ratio (BER) simulations with the Monte-Carlo method, we compare the required optical signal-to-noise ratio (OSNR in 0.1nm bandwidth) to achieve a BER of 5×10^{-4}, which is sufficient for error-free operation with forward error correction (FEC).

The results for OOK are shown in Fig. 2.11. The transmission distance is limited to less than 70km at a target OSNR penalty of 3dB and can be extended to about 100km by using either FFE[6] or FFE[4]-DFE[2]. However, both FFE and DFE show negligible performance improvement even with larger delay tap numbers. This

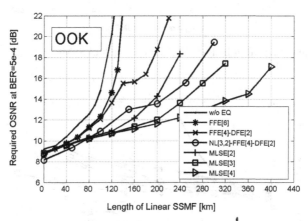

Fig. 2.11. Required OSNR at BER=$5*10^{-4}$ versus length of SSMF with and without using equalization for OOK.

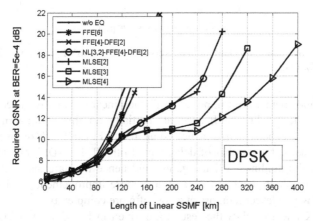

Fig. 2.12. Required OSNR at BER=$5*10^{-4}$ versus length of SSMF with and without using equalization for DPSK.

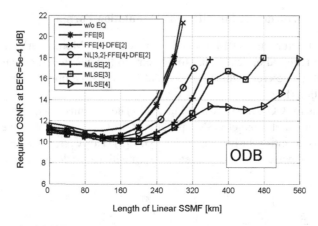

Fig. 2.13. Required OSNR at BER=5*10-4 versus length of SSMF with and without using equalization for ODB.

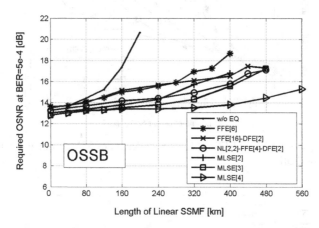

Fig. 2.14. Required OSNR at BER=5*10^{-4} versus length of SSMF with and without using equalization for OSSB.

shows the fundamental limit imposed by the nonlinear square-law detection of the photodiode. The NL-FFE-DFE accounts partially for this nonlinearity and can thus achieve much better performance compared to FFE-DFE. Moreover, we gain here by increasing the number of delay taps. Both NL-FFE-DFE and MLSE take into account the nonlinear ISI mitigation and hence they are less influenced by nonlinearity of photo detection. In addition, more improvement can be achieved for a MLSE with a larger memory at a cost of more complexity.

For DPSK and for ODB the results are given in Fig. 2.12 and Fig. 2.13 respectively. We confirm that DPSK based on balanced detection can achieve about 3dB sensitivity improvement compared to OOK for back-to-back (b2b) operation. As expected, ODB exhibits larger OSNR penalty for b2b. The optimum system performance is reached at transmission distance of around 100km. For DPSK and ODB, all EDC show relatively poor performance, especially at short transmission distances. Both FFE and DFE exhibit nearly no performance improvement. The main reason is the absence of the carrier [21] and thus enhanced nonlinearity of the whole system in ODB and DPSK system further degrade conventional equalization performance. Nevertheless, NL-FFE-DFE and MLSE both outperform FFE-DFE.

For OSSB, results are shown in Fig. 2.14. In order to obtain a good SSB signal, the extinction ratio should be chosen as a trade off between the single side band suppression and the b2b performance. For OSSB, chromatic dispersion results in basically linear distortions after detection and hence EDC performance is generally much less influenced by the square-law detection of photodiode. A simple linear FFE[6] can achieve good performance and extends the transmission distance from about 140km without equalization to about 300km at target of 3dB OSNR penalty. However, due to the non-ideal sideband suppression, the nonlinear distortions resulting from the interference between two sidebands in the presence of chromatic dispersion can not be mitigated efficiently by using linear equalizers. Therefore, best performance can only be achieved by using nonlinear equalizers including NL-FFE-DFE and MLSE.

We have investigated and compared NL-FFE-DFE performance for different modulation formats, respectively. We know that different equalizers show different EDC performance and the equalization performance is also strongly dependent on the modulation format. In order to show clearly these properties, we show in Fig. 2.15 the improved dispersion tolerance due to equalization at an OSNR of 16dB by using different equalizers for the four considered formats.

First of all, for each format, all equalizers show the similar trend: FFE<FFE-DFE<NL-FFE-DFE<MLSE (< means worse). That is to say, MLSE shows always the best performance if only a sufficient memory is assumed. NL-FFE-DFE, which takes into account nonlinear distortion, achieves sub-optimum performance and outperforms conventional FFE and FFE-DFE. FFE and FFE-DFE, which can only mitigate efficiently the linear distortion, shows worst performance, especially for carrier suppressed formats with enhanced nonlinearity such as DPSK and ODB. Secondly, for each equalizer, EDC performance shows the similar trend for the four formats: OSSB>OOK>DPSK>ODB (> means better). In the presence of chromatic dispersion, linear distortions dominate in OSSB while more nonlinear distortions occur for the other three formats, especially for carrier suppressed format DPSK and ODB. That is to say, enhanced nonlinear distortion in DPSK and ODB degrade EDC performance for all the equalizers.

On the one hand, different modulation formats have different dispersion tolerance as well as different receiver sensitivity without equalization. On the other hand, EDC shows different performance improvement for different formats. Therefore, we show in Fig. 2.16 the overall transmission distance at an OSNR of

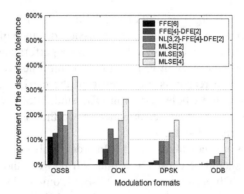

Fig. 2.15. The improvement of the dispersion tolerance due to the equalization for different modulation formats at an OSNR=16dB to achieve a BER of $5*10^{-4}$.

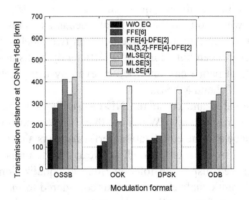

Fig. 2.16. The achievable transmission distance with and without using equalization for different modulation formats at an OSNR=16dB to achieve a BER of $5*10^{-4}$, SSMF-based 10Gb/s transmission.

16dB with equalization for the four formats. Fig. 2.16 shows that relatively, OSSB and ODB can achieve larger transmission distance compared to DPSK and ODB. This can be understood as follows: OSSB has large dispersion tolerance and EDC shows the largest performance improvement for OSSB. ODB has the largest dispersion tolerance without EDC despite of the worst EDC performance. DPSK and OOK have the comparable spectral width and hence the similar dispersion tolerance. DPSK has an advantage of 3dB receiver sensitivity compared to OOK. However, worse EDC performance is observed for DPSK than OOK. Based on these three aspects, OOK and DPSK shows the similar overall transmission distance at a same OSNR.

2.5   Optical Single Sideband Modulation

2.5.1   Compensation of Optical Dispersion

Optical single sideband modulation consists in the suppression of one sideband of any double sideband signal. However, such suppression leads to intensity distortion in the case of direct detection systems. As a consequence, not all modulation formats are suitable to be used together with OSSB. Two modulation formats (unipolar binary and AMI) are considered with nonreturn-to-zero and return-to-zero pulses. For simplicity, modulation formats using OSSB modulation with unipolar binary format and NRZ or RZ pulses are named OSSB-NRZ or OSSB-RZ, respectively. In the case of OSSB modulation with AMI format and RZ pulses, OSSB-AMI-RZ is used to identify such modulation format. Fig. 2.17 presents the eye opening penalty (EOP) as a function of the accumulated fibre dispersion considering OSSB signals with different signaling formats and EDC at the transmitter or receiver sides [25]. As can be seen from Fig. 2.17, using EDC at the transmitter side has a higher efficiency in the mitigation of the dispersion impairment compared to EDC at the receiver side as low increase of EOP is observed in the cases using EDC at the transmitter. It should be point out that EDC at the receiver side is not possible using the AMI format as it is a carrier-suppressed format. However, considering the AMI associated with OSSB modulation and RZ pulses (OSSB-AMI-RZ format), fibre dispersion can be efficiently mitigated after more than 1000 km of SSMF (equivalent to an accumulated dispersion of 17000 ps/nm) with less than 0.5 dB of increase of EOP compared to the back-to-back value.

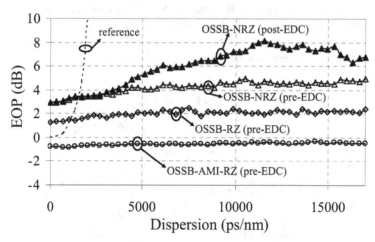

Fig. 2.17. EOP as a function of the accumulated dispersion at 10 Gb/s using electrical dispersion compensation at the transmitter (pre-EDC) or receiver side (post-EDC).

2.5.2   Reduction of Nonlinear Transmission Effects

Besides optical dispersion, optical modulations must also be robust to nonlinear transmission to avoid distortions due to SPM, XPM and FWM. Moreover, using signals with re-distortion leads to an increase of distortion due to nonlinear transmission. To reduce such impairments, OSSB signals using EDC at the transmitter side and concentrated optical dispersion compensation at the receiver side (either EDC after direct detection or a dispersion compensation fibre before direct detection) have been proposed to reduce the impact of the nonlinear effects [26], [27]. The basic ideal is to distribute the compensation of dispersion between the transmitter and receiver sides instead of compensating the accumulated dispersion only at the transmitter or receiver side. In the case of OSSB-AMI-RZ format, it has been found that 50% of the accumulated dispersion of the link should be compensated at the transmitter side (either using EDC or a DCF) whereas the remaining 50% of dispersion should be compensated at the receiver side (using DCF). Fig. 2.18 shows the Q-factor improvements using concentrated EDC at the transmitter side or distributing the dispersion compensation at the transmitter and receiver sides for a transmission system with 9 and 12 sections at 10 Gb/s and no inline dispersion compensation.

As can be seen from Fig. 2.18, the Q-factor obtained considering a distribution of dispersion compensation between the transmitter and receiver sides is always higher than the Q-factor obtained considering the compensation of dispersion concentrated at the transmitter side. In the case of the OSSB-RZ, the analysis has been done in the form of required OSNR to guarantee a specific bit-error-rate at the decision circuit. Fig. 2.19 presents the required OSNR for a transmission system with 8 sections and optimized distribution of dispersion compensation between the transmitter and receiver sides. As with the case of the OSSB-AMI-RZ, a significant improvement (around 3 dB of required OSNR) is observed between the values obtained with concentrated dispersion compensation at the transmitter side and the result obtained using compensation of dispersion at the transmitter and receiver sides.

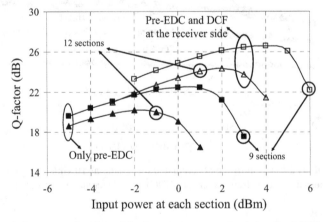

Fig. 2.18. Q-factor as a function of the input power at each section for OSSB-AMI-RZ.

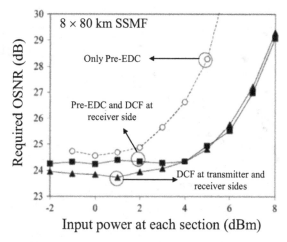

Fig. 2.19. Required OSNR as a function of the input power at each section for OSSB-RZ.

2.6 Enhancing the Performance DML Transmitters

As it was mentioned, in metro/access networks, the cost of transmission systems is a critical factor, closely related with the cost of terminal equipment. In this concept, directly modulated lasers (DMLs) are low cost transmitters that additionally offer low driving voltage, small size and high output power, compared to externally modulated lasers (EML). However, the frequency chirp characteristics of DMLs significantly limit the uncompensated transmission distance over SSMF to less than 10 Km at 10 Gb/s [28], and about 100 Km at 2.5 Gb/s. One of the proposed solutions for the mitigation of the chirp problem was the use of special fibre with negative characteristics [28]. However when SSMF is already deployed, other means of improving the transmission performance of DMLs should be used. Another promising approach was based on the development of special laser modules able even to reach 560 Km at 2.5 Gb/s by using either special DML designs, e.g. buried heterostructure gain coupled DFB semiconductor laser [29], or new techniques, e.g. chirped managed lasers (CMLs) able to achieve longer transmission distances [30], [31]. Research studies carried out within COST 291 have focused on the transmission performance enhancement of common low cost DMLs with the use of electrical processing at the receiver end similar to the use of EDC for EML based systems [32]-[34].

This research activity provided for the first time a complete study to evaluate the performance improvement when EDC is considered for DMLs having different chirp characteristics including both transient and adiabatic chirped DMLs at 2.5 Gb/s and 10 Gb/s [35]. It has been demonstrated for the first time to our knowledge the impact from the use of FFE and DFE on the performance of transient and adiabatic chirped DMLs at 2.5 Gb/s and 10 Gb/s that characterized by

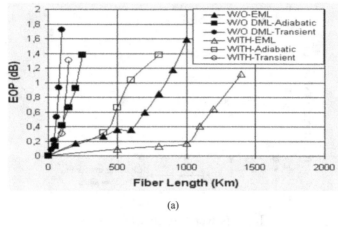

(a)

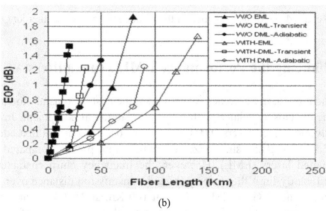

(b)

Fig. 2.20. EOP versus Fibre Length for 2.5 Gb/s (a) and 10 Gb/s (b) without and with equalizer DFE(5,1) [35].

the fact that their extracted parameters have been verified by simulations and experiments [35], [36]. Simulations predicted that 5-tap FFE and 1-stage DFE would be the optimum configuration in terms of performance against complexity. Fig. 2.11 shows the eye opening penalty (EOP) versus the transmission length for DMLs and EMLs at 2.5 Gb/s (Fig. 11(a)) and 10 Gb/s (Fig. 2.11(b)) respectively with and without equalizer of DFE (5,1) [35].

The transmission distance that can be obtained by using adiabatic dominated DMLs is better than transient dominated DMLs. Furthermore, by using EDC on adiabatic DMLs, the improvement on the performance of the system is more pronounced than transient chirp dominated DMLs. Finally, EDC equalized DMLs based systems are particularly attractive for metro/access networking applications at high data rates that require moderate signal improvement and low cost.

2.7 Conclusions

This chapter highlights the research efforts within the COST 291 action studying the effectiveness of electronic equalizers for the mitigation of optical transmission impairments. Different types of equalizers such as FFE, DFE, MLSE and nonlinear equalizers have been analyzed.

The effectiveness of MLSE has been investigated. More specifically, system optimization with the SQRT metric allowed us to find further evidence of the existence of the penalty saturation effect for increasing CD values. It also shown that, given proper filtering optimization, a theoretical penalty as low as 0.6 dB with respect to a back-to-back non-MLSE-RX can be achieved, even for large values of CD. Furthermore, our results demonstrated that MLSE is capable of dealing with the challenging combination of large SPM interacting with whole-link dispersion (13,360 ps/nm). The MLSE system does require about 3 dB higher OSNR than ODC and needs, at high power and accumulated dispersion, a large number of trellis states. On the other hand, the resulting non-linear threshold with MLSE is not far from the best simulated ODC scheme. The latter, in turn, represents an idealized best-case, difficult to implement in practice due to its extremely tight dispersion map tolerance and requiring DCUs and dual-stage EDFAs in each span.

Through the simulations of electrical dispersion compensation for different modulation formats in 10Gb/s optical fibre direct detection systems, we shown that nonlinear equalizers (NL-FFE-DFE) based on Volterra theory can achieve much better performance compared to conventional FFE-DFE and comparable performance to MLSE. NL-FFE-DFE takes into account the mitigation of nonlinear ISI and therefore it is less influenced by the nonlinearity of the square-law detection of the photodiode. Moreover, four kinds of equalizers including NL-FFE-DFE are compared comprehensively for four kinds of modulation formats. Different equalizers show different performance improvement and EDC performance is strongly dependent on the modulation format.

Furthermore, it was shown the effectiveness of EDC for the mitigation of chromatic dispersion and nonlinear transmission effects for optical single sideband modulation. Finally, DMLs have been primarily introduced for access network applications and their combination with electronic equalization has improved the performance of the system and makes them attractive for metro and access applications.

References

1. Proakis, J.G., Salehi, M.: Communication Systems Engineering. Prentice-Hall, Englewood Cliffs (1994)
2. Haykin, S.: Adaptive Filter Theory, 3rd edn. Prentice-Hall, Englewood Cliffs (1996)
3. Fonseca, D., Cartaxo, A., Monteiro, P.: On the use of electrical pre-compensation of dispersion in optical single sideband transmission systems. IEEE J. of Selected Topics in Quantum Electronics 12(4), 603–614 (2006)

4. Poggiolini, P., Bosco, G., Prat, J., Killey, R., Savory, S.: Branch Metrics for Effective Long-Haul MLSE. In: Proc. of ECOC 2006, Sept. 24-29, 2005, paper We2.5 (2006)
5. Bosco, G., Poggiolini, P.: Long-Distance Effectiveness of MLSE IMDD Receivers. IEEE Photon. Technol. Lett. 18(9), 1037–1039 (2006)
6. Franceschini, M., Bongiorni, G., Ferrari, G., Raheli, R., Meli, F., Castoldi, A.: Fundamental limits of electronic signal processing in direct-detection optical communications. J. Lightw. Technol. 25(7), 1742–1753 (2007)
7. Alic, N., et al.: Experimental demonstration of 10 Gb/s NRZ extended dispersion-limited reach over 600 km-SMF link without optical dispersion compensation. In: Proc. of OFC 2006, Mar. 5-10, 2007, paper OWB7 (2007)
8. Chandrasekhar, S., et al.: Chirp-Managed Laser and MLSE-RX enables Transmission Over 1200 km at 1550 nm in a DWDM Environment in NZDSF at 10 Gb/s Without Any Optical Dispersion Compensation. IEEE Phot. Technol. Lett. 18(14), 1560–1562 (2006)
9. Crivelli, D.E., Carrer, H.S., Hueda, M.R.: On the performance of reduced-state Viterbi receivers in IM/DD optical transmission systems. In: Proc. of ECOC 2004, Stockholm, Sweden, 5-9 Nov. 2004, We4.P.083 (2004)
10. Poggiolini, P., et al.: 1,040 km uncompensated IMDD transmission over G.652 fibre at 10 Gbit/s using a reduced-state SQRT-metric MLSE receiver. In: Proc. of ECOC 2006, Sep. 24-29, 2006, post-deadline paper Th4.4.6 (2006)
11. Otte, S.: Nachrichtentheoretische Modellierung und elektronische Entzerrung hochbitratiger optischer Uebertragungssysteme. Dissertation, University of Kiel (2003)
12. Xia, C., Rosenkranz, W.: Performance enhancement for duobinary modulation through nonlinear electrical equalization. In: ECOC 2005, Glasgow, paper Tu4.2.3, vol. 2, pp. 257–258 (2005)
13. Xia, C., Rosenkranz, W.: Mitigation of optical intrachannel nonlinearity using nonlinear electrical equalization. In: ECOC 2006, 24-28 September 2006, Cannes, French, paper We3.P.244 (2006)
14. Winzer, P.J., Chandrasekhar, S., Kim, H.: Impact of filtering on RZ-DPSK reception. IEEE Photonics Technology Letters 15(6), 840–842 (2003)
15. Gnauk, A.H., Winzer, P.J.: Optical phase-shift-keyed transmission. Journal of Lightwave Technology 23(1), 115–129 (2005)
16. Xie, C., Moeller, L., Ryf, R.: Improvement of optical NRZ- and RZ-duobinary transmission systems with narrow bandwidth optical filters. IEEE Photonics Technology Letters 16(9), 2162–2164 (2004)
17. Lyubomirsky, I., Pitchunani, B.: Impact of optical filtering on duobinary transmission. IEEE Photonics Technology Letters 16(8), 1969–1971 (2004)
18. Kim, H., Yu, C.X.: Optical duobinary transmission system featuring improved receiver sensitivity and reduced optical bandwidth. IEEE Photonics Technology Letters 14(8), 1205–1207 (2002)
19. Zaknich, A.: Principles of Adaptive Filters and Self-Learning Systems, pp. 257–265. Springer, Heidelberg (2005)
20. Proakis, J.G.: Digital Communications, 4th edn., pp. 660–708. McGraw-Hill, New York (1995)
21. Rosenkranz, W., Xia, C.: Electrical equalization for advanced optical communication systems. AEU - International Journal of Electronics and Communications 61(3), 153–157 (2007)

22. Xia, C., Rosenkranz, W.: Nonlinear electrical equalization for different modulation formats with optical filtering. IEEE/OSA Journal of Lightwave Technology 25(4), 996–1001 (2007)
23. Xia, C., Rosenkranz, W.: Electrical mitigation of penalties caused by group delay ripples for different modulation formats. IEEE Photonics Technology Letters 19(13), 954–956 (2007)
24. Sieben, M., Conradi, J., Dodds, D.E.: Optical single sideband transmission at 10Gb/s using only electrical dispersion compensation. Journal of Ligthwave Technology 17(10) (1999)
25. Fonseca, D., Cartaxo, A., Monteiro, P.: Highly efficient electrical dispersion compensation scheme for optical single sideband systems. In: Proc. IEEE Lasers and Electro Optics Society Annual Meeting (LEOS), Sydney, Australia, Oct. 2005, pp. 898–899 (2005)
26. Fonseca, D., Luis, R., Cartaxo, A., Monteiro, P.: Near pseudo-linear transmission regime in 10 Gb/s single sideband-alternate mark inversion systems using electrical dispersion pre-compensation. IEEE Photon. Technol. Lett. 19(15), 1127–1129 (2007)
27. Luis, R., Fonseca, D., Teixeira, A., Monteiro, P.: Dispersion management of electrically pre-compensated RZ single-sideband signals at 10 Gb/s without inline dispersion compensation. IEEE Photon. Technol. Lett. 19(14), 1039–1041 (2007)
28. Tomkos, I., et al.: 10-Gb/s Transmission of 1.55-μm Directly Modulated Signal over 100 Km of Negative Dispersion Fibre. IEEE Photon. Technol. Lett. 13(3), 735–737 (2001)
29. Nelson, L., Woods, I., White, J.K.: Transmission over 560 Km at 2. 5 Gb/s using a directly modulated buried heterostructure gain-coupled DFB semiconductor laser. In: OFC 2002, pp. 422–423.
30. Mahgerefteh, D., Cho, P.S., Goldhor, J., Mandelberg, H.I.: Penalty-free propagation over 600 Km of non-dispersion-shifted fibre at 2.5 Gb/s using a directly laser modulated transmitter. In: IEEE CLEO, May 1999, vol. 2, p. 182 (1999)
31. Mahgerefteh, D., Liao, C., Zheng, X., Matsui, Y., Johnson, B., Walker, D., Fan, Z.F., McCallion, K., Tayebati, P.: Error-free 250 Km transmission in standard fibre using compact 10 Gbit/s chirp-managed directly modulated lasers (CML) at 1550 nm. Electron. Lett. 41(9), 543–544 (2005)
32. Winters, J.H., Gitlin, R.D.: Electrical Signal processing techniques in long-haul fibre-optic systems. IEEE Trans. Comm. 38(6), 1439–1453 (1990)
33. Bulow, H.: PMD mitigation by optic and electronic signal processing. In: IEEE LEOS, Nov. 2001, vol. 2, pp. 602–603 (2001)
34. Watts, P.M., et al.: Performance of single mode fibre links using electronic feed forward and decision feedback equalizers. IEEE, Photon. Technol. Lett. 17(10) (2005)
35. Papagiannakis, I., et al.: Performance of 2.5 Gb/s and 10 Gb/s transient and adiabatic chirped directly modulated lasers using electronic dispersion compensation. In: Proc. ECOC 2007, Berlin, Germany (2007)
36. Tomkos, I., et al.: Extraction of laser rate equations parameters for representative simulations of metropolitan-area transmission systems and networks. Opt. Comm. 194, 109–129 (2001)

3 Optical Signal Processing Techniques for Signal Regeneration and Digital Logic

K. Ennser (chapter editor), S. Aleksic, F. Curti, D.M. Forin, M. Galili,
M. Karasek, L.K. Oxenløwe, F. Parmigiani, P. Petropoulos, R. Slavik,
M. Spyropoulou, S. Taccheo, A. Teixeira, I. Tomkos, G.M. Tosi Beleffi,
and C. Ware

Abstract. This chapter presents recent developments in optical signal
processing techniques and digital logic. The first section focuses on techniques to obtain key functionalities as signal regeneration and wavelength
conversion exploiting nonlinear effects in high nonlinear fibres and semiconductor optical amplifiers. The second section covers techniques for
clock recovery and retiming at high-speed transmission up to 320 Gb/s. In
addition a technique to obtain OTDM demultiplexing based on cross-phase
modulation is reported.

3.1 Optical Regeneration and Wavelength Conversion

3.1.1 640 Gbit/s Wavelength Conversion Based on XPM in HNLF

The single channel bit rate has continuously increased in deployed optical transmission systems and networks, reaching 10 – 40 Gbit/s in today's commercially
available systems. With the appearance of new technologies for optical transmitters and receivers operating near 100 Gbit/s [1], ultra fast signal processing becomes
increasingly relevant. At such high bit rates optical signal processing must be considered as a useful supplement to electronic processing. Several signal processing
tasks must be addressed in high speed communication systems and networks, including the indispensable wavelength conversion of data signals. Several approaches based on non-linear effects in optical fibres as well as semiconductor
structures have been investigated and two wavelength conversion set-ups have
been demonstrated up to 320 Gbaud symbol rates [2-3].

In this subsection, wavelength conversion by cross-phase modulation (XPM) in
highly non-linear fibre (HNLF) is demonstrated for a 640 Gbit/s (640 Gbaud
OOK) single channel, single polarization optical time division multiplexed (OTDM)
data signal. This constitutes the highest reported operating speed of a wavelength
converter to date. Error free conversion is achieved for all channels, and the best-
case penalty in receiver sensitivity is only 2.9 dB compared to the original 640
Gbit/s data signal.

I. Tomkos et al. (Eds.): COST 291 – Towards Digital Optical Networks, LNCS 5412, pp. 49–96, 2009.

3.1.1.1 Experimental Procedure

The experimental set-up is shown in Fig. 3.1. The optical signal is generated by an erbium glass oscillator pulse generating laser (ERGO-PGL) with a pulse repetition rate of 10 GHz and a wavelength of 1557 nm. The pulses are data modulated with a 2^7-1 PRBS in a Mach-Zehnder modulator (MZM) and subsequently multiplexed to 40 Gbit/s in a passive fibre delay 2^7-1 PRBS maintaining multiplexer (MUX). The 40 Gbit/s data pulses are then chirped and spectrally broadened by Self Phase Modulation (SPM) in 400 m of dispersion flattened highly non- linear fibre (DF-HNLF, $\gamma \sim 10$ W^{-1}km^{-1}, dispersion D = -1.2 ps/nm/km at 1550 nm and a dispersion slope of 0.003 ps/nm^{2}km – kindly provided by OFS Fitel Denmark). The positive dispersion in the remainder of the transmitter linearly compresses the data pulses to ~560 fs FWHM in the resulting 640 Gbit/s data signal. The signal is amplified by an EDFA to ~28 dBm and combined with a ~25 dBm CW at 1544 nm before injection into 200 m of HNLF ($\gamma \sim 10$ W^{-1}km^{-1}, zero dispersion at 1552 nm and a dispersion slope of 0.018 ps/nm^{2}km – kindly provided by OFS Fitel Denmark). The CW is phase modulated at 100 MHz to reduce Stimulated Brillouin Scattering (SBS). A counter-propagating 800 mW Raman pump enhances the wavelength conversion in the HNLF. The sidebands on either side of the CW are temporally off-set and it is thus necessary to select only one sideband to form the wavelength converted signal [2]. This is done using a Fibre Bragg Grating (FBG) as a notch filter to suppress the CW and part of one XPM sideband. A 9 nm band pass filter is used to further isolate the converter output, by suppressing the original data signal. The FBG has its centre wavelength at 1545.5 nm, a bandwidth of 3.2 nm and a suppression of ~40 dB.

The wavelength converted signal is demultiplexed to the 10 Gbit/s base rate in a non-linear optical loop mirror (NOLM) using 780 fs control pulses generated by adiabatic soliton compression of a 10 GHz pulse train (from a second ERGO PGL) in a dispersion decreasing fibre (DDF). The NOLM comprises 50 m of HNLF with the same fibre parameters as the one used for XPM wavelength conversion. Bit error rate (BER) measurements of the 10 Gbit/s OTDM tributaries are performed to evaluate the system performance.

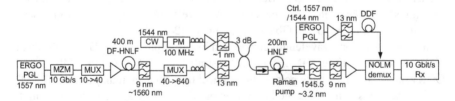

Fig. 3.1. Setup for 640 Gbit/s XPM wavelength conversion.

3.1.1.2 *Experimental Results*

Fig. 3.2(a) shows the optical spectrum of the wavelength converted data signal as well as the original 640 Gbit/s signal and the CW probe at the input to the HNLF. In the HNLF, the CW is spectrally broadened by the co-propagating data pulses in the original signal through XPM. In this way spectral sidebands are generated on the CW probe, which reflect the data logic of the original data signal. 640 GHz spectral components are clearly visible after conversion, as the wavelength converted signal has adopted the phase properties of the CW probe signal, giving a stable phase relationship between consecutive pulses in the wavelength converted OTDM data signal [4].

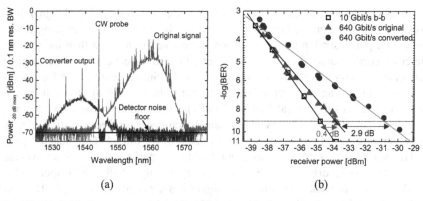

(a)   (b)

Fig. 3.2. (a) Optical spectrum of the 640 Gbit/s signal before and after wavelength conversion. (b) BER measurements for the converted and original 640 Gbit/s data signals.

Fig. 3.2(b) shows BER results for the 640 Gbit/s original data signal and for the wavelength converted signal when demultiplexed down to 10 Gbit/s. The 640 Gbit/s wavelength conversion is successful with error free performance. For both the original and the converted 640 Gbit/s signal, error free performance (defined as BER<10^{-9}) with no sign of an error floor and low penalty compared to the 10 Gbit/s back-to-back (b-b, measured straight out of MZM data modulator) is obtained. In Fig. 3.2(b), a typical channel is shown for the original 640 Gbit/s data signal (channel 64 in Fig. 3.4), whereas the BER curve for the converted 640 Gbit/s signal corresponds to one of the best performing channels in the converted signal (channel 23 in Fig. 3.4), having a conversion penalty of only 2.9 dB. In order to characterize all 64 channels in the OTDM signal, cross-correlations with a 500 fs sampling pulse were performed. In this way pulse amplitudes and pulse widths of each individual channel is investigated.

Fig. 3.3 shows cross-correlations of the 640 Gbit/s original data signal together with the converted 640 Gbit/s signal. The average pulse width of the two signals is measured on an auto-correlator to be ~560 fs before conversion and 660 fs after conversion. This increase in pulse width is seen to cause a reduction in contrast in

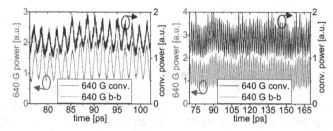

Fig. 3.3. Cross-correlations – Lower traces: 640 Gbit/s original data (FWHM ~560 fs). Upper traces: 640 Gbit/s converted data (FWHM ~660 fs). (a) zoom in on 16 channels (b) all 64 channels.

the cross-correlations of the converted signal compared to the original signal. It is expected that the contrast is mainly limited by the width of the sampling pulse, and actual pulse overlap after conversion is believed to me less than indicted by the cross-correlations. There is a ~1 dB amplitude difference among the original data channels and among the converted channels. Subsequently, all 64 channels are demultiplexed and subject to BER measurements. Fig. 3.4(b) shows the measured receiver sensitivities (at BER=10^{-9}) of all 64 channels in the converted and the original signals. All 64 converted channels achieve error free operation, clearly demonstrating successful wavelength conversion of the full 640 Gbit/s data signal.

The original OTDM signal has a good and even performance with an average sensitivity of -33.3 dBm, i.e. an average penalty of only 1.5 dB with respect to the 10 Gbit/s back-to-back. The converted signal has a penalty of only ~3 dB under optimised conditions with respect to the 10 Gbit/s back-to-back. This penalty is mainly believed to be caused by the pulse broadening associated with the wavelength conversion. The pulse broadening is in turn caused by the filter configuration used to extract the output signal from the wavelength converter. There is a ~ 6.5 dB sensitivity spread due to channel variations in the wavelength converted signal. On top of this an ambient drift affected the system for the duration of the meas-

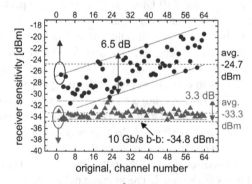

Fig. 3.4. Receiver sensitivity at a BER of 10^{-9} for all OTDM channels in the converted and original 640 Gbit/s signals.

urement, resulting in a slowly deteriorating sensitivity, yielding an average receiver sensitivity of -24.7 dBm. A significantly more stable performance of the converter is thus expected, if the impact of ambient variations can be reduced.

3.1.1.3 Conclusion

We demonstrated XPM based wavelength conversion of a 640 Gbit/s data signal. This constitutes the highest bit-rate reported in a wavelength conversion demonstration to date. The wavelength conversion allows error free operation of all tributary channels in the converted signal. Low penalty from wavelength conversion, compared to the input signal, is achieved in an optimized configuration of the converter.

3.1.2 Wavelength Conversion and Regeneration Based on Supercontinuum Generation

Nowadays that high capacity and flexible dense wavelength division multiplexing systems are commercially available. The evolution will probably be towards all-optical networking configuration where routing and switching will be performed in the optical domain [5-6]. Network scalability faces a major limitation due to noise accumulation from optical amplifiers, chromatic dispersion, crosstalk, jitter and nonlinearities. In fact the signal propagating in the All-Optical domain deteriorates determining a restriction to the maximum path length and the number of nodes through which signal can pass. This is to avoid the signals to suffer a fatal degradation in the all-optical network. In addition network aggregate capacity will increase demanding for more and more transmitted wavelength, also beyond the wavelength interval of standard C-band. The Network of the Future will need a key device: a λ-converter able to operate extremely wide conversion, 80 nm and more with limited power consumption and regeneration properties.

Several approaches have been propose, based on non linear effects in semiconductor optical amplifiers (SOA) [7] and in optical fibres (OF) [8-11]. All-fibre signal processing systems are very promising for their intrinsic robustness, format and bit rate independence, however some limitations are intrinsic: the use of four wave-mixing does not allow for carrier wavelength tuning far away from the zero dispersion wavelength where the phase matching between the interacting electromagnetic fields is maximum and is also polarization dependent [8]; the cross-phase-modulation (XPM) between signals and continuous wave (CW) carriers does not result directly in amplitude modulation [9] and the generation and subsequent slicing of SC [10] becomes critical for very large spectral bandwidth due to fibre non uniformity, is not efficient in the marks noise compression and requires power that scales as the square of the desired bandwidth [11-12].

On the other hand, increasing capabilities in the wavelength conversion and regeneration based on the adoption of auxiliary carrier has been demonstrated in a dispersion shifted fibre due to a non linear multi wave mixing effect [13]. XPM between a CW signal and a 10 GHz clock pulse train in a Dispersion Shifted fibre has been recognized as suitable solution for wavelength conversion implementing in WDM systems [14].

Adopting a high non linear dispersion flattened fibre (HNL-DFF), with constant normal dispersion, the position of AUX carrier wavelength in not important and the bandwidth is almost unlimited. The modulation of AUX carrier is made by phase-modulation induces by refractive index change due to strong signal field. The effectiveness of this process is enhanced by similar propagation velocity as in HNL-DFF. In addition in normal dispersion regime the signal pulse broadens in time and it may help to enhance conversion quality because it improves position synchronization. Furthermore, it will generate stable modulation avoiding modulation instability and random-shape output spectrum due by interplay between the coherent pulse and the amplified spontaneous emission background [15]. The converted signal will be consequently benefit for a 2R regeneration process.

3.1.2.1 Experimental Set-up

Fig. 3.5 shows the experimental set-up. The pulse train is modulated with a Lithium-Niobate Mach-Zehnder intensity modulator at 10 Gbit/s pseudo random bit sequence (PRBS) 2^{31}-1. The subsequent attenuator, which reduces the signal to noise ratio (SNR), simulates transmission degradations and allows us to verify the systems reshaping capabilities. This ps signal has been then processed by a block made up of an optical preamplifier (EDFA1), a polarization controller (PC1) and an high power optical amplifier (EDFA2, max output power 1W) that increases the power of the pulse up to the power necessary for the SC generation. A CW signal, the auxiliary carrier, generated by an external cavity laser (ECL) has been coupled with the pulse train and injected into the HNL-DF fibre. The output optical spectrum has been then filtered using a demultiplexer (DeMux) with optical bandwidth equal to 0.4 nm. As the HNL-DF fibre showed strong polarization dispersion, the PC1 has been necessary to select one of the fibre principal states of polarization (SOP). The PC2, otherwise, selected the correspondent output SOP subsequently filtered by means of a Polarizer (Pol). As a non linear medium we used in our experiments a 1 km long HNL-DF fibre with dispersion coefficient $D = -6$ psnm^{-1}km^{-1} in the 1500nm-1580nm range, non-linear coefficient $\gamma = 14.8$ W^{-1}km^{-1}, dispersion slope $S = 0.004$ pskm^{-1}nm^{-2} and attenuation $\alpha = 3.3$ dBkm^{-1} as illustrated in Fig. 3.5.

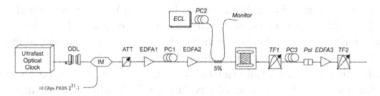

Fig. 3.5. Experimental set-up.

The extremely low S/D ratio guarantees spectrum symmetry [12] and the high D value help to avoid impact of fibre non uniformity [11]. The high γ value compensates for high D value and γ/D ratio allows for reasonable low power consumption. An ultra fast optical source generates a pulses train, with FWHM≈4–5 ps estimated from an autocorrelation trace, with a repetition rate of 10GHz.

3.1.2.2 Simulations

Fig. 3.6 shows several simulations. The numerical analysis has been performed with a commercial simulation package OptiSystem 4.0. The simulation relies on the split-step FFT solution of the non-linear Schrödinger equation and has been designed to correspond to the experimental setup presented in Fig. 3.5. The adopted pulse train has been a continuous first-order super Gaussian pulses train at λ_{signal} =1554.13 nm (ITU frequency grid channel 34) with FWHM = 5.5 ps and operating at 10 GHz repetition rate. As in the Fig. 3.5 this train has been amplified in a high-power EDFA, coupled with an auxiliary CW carrier and launched into the HNL-DF fibre. Amplified spontaneous emission generated by the EDFA has been suppressed in an OBPF (3-poles Gaussian filter with FWHM = 16 nm).

The spectrum at the HNLF output was sliced with an OBPF (fifth order super Gaussian filter, FWHM = 200GHz), detected by an PIN photodiode and the quality of the signal was evaluated by virtual OptiSystem electrical Eye Diagram Analyzer.

The CW carrier wavelength and power were set at 1570nm and 13dBm, respectively. The spectrum at the output of the HNL-DF fibre calculated for pulse train input power of 30 dBm is shown in Fig. 3.6(a). The inset, top right side, plots the zoomed part of the output spectrum resulting from the XPM between the SC

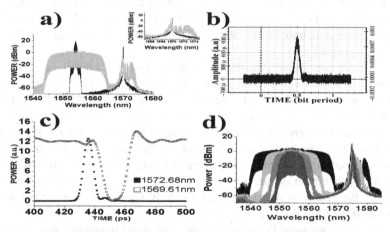

Fig. 3.6. Simulations results: a) HNL-DF input (black) and ouput spectra (grey) b) modulated Aux carrier at 1572.68 nm c) Phase modulation effect. Filled squares Aux detuned filtering, Not filled square Aux centered filtering d) Output spectrum for pulse train input power of 24 dBm, 27 dBm, 30 dBm and 32 dBm .

(grey) and CW Aux carrier (black). The modulation of the carrier, with the presence of the lateral bands [17], occurs in both the cases proposed (Fig. 3.7(a) 1535 nm and Fig. 3.7(b) 1570 nm). As it can be noticed, side-lobes of the XPM generated spectrum exhibit distinguished local maxima. When the OBPF was tuned at the wavelength of these maxima, e.g. at 1572.68 nm, clean eye diagram was obtained as is apparent from Fig. 3.6(b). The presence of a phase modulation effect, in our case XPM, is confirmed by the depletion of the carrier observed by centering the optical output filter at the auxiliary carrier wavelength, infact, when the filter is tuned close to the Aux CW original wavelength (e.g. at 1569.31 nm) an inverted signal is obtained, see Fig. 3.6(c) .Fig. 3.6(d) demonstrates the effect of the average signal power on the spectrum at the HNL-DF output for λ_{cw}= 1575 nm, and P_{cw}=21 dBm. The simulations results demonstrate that the power level of the side-lobes increases with P_{cw} increase, which is reflected in the eye noise level at marks and zeros, while favourable results, in terms of eye opening, have been obtained by setting the CW Aux carrier wavelength as far as ± 25 nm away from the signal wavelength (1530nm, 1580nm). It can be deduced that with an ideal HNL-DF fibre having zero dispersion slope, an infinite detuning can be achieved. In Fig. 3.7 we show the experimental optical spectra at the HNL-DF output. Note that the auxiliary carriers, 1535 nm and 1570 nm, are placed outside of the SC spectrum.

3.1.2.3 Experiments

Fig. 3.7(a) and 3.7(b) shows experimentally the output spectrum when the carrier is placed blue-or red-side with respect to signal wavelength. Note that the Auxiliary Carriers are placed outside of the SC spectrum. The modulation of the carriers, deduced by the presence of the lateral bands in the Fig. 3.6, occurs in both the case proposed tuning the aux carrier at 1535 and 1570 nm. In the experiment reported here the Auxiliary Carrier, tuned at 1535,9 nm, is filtered amplified and filtered again (Fig.3.5).

We see that the phase-modulation induced by the Super continuum generation enlarge the spectrum of the AUX, so as show in Fig. 3.7(d), filtering the signal at the wavelength of the AUX carrier we see an inverted modulation, Fig. 3.7(d) with a slightly detuning of the filtering, we archive a signal that reproduce the modulation of the original signal. Moreover employing an appropriate control of polarization for the input and for the output field, we can by means of the polarizer bring a further enhance in the quality of the signal and obtain the signal reported in Fig. 3.10(c). The optimum detuning for TF1 and TF2 were determined in order to maximize the quality of the regenerated signal. In Fig. 3.8 we can see the output spectrum of the filtered signal.

In Fig. 3.9(a) we show the eyes of the original signal modulated and attenuated entering the high non linear fibre input; In Fig. 3.9(b) we show the eye diagram of a slice of the Super-Continuum obtained by means of the Tunable filter; in Fig. 3.9(c) is reported the eyes of the Aux Modulated by the SC generation and filtered by two slightly detuned filter and by the Polarizer.

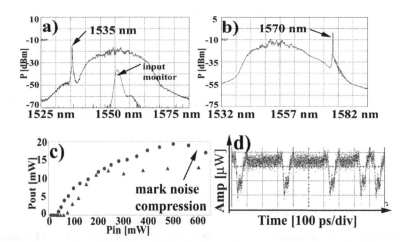

Fig. 3.7. Experimental results: a) Output spectrum with 1535 nm Aux carrier b) Output spectrum with 1570 nm Aux. c) Pin-Pout transfer function of SC slice (triangles) and Aux carrier (dots). d) Carrier depletion by Aux wavelength center filtering.

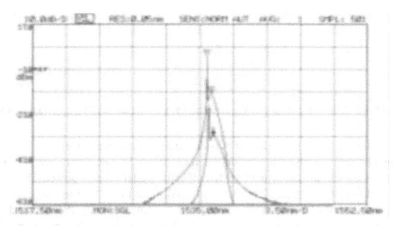

Fig. 3.8. Optical Spectrum of the Modulated Auxiliary carrier filtered by a slightly detuned filter.

As evident from Fig. 3.9(c), detuning TF1 and TF2 all optical reshaping by means of noise compression is demonstrated. The reshaping capabilities of this experimental device are confirmed by in-out power characteristics, due to the phase modulation effect that is the base of such kind of aux carrier based devices [18].

Fixing by means of EDFA1 and EDFA2 the mean peak power of the signal to set the working point near this maximum we obtain a reduction of the noise variance of the marks grater than the reduction achievable by the simple filtering from the Super Continuum.

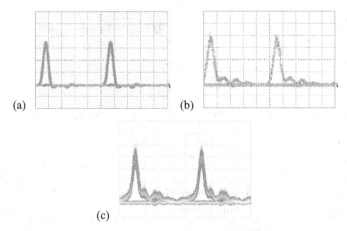

Fig. 3.9. Eyes of (a) the picosecond clean input signal, (b) the sliced SC, (c) the Modulated Aux Carrier (1535nm) Timescale is 20ps/Div for all the insets.

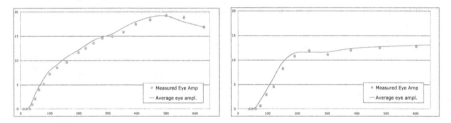

Fig. 3.10. Pin/Pout characteristics of the modulated aux carrier and the Super continuum slice.

It is possible to shift to wavelength outside the SC spectrum and therefore we have strong advantages with λ-conversion based on simply SC spectrum slicing. In fact the SC bandwidth increases as $P^{1/2}$, where P is the signal average power. Therefore to extend our SC up to 1557 nm we will need more than 10 times more power. Thus, the described wavelength converter has, compared with the standard SC spectral slicing approach, better regeneration transfer function, less power needed (20 times less), virtual infinite operational wavelength range and further two order of power reduction can be achieved by using optimize DF microstructure fibres with normal dispersion above 0.5 ps/nm/km and higher non linear coefficient.

3.1.3   Multi-Wavelength Conversion at 10 Gb/s and 40 GHz Using a Hybrid Integrated SOA Mach-Zehnder Interferometer

3.1.3.1   Experimental Setup and Results – 10GE

The experimental set-up of 10 Gb/s wavelength converter is shown in Fig. 3.11. Signal source was a 10 GE long-reach DWDM XFP transceiver. Wavelength of the transmitter was 1552.52 nm (channel #31 according to ITU 100 GHz grid), output power $P_s = 1.8$ dBm. The transceiver was inserted in bit-error ratio tester Sunrise Telecom STT Ethernet module, nominal receiver sensitivity was -24 dBm. Six CW wavelengths (1546.92 – 1550.92 nm, channels 38 -33) for the multi-cast channels were derived from ILX 7900 WDM tester, multiplexed in AWG multiplexer and launched in co-propagation direction with the 10 GE signal. SOA-MZI (CIP 40G-2R-ORP) was operated in standard configuration with signal launched to only one arm (A) which induced phase shift on the six CWs sent to arm C due to XPM. The Mach-Zehnder interferometer converts phase modulation to amplitude modulation.

The MZI was first balanced with CWs using the PC_2 and one of the thermo-optic phase shifters by monitoring output power at ports F and G. Extinction ratio better than 15 dB has been achieved. The hybrid SOA-MZI module was kept at constant temperature of 23°C while SOA_1 and SOA_2 were biased equally at 225 mA. Fig. 3.12 plots the optical spectrum at the input port C and the output port G of the SOA-MZI. Input power of CWs was -3 dBm/channel, optical spectrum analyzer (OSA) was connected to the destructive interference arm (G) of the MZI. Power of converted channels, $P_{con}(\lambda_{cw})$, at output port F was 15 dB higher than that shown in Fig. 3.12. Conversion gain defined as $P_{con}(\lambda_{cw})/P_s$ was wavelength independent and equal to ≈ 8.5 dB. Due to the SOA nonlinear effects, product of FWM appear on both sides of the converted channels. The strongest parasitic FWM products are 30 dB below the converted channels.

Output from arm F of the SOA-MZI was launched into 55 km of NZ-DSF fibre (zero dispersion wavelength 1458 nm, chromatic dispersion at 1550 nm $D = 4.2$ ps/nm/km). Chromatic dispersion of the link (231 ps/nm) has not been compensated. After demultiplexing (AWG designed for 100 GHz channel spacing) the FWM products are suppressed by additional 15 dB, as demonstrated in Fig. 3.13, where optical spectra recorded at individual output ports of the AWG

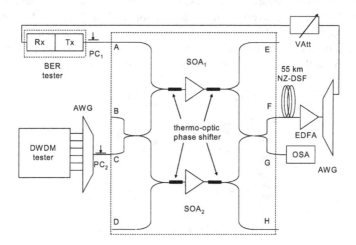

Fig. 3.11. Schematic diagram of 10 Gb/s experimental set-up.

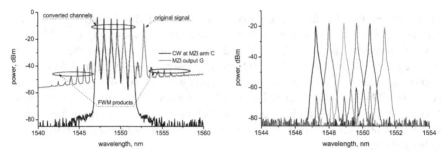

Fig. 3.12. Optical spectrum at the C port and the G port of SOA-MZI.

Fig. 3.13. Output spectrum recorded at individual output ports of the AWG demultiplexer.

demultiplexer are shown. Power of the 6 wavelength-converted signals was almost the same and equal to -17 dBm.

For the purpose of measuring the bit error ratio (BER) as a function of signal level for the wavelength-converted channels, the composite signal at the fibre end was amplified in an erbium-doped fibre amplifier (EDFA), demultiplexed and level was adjusted by variable optical attenuator, VAtt. Eye diagram of the converted signal at 1547.72 nm recorded at AWG output is plotted in Fig. 3.14.

Bit error ratio of individual converted channels was measured by the Sunrise Telecom STT Ethernet module as a function of receiver input power varied with variable attenuator VAtt. The BER tester was operated as Layer 2 tester in loop back regime. We used the 2^{31} test pattern with the longest frame size of 1518 bytes and performed each measurement for 20 minutes. This represents $\approx 10^9$ of transmitted frames. The tester detects the number of lost or badly received frames and their ratio to transmitted frames is evaluated as frame error.

Fig. 3.15 plots the logarithm of frame error ratio as a function of receiver input power of 4 of the 6 converted channels and for back to back configuration. It should be mentioned that lower than 10^{-10} frame error ration can not be achieved. Power penalty due to wavelength conversion and transmission over 55 km of NZ DSF is 2 dB for the best performing channel at 1550.92 nm.

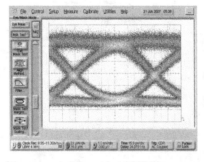

Fig. 3.14. Eye diagram of the converted channel at 1547.72 nm recorded at the receiver input.

3.1.3.2 Experimental Setup and Results – 40GHz

The experimental set-up of 40 Gb/s wavelength converter is shown in Fig. 3.16. A continuous pulse stream of sech2 shaped pulses with FWHM ≈ 3ps and repetition rate of 10 GHz generated in PMLLD is time division multiplexed in an OTDM to 40 Gb/s. The signal is then split in two by a 50:50 directional coupler to form control (or clock) pulses in push-pull (or differential) phase modulation configuration of the SOA MZI.

Variable optical delay line, VODL in arm A provides tight control over the differential time delay between the push and pull pulses distributed to arms A and D. The 2 CW channels (DFB LD_1, LD_2) were combined by a 50:50 directional coupler and launched in arm B of the SOA-MZI. SOA_1 and SOA_2 were biased at 250 mA. Composite signal from arm F was filtered by tunable optical band pass

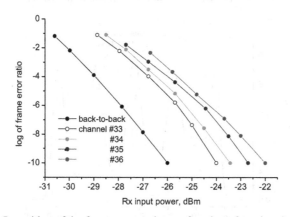

Fig. 3.15. Logarithm of the frame error ratio as a function of receiver input power.

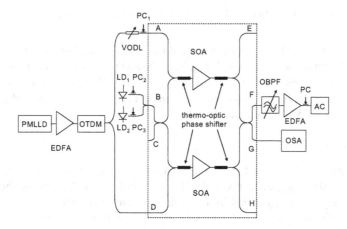

Fig. 3.16. Schematic diagram of 40 Gb/s experimental set-up.

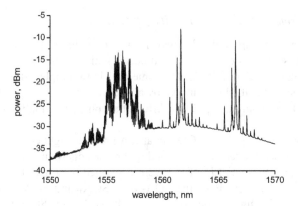

Fig. 3.17. Optical spectrum at the input of the OBPF.

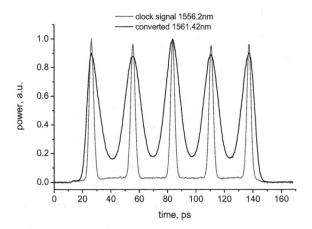

Fig. 3.18. Autocorrelation trace of the clock signal and of the converted signal at 1561.42 nm

filter OBPF (FWHM = 0.9 nm), amplified by an EDFA and autocorrelation trace recorded by an autocorrelator AC. Fig. 3.17 shows optical spectrum at the input of the OBPF when PMLLD was tuned at 1556.2nm, LD_1 and LD_2 at 1561.4 and 1566.3 nm, respectively. Autocorrelation trace of the clock signal and the converted signal at 1561.42 nm are shown in Fig. 3.18.

3.1.3.3 Conclusion

Using commercially available hybrid SOA-MZI we performed wavelength conversion of 10 GE signal to six 100 GHz spaced channels. Quality of the newly generated signals was verified by transmitting them over 55 km of NZ DSF and measuring the number of lost or badly transmitted frames by Sunrise Telecom STT Ethernet module. Power penalty of 2 dB for the best performing channel has been achieved at the end of the link. For 40 GHz wavelength conversion the SOA-

MZI was used in push-pull configuration. Continuous train of 3 ps pulses has been used as clock signal. One-to-two wavelength conversion has been demonstrated. Quality of the newly generated signals was checked by an autocorrelator.

3.1.4 All-Optical Multi-Wavelength Regeneration Based on Quantum-Dot Semiconductor Optical Amplifiers for High Bit Rates

Next generation optical communication networks require the advancement of many signal processing functionalities in the optical domain, bypassing the incumbent of electro-optical conversion. Furthermore, the necessity for hardware simplicity and cost effectiveness imposes the deployment of all-optical signal processing elements able to handle multiple input wavelengths simultaneously.

Quantum-Dot Semiconductor Optical Amplifiers (QD-SOAs) have demonstrated advanced properties such as their inhomogeneously broadened gain profile (~100nm) which can accommodate multiple input wavelengths, the size distribution of the quantum-dots which offers the categorization of dots in resonant groups acting as individual processing units for each input wavelength respectively and, their ultrafast carrier dynamics which makes them good candidates for high speed applications. In addition, they can be easily integrated due to their semiconductor nature. Apart from their use for linear broadband amplification [19] and regenerative multiwavelength amplification [20] they can be exploited for various nonlinear signal processing functions such as all-optical multiwavelength regeneration. The latter constitutes a challenging application which is still missing from the state-of-the-art. No experimental demonstration of multiwavelength all-optical regeneration has been reported yet in the literature. However, the physical mechanisms like spectral hole burning (SHB) within the quantum dots that dominate the carrier dynamics and are responsible for the sub-picosecond response time of QD-SOAs have been investigated in detail both numerically and experimentally [21-23]. In this short report, we present a summary of results illustrating the potential of QD-SOAs to be used in a 2R regenerative subsystem with the ability to process two input wavelengths simultaneously at 40Gb/s and 160Gb/s.

3.1.4.1 Setup Description

Fig. 3.19(a) illustrates the 2R regenerative subsystem which comprises of a cascade of two QD-SOAs each one followed by an optical passband filter. The two input data channels at the wavelengths λ_1 and λ_2 consist of 32% duty cycle RZ Gaussian pulses modulated by a 2^7-1 PRBS bit pattern. For every input data channel (hereinafter referred to as 'pump') a continuous working (cw) signal (hereinafter referred to as 'probe') co-propagates the device at a wavelength distance $\Delta\lambda$ defined such as to prevent spectral overlap of the two channels at the multi-gigabit line rate. As it is illustrated in Fig. 3.19(b), each pair of pump-probe signals is confined within the spectral broadening of a single-dot group bandwidth in order to isolate each channel from the other. Furthermore, channel regeneration is based

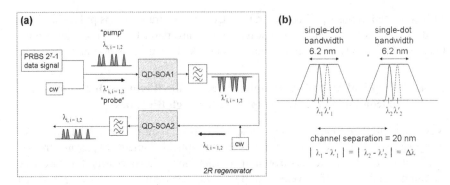

Fig. 3.19. (a) The 2R regeneration setup. (b) "pump" and "probe" wavelengths within adjacent single-dot homogeneous bandwidths under the QD-SOA inhomogeneously broadened gain.

on the cross-gain modulation (XGM) effect between the "pump" signal which modulates the carrier density and hence the gain of the QD-SOA and the cw "probe" signal which in turn acquires the same modulation. As a result, at the output of the first QD-SOA (QD-SOA1) the input data are wavelength converted from λ_1 to λ'_1 and from λ_2 to λ'_2, respectively. The respective "pump" signals after experiencing gain amplification at the output of the QD-SOA, they are cut-off by the passband filter. The modulated "probe" signals are fed as common input to the second QD-SOA (QD-SOA2) and play the role of the "pump" signals. Two new cw signals at the wavelengths λ_1 and λ_2 are introduced in QD-SOA2 as the 'probe' signals, facilitating wavelength conversion of the data to their initial wavelengths. At the output of QD-SOA2, the passband filter is again used to segregate the regenerated channels at wavelengths λ_1 and λ_2 as the only outputs. The channel spacing defined as the absolute spectral distance between λ_1 and λ_2 is 20nm suitable for CWDM scheme and the frequency detuning between the pump and probe signals is 200GHz at 40Gb/s and 600GHz at 160GHz.

The behavior of the QD-SOA devices is simulated based on a rate equation model which solves four rate equations concurrently each one describing the carrier density of a quantum-dot energy state: the ground state, the excited state, the continuum state and the wetting layer. The model has been implemented assuming that a) the dots are grouped together according to their size and resonant optical frequency b) the dots are spectrally and spatially isolated and communicate only via the wetting layer and c) carrier thermalization is highly improbable [24]. Carrier depletion at the ground state occurs at the incidence of optical power and the upper energy states act as reservoir of carriers for the lower energy states. Thus, it is important to improve the carrier refilling time for the upper energy levels (e.g. the wetting layer) [25]. The implemented model takes into account that the carrier relaxation time within the dots is 10ps and the carrier recombination lifetime within the dots and the wetting layer are 1ns and 0.4ns, respectively. Finally, the carrier capture time from the wetting layer to the continuum state is 1ps. It should be mentioned that the implemented model is suitable for system level studies and does not take into account spatiotemporal effects that would arise from the non-linear interaction of light with matter combined with the spatial dispersion of the gain and the refractive index, which is expected to further limit the quality of the output signals [26].

3.1.4.2 Results

The performance evaluation of the 2R regenerative subsystem illustrated in Fig 19(a) has been carried out based on extensive numerical simulations by optimizing the operating parameters of the QD-SOA cascade such as the power levels of the cw input signals and the device parameters such as the homogeneous broadening of the single dot-group gain, respectively. The optimization process relies on the measurement of the extinction ratio and Q-factor depicting the quality of the output signals. Detailed simulation results can be found in [27]. The input power levels of the data signals are such that operation at the nonlinear regime beyond the saturation power of a 10mm QD-SOA is ensured. Furthermore, the power levels of the cw probe signals into QD-SOA1 and QD-SOA2 play a significant role in the carrier dynamics of the devices. The increase of the probe signal in QD-SOA2 causes the device to perform faster hence Q-factor enhancement is observed. However, the amplitude jitter suppression at the marks is accompanied by extinction ratio deterioration which is a well studied problem in QD-SOA devices under XGM operation [28]. It is thus critical to trade off between Q-factor improvement and limitations of the signal quality introduced by the degradation of the extinction ratio.

A second optimization study refers to the interplay between the homogeneous single dot-group bandwidth and the spacing of adjacent channels. It has been shown that, the narrower the homogeneous broadening is, the better isolated the channels are, thus preventing interference. On the other hand, if the homogeneous bandwidths (in which the input channels are confined) overlap to a great extent which is comparable to the channel spacing, the quality of the output signals degrades. Numerical results at 40Gb/s have illustrated that effective suppression of the amplitude jitter at the marks is observed when the homogenous broadening is 6.2nm and the channel spacing is minimum 10nm, whereas similar performance is observed for homogeneous broadening 12.4nm at a minimum 17nm channel spacing. At 160Gb/s regenerative performance has been observed only for 6.2nm homogeneous bandwidth at a 15nm channel spacing the minimum. Fig. 3.20 illustrates the eye diagrams of the input signals and the eye diagrams of the regenerated signals both at 40Gb/s and 160Gb/s. Q-factor improvement is evident for both channels at both bit rates indicated by the amplitude jitter suppression at the marks. However, it is obvious that the power level of the spaces has risen causing the extinction ratio degradation. A saturable absorber may be used appropriately to combat this degradation [24].

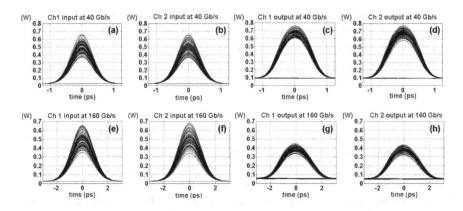

Fig. 3.20. Input – Output eye diagrams (a-d) at 40 Gb/s and (e-h) at 160 Gb/s.

Finally, the possibility of 2R regeneration of more than two input channels has been investigated with detailed results presented in [29]. It has been shown that four-channel regeneration is limited by the low Q-factor performance of the two "outer" channels with respect to the center of the inhomogeneously broadened gain of the QD-SOAs.

3.2 Optoelectronic Clock Recovery, Retiming and OTDM Demultiplexing

In this section several techniques for clock recovery, retiming and optical time division multiplexing (OTDM) demultiplexing are presented and discussed. In particular ultra-high speed transmission is achieved.

3.2.1 320 Gbit/s Clock Transmission and Channel Identification

One of the main issues to be addressed in order for Optical Time Division Multiplexing (OTDM) to become a feasible network technology is the ability to unambiguously identify the individual tributary channels. This issue originates from the lack of an absolute clock reference throughout a transmission system. A few schemes have been proposed for channel identification (ID) e.g. [30-31]. A second issue to be addressed in high-speed communication systems is clock recovery to synchronise a receiver to an incoming signal. The two main approaches have been PLL-based [32] or based on injection locking [33].

In this paper, we demonstrate 320 Gbit/s operation of a clock recovery scheme where a clock signal is added as a phase modulation to one of the tributary OTDM channels in a high-speed on/off keying data signal. In this way by detecting the phase modulation the clock can be extracted and the clock modulated channel is uniquely identified. Adding the clock modulation, is virtually penalty-free. We present detailed characterizations at 160 Gbit/s, and further demonstrate error free, low-penalty performance at 320 Gbit/s.

3.2.1.1 Experimental Procedure

The experimental set-up is shown in Fig. 3.21.

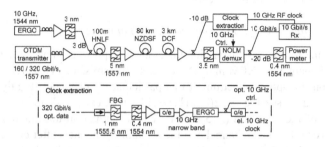

Fig. 3.21. XPM modulation and transmission set-up.

A high-speed optical signal is generated in a standard OTDM transmitter comprising a 10 GHz pulse source generating 1.3 ps pulses (ERGO) at 1557 nm, which are data modulated in a Mach-Zender modulator (2^7-1 PRBS) and multiplexed to 160 or 320 Gbit/s in a fibre delay multiplexer. The high-speed data signal is injected into 100 m of highly non-linear fibre (HNLF) along with a 10 GHz pulse train at 1544 nm from a second pulse source. The two signals are temporally aligned for overlap between one of the 10 Gbit/s channels in the high-speed signal and the 10 GHz pulse train giving rise to cross phase modulation (XPM) between the two signals. At the output of the HNLF the pulses at 1544 nm are suppressed by a band pass filter and the 320 Gbit/s data signal is transmitted through 80 km of non-zero dispersion shifted fibre (NZDSF) compensated with 3 km of dispersion compensating fibre (DCF). After transmission, the signal is split in a 10 dB coupler. The weak output is used for clock extraction while the strong output is injected into a non-linear optical loop mirror (NOLM) for demultiplexing. Clock extraction is performed by filtering on the low wavelength edge of the data spectrum using a combination of a fibre bragg grating (FBG) and a 0.4 nm band pass filter to isolate the sideband created on one data channel by XPM in the transmitter. The filtered sideband is then detected in a photo detector and the electrical signal is filtered to enhance the 10 GHz component. An ERGO optical pulse source is synchronized to the 10 GHz component in the RF signal using its internal PLL. In this way the transmitted clock is extracted to synchronize the demultiplexer and the receiver. The optical output of the pulse source then represents a 10 GHz clock signal which is synchronized to one channel in the input data signal. Clock extraction is achieved for PRBS lengths up to 2^{15}-1, limited by the particular PLL used in the ERGO. The optical pulses are then split into two arms. One is used as control pulses for the demultiplexer and the other is o/e converted to generate a 10 GHz RF clock to synchronize the base rate receiver.

After demultiplexing of the 320 Gbit/s data signal a 20 dB coupler is used to tap the demultiplexed signal for channel identification. A narrow filter at the edge of the signal spectrum is used to identify the channel carrying the clock similar to what is done for clock extraction. The power through the filter is measured using a sensitive low bandwidth power meter. The strong output form the demultiplexer is used for bit error rate measurements (BER).

3.2.1.2 Principle of Clock Transmission and Channel ID

In this scheme for clock transmission the clock is applied as a phase modulation to one OTDM channel by XPM in a HNLF before transmission. This creates spectral sidebands on the optical spectrum of that particular data channel slightly broadening its spectrum – see Fig. 3.22(b). These sidebands can be identified by filtering at the edge of the spectrum (~1554 nm). The difference in this filtered sideband power between the clock channel and the remaining channels after demultiplexing can be seen in Fig. 3.22(a). A 6 dB power contrast between the clock channel and the remaining channels is achieved in this way (Fig. 3.22(a)), yielding a clear channel identification.

In this scheme for clock transmission the clock is applied as a phase modulation to one OTDM channel by XPM in a HNLF before transmission. This creates spectral sidebands on the optical spectrum of that particular data channel slightly broadening its spectrum – see Fig. 3.22(b). These sidebands can be identified by filtering at the edge of the spectrum (~1554 nm). The difference in this filtered sideband power between the clock channel and the remaining channels after demultiplexing can be seen in Fig. 3.22(a). A 6 dB power contrast between the clock channel and the remaining channels is achieved in this way (Fig. 3.22(a)), yielding a clear channel identification.

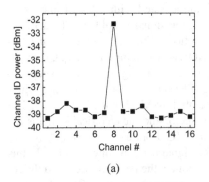 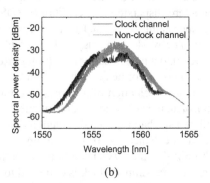

(a)   (b)

Fig. 3.22. (a) measured power for identifying the clock channel in the receiver (b) spectrum of demultiplexed data signal for the clock channel and for one other channel.

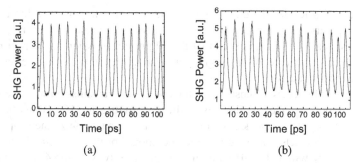

Fig. 3.23. (a)160 Gbit/s data with phase modulated clock before transmission and (b) after transmission.

The slight spectral broadening of the clock channel could be feared to make this channel more susceptible to dispersion in the transmission span. However, this is easily overcome by standard passive dispersion compensation. Fig. 3.23 shows the cross correlations of a 160 Gbit/s data signal with phase modulated clock channel before transmission (left) and after transmission (right). In both cases no discernible pulse width difference among the 16 channels is observed indicating sufficient dispersion compensation in the transmission line.

3.2.1.3 Bit Error Rate Characterizations

Fig. 3.24 shows BER measurements for the 320 Gbit/s data signal transmitted 80 km with a phase modulated transmitted clock.

BER curves for the clock modulated channel and for one other channel are shown. Both are error free with no error floor. The clock channel has a penalty in receiver sensitivity of only 1.2 dB, demonstrating that this scheme works well for 320 Gbit/s data. For a 160 Gbit/s data signal, BER curves are shown with and

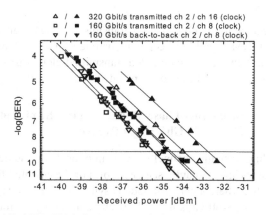

Fig. 3.24. BER: 160 / 320 Gbit/s 80 km transmission and back-to-back with phase modulated clock.

without transmission for the clock channel and for one other channel. In all cases error free operation is achieved. The clock channel has a penalty in receiver sensitivity of less than 0.8 dB. It is seen that the 80 km transmission has almost no effect on the performance of the signal, indicating that the clock is effectively transmitted to the receiver.

In Fig. 3.25 the receiver sensitivities for 160 Gbit/s transmission with the phase modulated clock are shown, along with the sensitivities for 160 Gbit/s back-to-back using the electrical clock from the transmitter to synchronize the receiver. In both cases all channels achieve error free operation. For the transmitted signal the sensitivities vary within 1.3 dB with an average sensitivity of -35.4 dBm, i.e. the clock channel lies within the sensitivity variation, rendering it virtually penalty-free. In the back-to-back case the variation is 0.7 dB and the average sensitivity is -35.7 dBm. The increase in channel variation is expected to be due to imperfect multiplexing of the data signal resulting in varying performance after transmission.

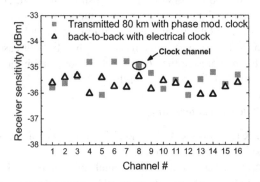

Fig. 3.25. Receiver sensitivities for 160 Gbit/s data signal transmitted 80 km with phase modulated clock and back-to-back with electrical clock..

3.2.1.4 Conclusion

In this paper, a scheme for combined clock transmission and channel identification in a high-speed OTDM signal has been demonstrated for a 320 Gbit/s data signal transmitted 80 km with a phase modulated transmitted clock. Error free performance and very low penalty has been demonstrated.

3.2.2 Filtering-Assisted Cross-Phase Modulation in a Semiconductor Optical Amplifier Enabling 320 Gb/s Clock Recovery

In the pursuit of higher single-channel bit rates in optical communication systems, new fast technologies are constantly searched for. For the essential functionality of clock recovery (CR), where a network node is synchronized to a data signal, locking to data rates above 160 Gb/s has only been achieved by 2 groups worldwide [34,35]. In [34], a rather extensive set-up using four wave mixing (FWM) was used to lock to 400 Gb/s, and in [35], a 320 Gb/s RZ-DPSK was locked to using very fast electroabsorption. We have previously shown 160 Gb/s CR using FWM [36] in a potentially very compact set-up. To upgrade from 160 to 320 Gb/s requires a very high timing resolution based on narrow control pulses and a fast and efficient physical effect. In this paper, we report on a novel scheme for 320 Gb/s CR. A new narrow-pulse control laser is introduced and filtering-assisted cross-phase modulation (f-a XPM) in a semiconductor optical amplifier (SOA) enables locking to 320 Gb/s. All active components are semiconductor-based, paving the way for future monolithic integration.

3.2.2.1 Experimental Procedure

The experimental set-up is shown in Fig. 3.26. To generate the 320 Gb/s data signal, an erbium glass-crystal-based external cavity mode-locked laser (ERGO PGL) is used as pulse source. It produces 2 ps pulses, which are further compressed to 1.3 ps in a fibre-based pulse compressor. The pulses, running at a 10 GHz repetition rate at 1555 nm, are data modulated before compression with a 2^7-1 PRBS sequence and optically time division multiplexed (OTDM) to 320 Gb/s in a PRBS and polarization maintaining fibre-based multiplexer. The data signal is amplified and spectrally filtered before injected into the clock recovery (CR) circuit. The data pulse shape at the input to the CR circuit has maintained its narrow pulse width, which is adequate for 320 Gb/s RZ OTDM data.

The clock recovery part of the set-up is based on compact semiconductor components. The data signal is injected into a semiconductor optical amplifier, which acts as a phase comparator between the data and a local clock signal, thus generating an error signal, proportional to the cosine of the phase difference between the clock and data. The local clock signal consists of 10 GHz pulses from a commercial semiconductor tuneable mode-locked laser (TMLL), which is driven by a voltage controlled oscillator (VCO), which in turn is tuned by the generated error signal. The overall bit rate limitation of this set-up is determined by the pulse width of the local clock pulse source (in this case a TMLL) together with the intrinsic response time of the mixing process that takes place in the phase comparator.

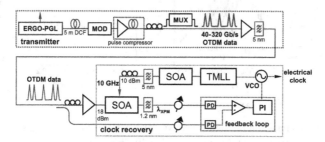

Fig. 3.26. High-speed clock recovery set-up.

Therefore, it is imperative to have a pulse source with very narrow pulses and to use a phase comparator with an ultra-fast response. In this set-up the TMLL was running at 1560 nm with a pulse width of 2.2 ps. In [36], four wave mixing between the data and control was used as mixing process, but with the specific SOA this was too inefficient for 320 Gb/s. Instead, filtering-assisted cross-phase modulation of the clock pulses by the data pulses is used. As the data pulses travel through the SOA, they give rise to changes in the carrier density due to ultra-fast carrier dynamics such as spectral hole burning and carrier heating. This results in gain as well as refractive index changes, which in turn results in an amplitude and phase modulation of the clock pulses – they get chirped. As shown in [37] this chirp may be used to enhance the speed response of an SOA if a narrow filter is placed at the SOA output, such that the slope of the filter is tuned to the centre of the clock spectrum. In this experiment, a 1.2 nm filter is placed with its slope on the clock wavelength and its peak on the red shifted side of the clock ~ 1562 nm, see Fig. 3.27. A phase modulation of the clock will thus be transferred into an intensity modulation, which can be directly detected. The filtered signal will only occur when a clock and data pulse overlap in time, and hence, if they are not synchronized and therefore move across each other at the difference frequency, a slowly varying error signal is generated. Fig. 3.27 shows the input and output spectra from the SOA together with an error signal from a 40 Gb/s data signal. The filter is tuned so that both an amplitude and phase change is allowed through, yielding a dispersion shaped 3-level curve [37]. This contains a very sharp rising feature, which can be used to lock to.

The phase comparator SOA is a 2 mm long InGaAsP multiple quantum well (MQW) device with 8 wells, designed for signal processing with a high differential gain and a high optical confinement. An additional SOA is used to boost the pulses from the TMLL. This booster SOA is designed with a low differential gain to act as a linear amplifier with high output power. It has 3 quantum wells, is 2 mm long and provides 15 dBm output power (10 dBm after filter).

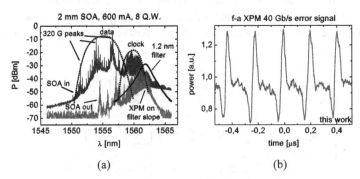

Fig. 3.27. Phase comparison through f-a XPM. (a) spectra of the incoming signals, the 1.2 nm filter and the filtered output. (b) error signal at 40 Gb/s.

3.2.2.2 Temporal Resolution and Locking Results

Fig. 3.28 shows results on the timing resolution. The local clock has pulses that are sufficiently narrow to partly resolve the data signal, Fig. 3.28(a). The produced error signals at 80, 160 and 320 Gb/s, as portrayed in Fig. 3.28(b) show clear traces of the high-speed data signals, although the 320 Gb/s signal is quite small. This is due to the slightly too wide clock signal and a slow gain recovery tail in the SOA.

Fig. 3.29 shows the locking performance of this scheme in terms of electrical power spectra. Fig. 3.29(a) shows locking at 80, 160 and 320 Gb/s. In all cases, successful locking is achieved. There is no significant difference in the performance at the different bit rates. Fig. 3.29(b) shows how the sideband noise can be minimized by altering the PLL gain. For low gain the CR will only be able to follow the data signal very close to the carrier frequency, and PLL-noise (from the VCO, the TMLL, the mixer, and the active components in the loop filter) will dominate the spectrum even close to the carrier. As the gain increases, the CR will be able to follow the data frequency to higher offset frequencies from the central carrier and thus push the PLL's own phase noise further down and away from the carrier, yielding lower timing jitter. The best performance obtained for this set-up corresponds to a timing jitter of 800 fs (measured on a time precision oscilloscope).

This jitter value can be further minimized by optimizing the PLL loop in terms of length (due to a number of fibre pigtails, the loop runs a total length of about 20 m) [38], in terms of narrower clock pulses enabling a bigger error signal, and generally using low-noise components in the loop. For instance, in [36] a low-noise monolithically mode-locked laser was used as clock signal yielding overall lower phase noise of the CR.

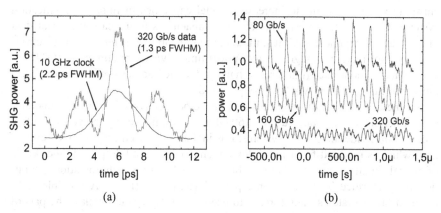

Fig. 3.28. Temporal resolution. (a) autocorrelation traces of 320 Gb/s data and local clock. (b) error signals at various bit rates up to 320 Gb/s.

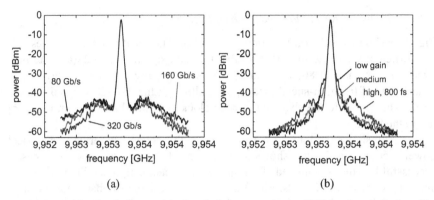

(a)   (b)

Fig. 3.29. Locking performance (a) electrical power spectra of VCO when locked to 80, 160 and 320 Gb/s (b) PLL-gain optimization reduces phase noise.

3.2.2.3 Conclusions

We have reported on a novel clock recovery scheme, relying on ultra-fast filtering-assisted cross-phase modulation in a semiconductor optical amplifier. The temporal resolution of the set-up was sufficient to resolve a 320 Gb/s OTDM data signal, and locking at bit rates up to 320 Gb/s was successfully achieved. With additional optimizations of the set-up, it looks very promising for high-speed clock recovery.

3.2.3 640 Gbit/s Data Transmission and Clock Recovery Using an Ultra-Fast Periodically Poled Lithium Niobate Device

The demand for higher telecom bandwidth is steadily increasing, and there are currently various pursuits to increase channel capacities in fibre communication systems. One way to increase the channel capacity is to increase the serial data rate. To go far beyond 100 Gbit/s, though, requires optical techniques, as electronics is not yet mature enough. Using optical time division multiplexing (OTDM), 640 Gbaud symbol rates (pulse rates) have so far been demonstrated as the highest pulse rate carrying data by a few groups worldwide.

For these very-high-bit-rate transmissions, clock recovery is a critical functionality. Locking to data rates above 160 Gbit/s has only been achieved by a few groups worldwide. In OTDM systems, this includes the need for sub-clock extraction, which is required for demultiplexing, channel add-and-drop, and other kinds of optical signal processing. At lower bit rates, such clock synchronization is traditionally performed by an electronic phase-locked loop (PLL). A possible technique in faster systems, therefore, is to use optical devices to emulate the part of the PLL which requires the highest speed.

Optical nonlinearities in semiconductor optical amplifiers (SOAs) have been successfully used to that end. Very recently, a SOA-based opto-electronic PLL was demonstrated at 640 Gbit/s by the COBRA Research Institute and COM•DTU. SOAs can thus clearly be very fast, but they also have inherent slow recovery times, which will inevitably lead to patterning effects for OOK modulation formats. To avoid this, non-linear three-wave mixing (TWM) in Periodically-Poled Lithium Niobate (PPLN) was proposed and demonstrated in a PLL at 40 Gbit/s by GET/ENST. Such devices being passive, they do not exhibit amplified spontaneous emission noise or patterning effects. Furthermore, the resulting TWM is in the visible-light range, hence easily isolated from near-infrared input signals.

This collaboration leveraged the competences of GET/ENST and COM•DTU, as well as the National Institute for Materials Science (NIMS) in Japan. Over October and November 2007, a series of joint experiments demonstrated clock recovery and OTDM demultiplexing up to 640 Gbit/s. We used efficient three-wave mixing in an adhered-ridge waveguide PPLN module prototype from NIMS; an electronic loop filter designed and realized by GET/ENST; and COM•DTU's OTDM test bench and expertise.

We have succeeded in extracting a 10-GHz clock from RZ-DPSK data streams at 160 and 320 Gbit/s, and RZ-OOK data streams at 320 and 640 Gbit/s after transmission over a 50-km fibre. We were able to optimize the setup to obtain a clock with a time jitter less than 100 fs. This clock was used in a full error-free OTDM demultiplexing, albeit with a 3-dB penalty.

At the time of this writing, we have submitted two articles [39-40] resulting directly from this joint action, and two more such are under preparation.

3.2.3.1 Main Technical Results

The clock recovery system comprises the three same basic building blocks of a conventional PLL: voltage-controlled oscillator (VCO); mixer/phase comparator; and loop filter. Unlike a conventional PLL, however, the input signal and local clock are optical, and the phase comparator is based on a nonlinear optical effect. Fig. 3.30 describes the detailed setup for 640 Gbit/s clock recovery [40].

The VCO is a standard electronic oscillator, which drives a tunable mode-locked laser (TMLL) to form an optical pulse train of repetition rate $f_c \approx 10$ GHz

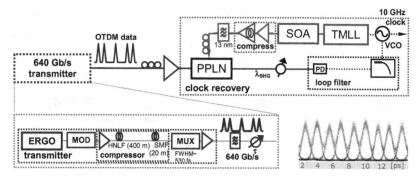

Fig. 3.30. OTDM joint experiment setup.

at wavelength $\lambda_c = 1557$ nm. The pulse width is 2.7 ps, which proved sufficiently short for clock recovery up to 320 Gbps, but not for 640 Gbps. For the latter, soliton compression in a high-power EDFA was used to bring the pulse width down to 730 fs. This forms the clock signal.

It is inserted into a 3-dB optical coupler along with the input data signal whose clock is to be extracted. The latter is an OTDM signal around $f_s \approx N \times 9.95328$ GHz with N being 16, 32 or 64, which yields a 160, 320 or 640 Gbit/s data stream at wavelength $\lambda_s = 1557$ nm.

Upon injection into the PPLN device, these optical signals generate a TWM beam at $[\lambda_s^{-1} + \lambda_c^{-1}]^{-1} \approx 781$ nm $= \lambda_{TWM}$, which is detected by a silicon avalanche photodetector. Being insensitive to the infrared input signals, and having a limited electrical bandwidth, the detector itself acts as a filter and isolates the TWM signal's low-frequency components. This is equivalent to a low-pass-filtered mixer, which can act as a phase comparator. It is to be noted that this operation is not limited by the PPLN's three-wave mixing bandwidth, which is narrower than the signals [39] This phase comparator's output is then electrically low-pass filtered and shifted to drive the VCO, thus closing the loop.

Successful clock recovery at both 320 and 640 Gbit/s is achieved. Fig. 3.31 and Fig. 3.32 show the 640 Gbit/s locking performance of this scheme in terms of an electrical power spectrum and a single sideband to carrier ratio (SSCR) phase noise curve plotted together with the integrated timing jitter. Fig. 3.31 shows a clean frequency peak with low noise. There is a noise peak at the PLL-bandwidth of 200 kHz where the CR phase noise shifts from being dominated by the data noise to the PLL noise as expected. Fig. 3.32 shows the SSCR and the integrated phase noise (from 1 kHz and upwards) yielding an rms timing jitter value depending on the upper integration limit. Beyond the PLL bandwidth, where the phase noise gets very low, the timing jitter remains fairly constant at ~250 fs. This value is very good considering the large phase noise contribution from the TMLL, which has a free-running rms timing jitter of ~400 fs.

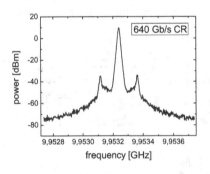
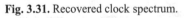

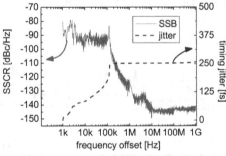

Fig. 3.31. Recovered clock spectrum.

Fig. 3.32. Recovered clock phase noise: single sideband to carrier ratio and integrated jitter.

OTDM demultiplexing was also achieved, extracting a 10 Gbit/s OTDM channel from a 640 Gbit/s data stream transmitted over 50 km of SMF-IDF fibre. We succeeded in reaching error-free, although not penalty-free, performance: the penalty from the clock recovery is 3 dB, and that from the 50-km transmission span is 0.5 dB.

3.2.3.2 Conclusions

We have reported on a novel clock recovery scheme, relying on a truly ultra-fast sum-frequency generation in a PPLN. The temporal resolution of the set-up was sufficient to resolve a 640 Gbit/s OTDM data signal; locking at bit rates up to 640 Gbit/s was successfully achieved, and the clock was used to achieve error-free OTDM demultiplexing. Submitted paper [40] constitutes the first report on the use of a PPLN at such high bit rates and is only the second demonstration of 640 Gbit/s clock recovery.

3.2.4 All-Optical Clock Extraction Circuit Based on a Mode-Locked Ring Laser Comprising SOA and FP Filter

An accurate timing extraction is required by many transmission subsystems such as optical receivers, 3R regenerators, cross-connects, and OTDM demultiplexers. Various methods for optical clock recovery at high data rates have been studied recently [34, 41-46].

The method we investigated within the duration of the COST291 action uses a semiconductor optical amplifier (SOA) as gain element of an optically injected fibre ring laser and a Fabry-Pérot (FP) filter at the input of the structure for reducing the pattern effect. Our results regarding the influence of the signal power, pulse width, mark probability and length of sequences with consecutive "zeros" on the quality of the recovered clock signal have shown that this scheme can produce a high quality optical clock for a relatively wide range of the considered parameters.

3.2.4.1 Injected Mode-Locked Ring Laser with FP Filter

The schematic of the clock recovery based on a mode-locked ring laser (MLRL) with a semiconductor optical amplifier (SOA) in its cavity is shown in Fig. 3.33. The SOA with sufficient gain to ensure lasing in the ring cavity at wavelength λ_2 is used as a gain element. The incoming data signal at λ_1 modulates the gain of the SOA. The optical input signal is injected into the MLRL through a Fabry-Pérot filter and a coupler. Consequently, the signal in the loop cavity is modulated according to the incoming data pattern by exploiting the cross gain modulation (XGM) in SOA. The length of the loop can be adjusted precisely by the tunable optical delay element τ. The Fabry-Pérot filter with a free spectral range equal to the data rate of the incoming signal is placed at the input of the structure in order to compensate for the pattern effect in SOA [47]. An isolator within the ring cavity ensures unidirectional operation.

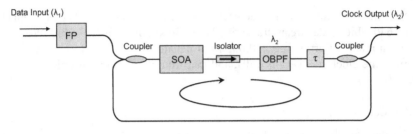

Data Input (λ_1)   Clock Output (λ_2)

FP: Fabry-Perot Filter   SOA: Semiconductor Optical Amplifier   OBPF: Optical Band Pass Filter

Fig. 3.33. All-optical clock extraction circuit based on a SOA-based mode-locked ring laser.

In our simulation study, we observed the relative variation of peak power (i.e., amplitude jitter) and the timing jitter of generated clock pulses at the output. The relative variation of peak power is defined as $v = (P_{max} - P_{min})/(P_{max} + P_{min}) \cdot 100\%$. The results we obtained for 40 Gbit/s return-to-zero (RZ) data and by using numerical simulations are shown in Fig. 3.34 – Fig. 3.37. The main simulation parameters can be found in [48].

The influence of occurring data sequences with many consecutive "zeros" is shown in Fig. 3.34. It can be seen that both the relative variation of peak pulse power and the peak-peak timing jitter increase with increasing the number of consecutive "zeros". A power variation lower than 6.5 % and peak-peak timing jitter up to 1.2 ps have been obtained for data sequences containing up to 17 consecutive "zeros". Fig. 3.35 shows the quality of generated optical clock versus mark probability of the input data. Both the timing jitter and the relative peak power variation increase with decreasing the mark probability of PRBS data sequences. A maximum timing jitter value of 1.83 ps and a power variation of $v = 3.58$ % have been obtained for a mark probability of 0.3.

Further, we investigated the influence of pulse power of data signal on timing jitter and amplitude fluctuation of the generated clock. As it can be seen in Fig. 3.36, the structure produces high quality optical clock for a wide range of power of data signal. The best result we obtained for a peak power of 10 mW. At this point, the timing jitter and the peak power variation of clock pulses are 0.5 ps and

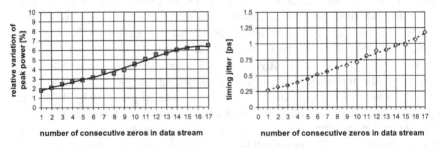

Fig. 3.34. Relative variation of peak power and timing jitter vs. number of consecutive "zeros".

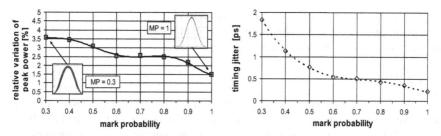

Fig. 3.35. Relative variation of peak power and timing jitter versus mark probability of PRBS data.

1.3 %, respectively. For an increase of the peak power up to 100 mW, the timing jitter remains almost constant, while the relative variation of the clock amplitude increases to slightly more than 3 %. If the power of the data signal is lower than 5 mW, the modulation of SOA is not deep enough, and consequently, both timing jitter and amplitude jitter become large.

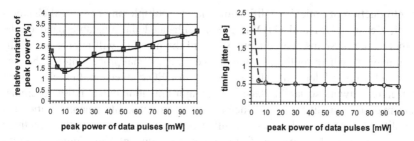

Fig. 3.36. Influence of data pulse power on clock quality.

We also studied the effect of different data pulse widths on generated clock. The best results were obtained for a pulse width of 2.5 ps (see Fig. 3.37).

A clock signal with a timing jitter below 0.6 ps and a relative variation of peak power less than 3 % can be generated for data pulse widths within the range from 2 ps to 6 ps. When the pulse width exceeds 6 ps, both timing jitter and especially

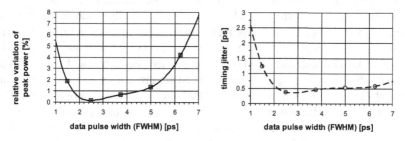

Fig. 3.37. Influence of data pulse width on clock quality.

amplitude modulation of clock are strongly impacted by overlapping of neighbouring data pulses. If the pulse width becomes shorter than 2 ps, both timing jitter and relative variation of peak power increase.

3.2.4.2 Conclusion

An optical clock recovery method based on a mode-locked ring laser (MLRL) comprising a semiconductor optical amplifier (SOA) as gain element and a Fabry-Pérot (FP) filter to reduce the pattern effect was investigated by extensive numerical simulations. Our results concerning different number of consecutive "zeros" within the data stream, different mark probabilities of pseudorandom data sequences, as well as various signal powers and pulse widths have shown that this scheme is able to produce a high quality optical clock for a relatively wide range of considered parameters.

3.2.5 OTDM Demux Based on Induced Modulation on an Auxiliary Carrier by Means of Super-Continuum Generation

The introduction of optical amplification and the multiplexing of several channels (WDM, Dense WDM and OTDM) greatly enhanced the available band of the communication systems. The direction toward the network will evolve is the all-optical network where routing, switching and all the signal processing will be performed in the optical domain with no necessity of OEO conversion. In such network architecture, network scalability faces some major limitation due to noise accumulation from optical amplifiers, chromatic dispersion, crosstalk, jitter, nonlinearities and possibility of congestion due to the conflict between used wavelengths. For these reasons All-Optical Regeneration, format conversion, optical Time domain demultiplexing, switching, logical functions and wavelength conversion over a wide spectral range are keys issue for all-optical transmission system [49].

The WDM signals are preferred to the OTDM for several reasons; one of them is that nowadays all the signal processing in the network node is operated in the electrical domain and the limitation of the capability of the electrical signal processing impose a data rate limit to the signal elaboration (40 Gbps). Therefore each single data channel has to be separated (Demux function) from incoming very high bit-rate signal. A WDM Demux is a passive, commercially available object. On the other hand the optical signal processing is generally much faster than the electrical, so is acceptable to think that is convenient to have in an All-Optical Network fewer channel at higher bit rate (180, 320 Gbps and more). In order to manage an OTDM channel an OTDM Demux technique has to be developed. Generally, the ultra fast response time we need are granted by non-linear effect in fibre and several fibre-based optical OTDM Demux were proposed [50]. The intergraded optics OTDM demux proposed are not so fast, under 40 Gbps [51].

We present a new set-up based on SC generation and XPM that is exploitable to achieve time domain demultiplexing; by simple mixing the modulated signal, a clock and a CW auxiliary carrier we manage to transfer the data stream of one of the tributary to the new wavelength with simultaneous 2R regeneration effect.

The modulate signal and a clock synchronized to the signal we want to extract are mixed in polarization difference; this solution requires generating a very limited SC spectrum while allows to tune the carrier over 30 nm away.

3.2.5.1 Proposed Experimental Set-up

Fig. 3.38 shows the experimental set-up. An ultra fast optical source generates a pulses train FWHM~1.5ps, repetition rate 40GHz The pulse train is modulated and multiplexed to obtain a sequence with 80/160Ghz Repetition Rate.

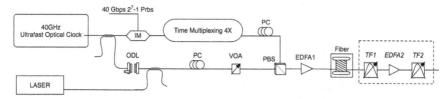

Fig. 3.38. Experimental set-up of the all-optical wavelength converter-2R scheme.

This ps signal has been then processed by a polarization controller (PC) and an high power optical amplifier (EDFA1) that increases the power of the pulse up to the power necessary for the SC generation. An auxiliary CW carrier, tuned outside the SC spectrum is generated by a tunable ECL and is coupled into the fibre with the pulse train.

The 80 Gbps signal is combined with an orthogonally polarized 40 Gbps pulse train and the continuous wave (CW). The 40GHz clock is variably delayed to synchronize it with the signal to extract. The resulting signal is injected in a highly one of our non-linear fibre; the first fibre we tested was 1 km long with dispersion coefficient D=-6 ps nm^{-1}km^{-1} in the 1500nm-1580nm range, γ=14.8 W^{-1}km^{-1}, S= 0.004 ps km^{-1}nm^{-2} and α=3.3 dB/km. The extremely low S/D ratio guarantees spectrum symmetry and the high D value helps to avoid impact of fibre non-uniformity. The γ/D ratio allows for reasonable low power to obtain the non-linear effect. The second fibre is a photonic crystal optical fibre (PCF). The PCF is constituted from pure silicon, has triangular core of 2.1 µm and a cladding diameter of 128µm. 50 mt length, α=9 dB/km, numerical opening NA=0.4 and $\gamma = 11(wkm)^{-1}$.

The dispersion of the HNLF is reported in Fig. 3.39(a); the section and dispersion of the PCF are reported in Fig. 3.39(b).

The modulated signal generates supercontinuum (SC) in the normal dispersion regime. The AUX will suffer the XPM obtained via the index refraction modulation induced by the Super continuum generation [10, 18, 52]. This Phase modulation will follow the Amplitude modulation of the original Signal so, filtering the output of the fibre using a cascade of two optical filters and an amplifier we obtain

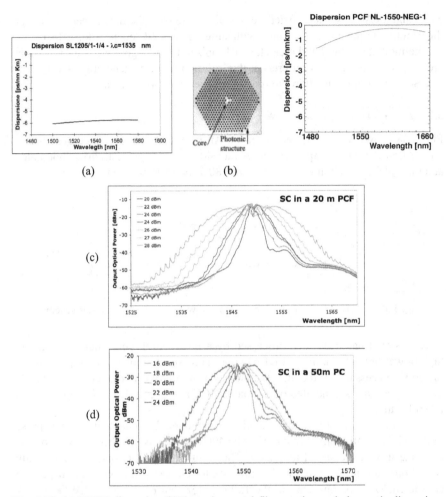

Fig. 3.39. (a) HNLF dispersion (b)Photonic crystal fibre section and chromatic dispersion (on the right). SC generation in two PCF fibres (c) of 20 m and (d) 50 m.

a replica of the original signal. This solution requires the generation of a very limited SC spectrum while allows to tune the carrier over 30 nm away with no regard to the wavelength of the original signal [11].

3.2.5.2 Technique Description

Preliminary studies of the technique has been done using a modulated Pulse train with repetition rate 10 GHz and FWHM~4 ps. In Fig. 3.40 and Fig. 3.41 we show the experimental output optical spectrum obtained by the experiment described above and the simulation of the output optical spectrum. Note that the Auxiliary Carrier is placed outside of the SC spectrum. Fig. 3.40(a) refers to the simulation

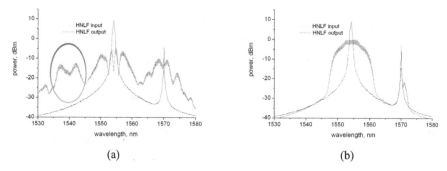

(a)   (b)

Fig. 3.40. Simulated Output Optical Spectrum obtained simulating two different dispersion flattened fibre; D=0 in A), D=-6 in B).

of a fibre with Dispersion Flattened, D=0, a dispersion behaviour similar to the one we have in the PCF; while Fig. 3.40(b) refers to a Dispersion Flattened and D=6 ps/nm/km, a dispersion behaviour similar to the one we have in the HNLF. The modulation of the Auxiliary carriers, deduced by the presence of the lateral bands in the inserts, occurs in both the case proposed tuning the Auxiliary Carrier at both 1535 and 1570nm [11]. However in the HNLF (D=-6) the dispersion reducing the phase matching between the interacting wave prevents the generation of the FWM replica of the modulated auxiliary carrier; the replica circled in Fig. 3.40(a) is present in the case of low dispersion (D~0) [12].

Moreover, because of the dispersion we observe in Fig. 3.40(b), in the non-linear regime, the reduction of one of the side band due to the interaction between the amplitude modulation and the phase modulation.

The XPM obtained via the index refraction modulation induced by the Super continuum generation, causes the strong enlargement of the Aux spectrum. The presence of a phase modulation effect (XPM) is confirmed by the depletion of the carrier observed by centering the optical output filter at the auxiliary carrier wave-

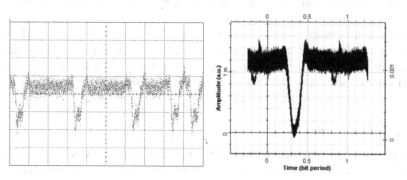

Fig. 3.41. Measured and simulated inverted signal obtained filtering the output at the AUX wavelength.

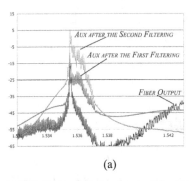 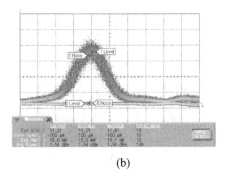

(a)   (b)

Fig. 3.42. Optical Spectrum of the Modulated Auxiliary carrier at the output of the fibre (downwards), after the first filtering and after the second filtering (upwards).

length, in fact, when the filter is tuned to the Aux CW original wavelength (e.g. at 1569.6 nm) an inverted signal is obtained.

The output of the Fibre the Auxiliary Carrier is filtered amplified and filtered again. We filter a band interested by the modulation induced spectrum enlargement of the auxiliary carrier.

The optimum detuning for the double filtering was determined in order to maximize the quality of the converted signal. In Fig. 3.42 we can see the HNLF output spectrum (downwards in Fig. 3.42(a)); the spectrum of the signal filtered amplified and further filtered (upwards in Fig. 3.42(a)). It can be noticed that the SC spectrum and the XPM generated spectrum broaden with increasing P. Furthermore, we have studied the effect of the Aux CW carrier input power (Pcw) and the frequency detuning between the signal and the Aux CW carrier. The power level of the side-lobes increases with Pcw increase, which is reflected in the eye noise level at marks and zeros the Q factor In-Out experimental results has been reported; results and simulations demonstrate the noise compression via aux carrier adoption. The Aux pulse show a good 2R regeneration effect, in particular on marks and 2Rregeneration is demonstrated by Q factor measurement with an increase from 6.5 to over 11 (Fig. 3.42(b)).

This demonstrates reshaping capabilities of this experimental device. The described wavelength converter has, compared with the standard SC spectral slicing approach, better regeneration transfer function, less power needed (20 times less). Furthermore because of the very good behaviour of the PCF a very large operational wavelength range is achievable.

3.2.5.3 Time Domain Demux

In Fig. 3.43(a) are reported the input 80Gbps PRBS signal (upwards Fig. 3.43(a)), the 40 GHz clock (downwards Fig. 3.43(a)) and the resulting signal generating the SC. In Fig. 3.43(b)) is reported the output optical spectrum, the 80 GHz modulation of the Aux Carrier is evident. Also in Fig 3.44 where is reported the output optical spectrum, the 160 GHz modulation of the Aux Carrier is evident.

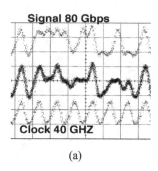

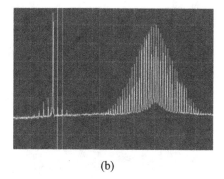

(a)   (b)

Fig. 3.43. (a) SC generating Signal and (b) Optical Spectrum at the output of the fibre (80 Gbps).

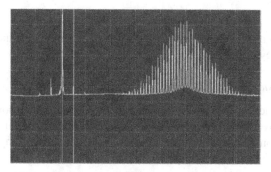

Fig. 3.44. Optical Spectrum at the output of the fibre (160 Gbps Signal).

The alignment of an information impulse with a control impulse induces a particularly high phase shift; a frequency shift significantly stronger that the phase shifts induced by the 40GHz clock or information signal alone. As such, filtering the aux carrier with the correct detuning it is possible to define a range of wavelength within which the Power of the modulated CW is positioned only when the control impulses superimpose the information impulses. Tuning the Tunable Filter, TF1 and TF2, in this frequency range allows extracting the impulses from the original information signal. Fig 3.45.

The concept behind the demultiplexing achieved with this setup, is the dependency of the nonlinear phase on the power of the XPM inducing signal. This higher power of the latter, the higher is the nonlinear phase shift suffered by the CW. Also, since coherence is a problem that could endanger the process, two polarizations were used. By formatting the filter shape and central position, is possible to control the XPM part that is extracted. In the experiments a simple off the shelf filter was used, therefore limiting the overall result. Another effect which will certainly be of interest are the time delays and amplitude difference between the demultiplexing stream and the stream to be demultiplexed. The advantage of having

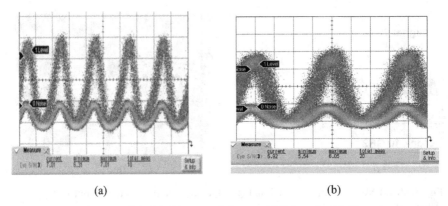

(a)   (b)

Fig. 3.45. Eye diagram of the 40 Gbps signal demuxed (a) from the 80 Gbps signal and (b) from the 160 Gbps.

the two in the same wavelength is that the walk off is minimized, increasing therefore the efficiency of the process.

3.2.5.4 Conclusions

We proposed and experimentally verified an all-optical Time domain demux - wavelength converter scheme based on supercontinuum generation and cross-phase modulation that offers simultaneously a signal quality improvement and broad wavelength conversion range.

3.2.6 160 Gb/s Retiming Using Rectangular Pulses Generated Using a Superstructured Fibre Bragg Grating

At high-speed serial data rates, timing jitter becomes a serious detrimental factor. Timing jitter of data pulses should in general be less than about 5% of the high repetition rate timeslot, but at bit rates exceeding 160 Gbit/s (timeslot <6.25 ps), it gets increasingly difficult to find pulse sources that can fulfill these requirements, which become tighter if transmission is required. Therefore, retiming in a regenerator is a crucial functionality for high-speed systems. Most optical regenerator schemes also imply wavelength conversion, though it is often desirable to maintain the wavelength, without extra wavelength conversion [53].

In this section, we demonstrate a 160 Gbit/s pulse retiming scheme, while maintaining the original wavelength of the data signal. This scheme contains a polarization-insensitive superstructured fibre Bragg grating (SSFBG) [54-55] with a sinc-shaped transfer function, to temporally shape the incoming data pulses into rectangular pulses, and a polarization-rotating Kerr switch based on 200 m highly non-linear fibre (HNLF). Using this scheme, a 160 Gbit/s data signal, with an error floor mainly limited by the data signal timing jitter, is successfully retimed to become error free when incorporating the SSFBG and the Kerr switch.

3.2.6.1 Experimental Set-up

The experimental set-up is shown in Fig. 3.46. The heart of the set-up is the SSFBG and the Kerr switch. The SSFBG is used to shape the 160 Gbit/s data pulses into ~5 ps rectangular pulses [54]. This waveform provides optimal resilience to timing jitter induced errors, and also reduces the absolute accuracy for temporal bit alignment. The shaped pulses are aligned to the Kerr switch polarizer so as to be attenuated in the Kerr switch when the control signal is absent. A short clock pulse then rotates the polarization of the part of the rectangular data pulse that overlaps in time with the clock pulse due to cross-phase modulation in the HNLF, thereby allowing this part of the original data pulse to be transmitted through the polarizer. This configuration ensures that the switched retimed pulse maintains its original wavelength, since it is actually part of the original data pulse. That is to say, the retimed pulses are carved out by the short clock pulses, which have low timing jitter, and hence, the retimed data pulses adopt the low jitter of the clock pulses. The clock pulse being narrower than the flat-topped signal generated by the SSFBG ensures that the switching is insensitive to mistiming in the data signal.

For the data signal, a 10 GHz semiconductor tunable mode-locked laser (TMLL) is used – the wavelength is set to 1557 nm, the FWHM pulse width is 1.8 ps and the root mean square (rms) timing jitter is ~410 fs. These pulses are amplitude modulated (MOD) with a 2^7-1 PRBS and subsequently multiplexed in a PRBS and polarization maintaining fibre-based pulse-interleaving multiplexer (MUX). The data is amplified, injected into the SSFBG generating the rectangular pulses and, through a polarization controller (PC), aligned at 90° to the Kerr switch polarizer. For the clock pulses, an Erbium-glass mode-locked laser (ERGO) with low rms timing jitter (~210 fs) is used – the wavelength is 1544 nm and the FWHM is 1.3 ps when linearly compressed in dispersion compensating fibre (DCF). The rms timing jitter is measured at the FWHM point on the trailing edge of the pulse on a Digital Communication Analyser at the repetition rate of 40 Gbit/s.

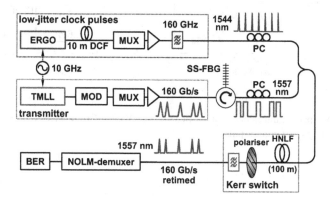

Fig. 3.46. 160 Gbit/s retiming set-up.

The pulses are synchronized to the data pulses via the same synthesizer and they are multiplexed in a second fibre-based polarization maintaining multiplexer. The 160 GHz clean pulse train is amplified, filtered and its state of polarization is aligned at 45° to the Kerr switch polarizer.

The Kerr switch contains a 200m HNLF with dispersion slope of ~ 0.017 ps/nm²km, zero dispersion at 1551 nm, and the non-linear coefficient of $\gamma \sim 10.5$ W⁻¹km⁻¹. These fibre properties ensure negligible pulse-to-pulse walk-off and pulse broadening. The 160 Gb/s retimed data pulses have adopted the clock FWHM and jitter (i.e. ~1.2 ps and ~250 fs, respectively). In order to perform BER characterization, the retimed data signal is demultiplexed down to 10 Gbit/s in a non-linear optical loop mirror (NOLM) demultiplexer with a 50 m HNLF (slope 0.018 ps/nm²km, zero dispersion wavelength at 1554 nm, $\gamma \sim 10.5$ W⁻¹km⁻¹).

3.2.6.2 Experimental Results

Fig. 3.47 shows the response of the SSFBG to the input pulse from the TMLL – in this case at 10 GHz. The -3dB bandwidth of the input spectrum is ~2 nm, while the spectrum at the output of the grating clearly shows sinc-like features corresponding to rectangular waveforms. Note that the slight spectral asymmetry originates from a corresponding slight asymmetry in the SSFBG spectral response. The FWHM of the measured temporal pulse is ~5.7 ps, though the flanks are not as steep as expected (due to the limited bandwidth of the input spectrum). The temporal width is measured with a cross-correlator with a 600 fs sampling pulse, so the de-convoluted trace has slightly steeper flanks. The flat top of the pulse extends over 2 ps, which is enough to eliminate most of the ~410 fs rms timing jitter of the data pulses.

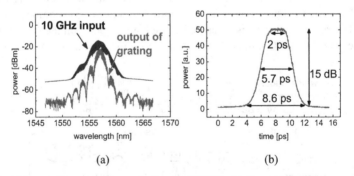

(a)   (b)

Fig. 3.47. (a) Spectral traces before and after the SSFBG. (b) Cross-correlation measurement of the shaped 5ps rectangular pulse.

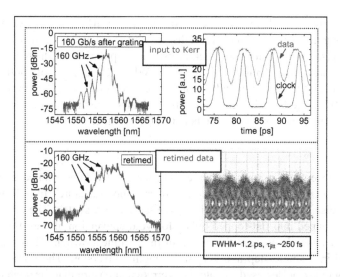

Fig. 3.48. Switching results. Top: Input spectrum to Kerr switch – cross-correlation measurements of the shaped data and of the clock pulses at 160Gbit/s. Bottom: Output Spectrum of Kerr switch (retimed data signal) – Corresponding eye diagrams.

Fig. 3.48 shows results before and after the retiming system at 160 Gb/s. The spectrum shows the 160 GHz tones on the sinc-like spectrum. The clock pulses are temporally aligned to the nominal centre of the data pulses, which is thus sampled by the short clock pulses, allowing the retimed pulses to be only ~1.2 ps wide with a rms timing jitter drastically reduced to ~250 fs. Accordingly, the sampled data pulses have a broader spectrum, but still centered at the original wavelength.

Fig. 3.49 shows the BER characteristics with and without retiming. When the retimer is applied, all 16 channels are successfully retimed and error free operation

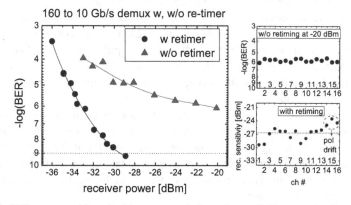

Fig. 3.49. BER curves with and without any retiming scheme (left) corresponding receiver sensitivity for all 16 channels with and without any retiming scheme (right).

is achieved for all of them (average sensitivity of -27 dBm). The spread in sensitivity among the retimed channels (~ 5dB) is due to uneven amplitudes in the multiplexed pulses in the data and clock arms, and to polarisation drifts in the Kerr switch (see last three channels in Fig. 3.49). When retiming is not applied to the system, the 410 fs rms jitter induces a severe error floor for all channels, and a BER of 10^{-6} is the best that can be obtained (at the maximum receiver power of -20 dBm), which clearly reveals the benefit of the retimer.

3.2.6.3 Conclusions

We have presented a retiming scheme and successfully demonstrated its use in a 160 Gb/s experiment. The system incorporates a superstructured fibre Bragg grating, as a pulse shaping element, and a HNLF-based Kerr switch. The original data pulses are shaped into flat-topped pulses to avoid conversion of their timing jitter into pulse amplitude noise at the output of the nonlinear switch. Thus retiming is performed in a single step avoiding wavelength conversion. The ultra-short and low jittery clock pulses used to gate the Kerr switch ensure that the switched data have low timing jitter. This allows error free operation, which would not be possible without the retimer.

3.2.7 Timing Jitter Tolerant 640 Gb/s Demultiplexing Using a Long-Period Fibre Grating-Based Flat-Top Pulse Shaper

For high-speed serial data operating at rates above 160 Gb/s, it becomes increasingly challenging to find pulse sources with sufficiently low timing jitter. Therefore it can be very advantageous to have a switch with a high tolerance to timing jitter. Such a switch may be obtained by generating a flat-top (FT) switching window, e.g. by the use of FT control pulses in a fibre-based switch relying on the ultra-fast (fs-response) Kerr-effect in non-linear fibres. For 160 Gb/s, various approaches have been reported [53,55-56], but above 160 Gb/s, a more promising scheme seems to be the flat-top pulse shaper (FTPS) based on optical differentiation in a long-period fibre grating (LPG) [57]. This scheme has the great advantage of being able to handle arbitrary pulse durations. We very recently reported a proof of principle experiment at 640 Gb/s [58], in which we demonstrated error-free demultiplexing in an otherwise jitter-limited system with the use of this technique.

In this paper, we elaborate on this new switch configuration and demonstrate that it allows not only for error-free but also jitter tolerant operation at 640 Gb/s. We show a 350 fs timing tolerance, corresponding to 20% of the bit duration, and present results on the demultiplexing of all 64 tributary channels, and also show a 14 dB improvement in receiver sensitivity when employing the FT rather than a Gaussian shaped control pulse.

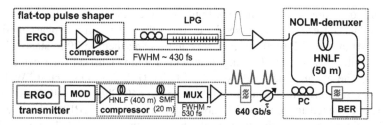

Fig. 3.50. Schematic set-up for jitter tolerant demux.

3.2.7.1 Principle and Experimental Set-up

Fig. 3.50 shows the experimental set-up. A 2 ps pulse at 1543 nm (derived from an Erbium glass oscillating pulse source, ERGO) is soliton compressed through a fibre with gain (here an EDFA) to a sech-pulse with a FWHM of 430 fs. The pulse is sent through the LPG which has a characteristic transfer function with a strong dip (see Fig. 3.51, right) and a □-phase jump across it. The dip divides the spectrum of the input pulse into to two uneven parts and, at the same time, inverts their relative phases. The effect of this is to create a superposition of the original pulse shape with its differential, which consists of a temporal double-pulse [57].

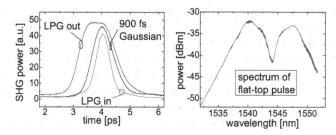

Fig. 3.51. Cross-correlation traces of the input and output of the LPG and a reference 900 fs Gaussian pulse (left) output spectrum of the LPG with the characteristic dip (right).

The original pulse can fill the gap between the double-pulse. The amount of this filling is determined by the detuning of the LPG dip with respect to the central spectral carrier. For a certain detuning, the gap disappears completely forming a perfect FT waveform. Here, the detuning is about 2 nm and the resulting FT pulse has a FWHM of 1.4 ps and a flat top part of 400 fs (within 1% of the peak intensity), see Fig. 3.51. This pulse is subsequently used as control in a non-linear optical loop mirror (NOLM) with only 50 m of highly non-linear fibre (HNLF) (dispersion slope ~ 0.018 ps/nm^{2}km, zero dispersion at 1554 nm, and non-linear coefficient of γ ~ 10.5 W^{-1}km^{-1}). A second ERGO at 1557 nm is used in an OTDM transmitter producing a 640 Gb/s serial data signal (2^7-1 PRBS, single-polarization). This ERGO runs at ~10 GHz (as the first ERGO) but is data modulated and multiplexed to 40 Gb/s before pulse compression – this procedure reduces individual channel broadening of the compressed pulses in the first stages of the MUX. The 40 Gb/s data pulses are compressed by chirping them through self-phase modulation (SPM) in a HNLF (400 m), linearly compressing them in 20 m standard single-mode fibre (SMF), and filtering them with a 9 nm Gaussian filter centered at 1560 nm. The obtained data pulses are 530 fs FWHM (7.5 nm spectral FWHM ~ 1.14 times transform limit), and are multiplexed to 640 Gb/s. The 640 Gb/s data is then demultiplexed to 10 Gb/s for subsequent bit error rate (BER) characterization. To minimize the relative timing jitter between the two ERGO lasers, the data ERGO is free running (acting as a local low-noise oscillator with 70 fs rms timing jitter), and the control ERGO together with the rest of the whole system is locked and synchronized to the data ERGO.

3.2.7.2 Dynamic Characterization – BER Performance

Fig. 3.52 shows the BER curves for 640 to 10 Gb/s demultiplexing with the use of the FT and 900 fs Gaussian control pulses. Both measurements are made using the same best-performing channel, number 31. Error-free performance is obtained in both cases, but the receiver sensitivity is greatly improved by 14 dB when using the FT pulses. The FT-demultiplexed channel 31 has no error floor and a sensitivity of -32.5 dBm, i.e. an only 3.4 dB penalty with respect to the 10 Gb/s back-to-back. In sharp contrast to this, is the Gaussian-demultiplexed channel 31 with its sensitivity of -18.5 dBm and clear error floor. Furthermore, as follows from the demultiplexed eye diagrams in Fig. 3.52, the FT pulse converts considerably less timing jitter (phase noise) into amplitude noise compared to the Gaussian control pulse.

To characterize the system tolerance to timing jitter, the data signal is gradually displaced in time relative to the position of the FT pulse and hence the switching window. Fig. 3.53 shows the results of this characterization, where channel 1 is demultiplexed with the FT pulse. The measurements are carried out at a receiver power of 5 dB above the receiver sensitivity to allow for some measurement margin.

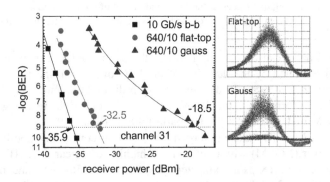

Fig. 3.52. Square pulse demultiplexing (channel 31). Left: BER of 10 Gb/s back-to-back, 640 to 10 Gb/s demultiplexing with LPG-based or Gaussian pulse. Right: Eyes for flat-top or Gaussian control.

The FT pulse is seen to maintain error-free (BER < 10^{-9}) performance with a 350 fs tolerance to temporal displacement, which roughly corresponds to the flat part of the pulse top. The BER increases rapidly for larger data-control displacements, due to the presence of the neighbouring channels. Using the Gaussian control, it is not possible to measure the timing tolerance, as the quality of this signal is too marginal. Also shown in Fig. 3.53, are the results of demultiplexing all 64 channels using the FT control pulses, in terms of receiver sensitivities. The FT pulse is clearly able to resolve all 64 channels. The relatively large spread in the measured sensitivities, is attributed to suboptimum alignment of the multiplexer, rather than the performance of the FTPS. Most channels (54) are error free. The remaining 10 channels are not error-free, due to their partial overlap with a neighbouring channel (the non-error-free channels always come in pairs – their lowest BER are listed in the inset of Fig. 3.53), something which can be avoided by careful optimization of the multiplexer. Despite these odd 10 channels, the FT control pulses are seen to clearly distinguish all 64 channels, in spite of some of them being impaired by overlap with neighbours.

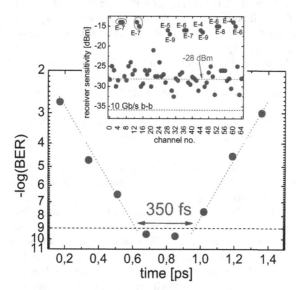

Fig. 3.53. BER at different control-data displacements (power=5 dB + receiver sensitivity). Inset: receiver sensitivities for all 64 channels.

3.2.7.3 Conclusions

We demonstrate the use of flat-top pulses for 640 Gb/s switching, with a significant improvement of the tolerance to timing jitter, enabling a strong reduction (14 dB) in penalty for 640 to 10 Gb/s demultiplexing. The pulses are created by a pulse shaping technique based on optical differentiation by an LPG, and we present a 350 fs timing tolerance, and demonstrate that all 64 OTDM channels can be distinguished.

References

1. Winzer, P., et al.: 107-Gb/s optical ETDM transmitter for 100G Ethernet transport. In: Proc. of ECOC, paper Th4.1.1 (2005)
2. Galili, M., et al.: Low-penalty Raman-Assisted XPM Wavelength Conversion at 320 Gb/s. In: Proc. of CLEO US, paper CThF4 (2007)
3. Liu, Y., et al.: Error-Free 320 Gb/s SOA-Based Wavelength Conversion Using Optical Filtering. In: Proc. of OFC, paper PDP28 (2006)
4. Rau, L., et al.: All-optical 160-Gb/s phase reconstructing wavelength conversion using cross-phase modulation (XPM) in dispersion-shifted fibre. IEEE Photonics Technology Letters 16(11), 2520–2522 (2004)
5. Bergano, N.S.: WDM long haul transmission networks. In: Proc. ECOC96, Oslo, pp. 65–71 (1996)
6. Agrawal, G.P.: Non Linear Fibre Optics, 4th edn. Academic Press, London (2006)
7. Ueno, Y., et al.: IEEE Photon. Tech. Lett. 13(5), 469–471 (2001)
8. Murata, S., et al.: IEEE Photon. Tech. Lett. 3, 1021–1023 (1991)
9. Olsson, B.-E., Blumenthal, D.J., et al.: IEEE Phot. Technol. Lett 12(7) (2000)
10. Mamyshev, P.V.: All-optical data regeneration based on self-phase modulation effect. In: Proc. Eur. Conf. Optical Communication (ECOC'98), Madrid, Spain, Sep. 20–24, 1998, pp. 475–476 (1998)
11. Taccheo, S., Vavassori, P.: paper ThU5. In: OFC 2002, Anaheim, CA, USA (2002)
12. Taccheo, S., Boivin, L.: paper ThA1. In: OFC 2000, Baltimore, USA (2000)
13. Forin, D.M., Curti, F., Tosi Beleffi, G.M., et al.: IEEE Phot. Tech. Lett. 17(2), 429–431 (2005)
14. Patrick, D.M., Ellis, A.D.: IEE Elec. Lett. 29(15), 1391–1392 (1993)
15. Nakazawa, M., Tamura, K., Kubota, H., Yoshida, E.: Opt. Fibre Tech. 4, 215–223 (1998)
16. Nakazawa, M., Kubota, H., Tamura, K.: Opt. Lett. 24, 318–320 (1999)
17. Schwartz, M., Bennet, W.R., Stein, S.: Communication Systems and Techniques. McGraw-Hill, New York (1966)
18. Forin, D.M., Curti, F., Tosi Beleffi, G.M., Matera, F.: All Optical Fibre 2+1 Auxiliary Carrier Transponder-Regenerator. Photonics Technology Letters 17(2), 429–431 (2005)
19. Wong, H.C., Ren, G.B., Rorison, J.M.: The constraints on Quantum-dot semiconductor optical amplifiers for multichannel amplification. IEEE PTL, vol 18 (2006)
20. Sygletos, S., et al.: Multi-wavelength regenerative amplification based on quantum-dot semiconductor optical amplifiers. In: 9th Intern. Conf. on Transparent Optical Networks (ICTON'07), Rome, Italy, July 1–5, 2007, Paper We.D2.5 (invited), pp. 234-237 (2007)
21. Uskov, A.V., Berg, T.W., Mørk, J.: Theory of pulse-train amplification without patterning effects in Quantum-Dot Semiconductor Optical Amplifiers. IEEE Journal of Quantum Electronics 40(3), 306–320 (2004)
22. Borri, P., et al.: Spectral Hole Burning and carrier heating dynamics in InGaAs Quantum-dot amplifiers. IEEE Journal of selected topics in Quantum electronics 6(3), 544–551 (2000)
23. Akiyama, T., et al.: Application of spectral-hole burning in the inhomogeneously broadened gain of self-assembled quantum dots to a multiwavelength-channel nonlinear optical device. IEEE PTL 12(10), 1301–1303 (2000)

24. Sugawara, M., et al.: Quantum-dot semiconductor optical amplifiers for high-bit-rate signal processing up to 160 Gb s-1 and a new scheme of 3R regenerators. Meas. Sci. Technol. 13, 1683–1691 (2002)
25. Berg, T.W., et al.: Ultrafast gain recovery and modulation limitations in self-assembled Quantum-dot devices. IEEE PTL 13(6), 541–543 (2001)
26. Gehring, E., et al.: Dynamic spatiotemporal speed control of ultrashort pulses in quantum-dot SOAs. IEEE journal of quantum electronics 42(10), 1047–1054 (2006)
27. Spyropoulou, M., Sygletos, S., Tomkos, I.: Simulation of multi-wavelength regeneration based on QD semiconductor optical amplifiers. IEEE PTL 19(20), 1577–1579 (2007)
28. Lee, H., et al.: Theoretical study of frequency chirping and extinction ratio of wavelength-converted optical signals by XGMand XPM using SOA's. IEEE Journal of Quantum Electronics 35(8) (1999)
29. Spyropoulou, M., Sygletos, S., Tomkos, I.: Investigation of multiwavelength regeneration employing Quantum-dot semiconductor optical amplifiers beyond 40Gb/s. In: Proc. of the International Conference on Transparent Optical Networks (ICTON 2007), Rome, pp. 102–105 (2007)
30. Kagawa, M., et al.: ECOC, paper We3.2.4 (2005)
31. Clausen, A.T., et al.: CLEO, paper CThQ7 (2004)
32. Boerner, C., et al.: OFC, paper OTuO3 (2003)
33. Lach, E., et al.: OFC, paper TuA2 (2002)
34. Kamatani, O., Kawanishi, S.: Prescaled Timing Extraction From 400 Gbit/s Optical Signal Using an Phase Lock Loop Based on Four-Wave-Mixing in a Laser Diode Amplifier. IEEE Photonics Technology Letters 8(8), 1094–1096 (1996)
35. Marembert, V., et al.: ECOC, paper Th4.4.1 (2004)
36. Oxenløwe, L.K., et al.: ECOC, paper We3.5.2 (2004)
37. Nielsen, M.L., et al.: Electron. Lett. 39(18), 1334–1335 (2003)
38. Zibar, D., et al.: CLEO, paper CMZ4 (2005)
39. Ware, C., Oxenløwe, L.K., Agis, F.G., Mulvad, H.C.H., Galili, M., Kurimura, S., Nakajima, H., Ichikawa, J., Erasme, D., Clausen, A.T., Jeppesen, P.: 320 Gbps to 10 GHz sub-clock recovery using a PPLN-based opto-electronic phase-locked loop. Submitted to Optics Express (November 2007)
40. Oxenløwe, L.K., Gomez Agis, F., Ware, C., Kurimura, S., Mulvad, H.C.H., Galili, M., Kitamura, K., Nakajima, H., Ichikawa, J., Erasme, D., Clausen, A.T., Jeppesen, P.: 640 Gbit/s clock recovery using periodically poled Lithium Niobate. Submitted to Electronics Letters (December 2007)
41. Sartorius, B., Bornholdt, C., Brox, O., Ehrke, H.J., Hoffman, D., Ludwig, R., Möhrle, M.: All-Optical Clock Recovery Module Based on Self-Pulsating DFB Laser. IEE Electronics Letters 34(17), 1664–1665 (1998)
42. Jinno, M., Matsumoto, T.: Optical Tank Circuits Used for All-Optical Timing Recovery. IEEE Journal of Quantum Electronics 28(4), 895–900 (1992)
43. Barnsley, P.: All-Optical Clock Extraction Using Two-Contact Devices. IEE Proceedings – Photonics Journal 140(5), 325–336 (1993)
44. Yamamoto, T., Oxenløwe, L.K., Schmidt, C., Schubert, C., Hilliger, E., Feiste, U., Berger, J., Ludwig, R., Weber, H.G.: Clock Recovery from 160 Gbit/s Data Signals Using Phase-Locked Loop with Interferometric Optical Switch Based on Semiconductor Optical Amplifier. IEE Electronics Letters 37(8), 509–510 (2001)
45. Vlachos, K., Theophilopoulos, G., Hatziefremidis, A., Avramopoulos, H.: 30 Gb/s all-optical clock recovery circuit. IEEE Photon. Techonol. Lett. 12, 705–707 (2000)

46. Carruthers, T.F., Lou, J.W.: 80 to 10 Gbit/s Clock Recovery Using Phase Detection with Mach-Zehnder Modulator. IEE Electronics Letters 37(14), 906–907 (2001)

47. Wang, T., Li, Z., Lou, C., Wu, Y., Gao, Y.: Comb-Like Pre-processing to Reduce the Pattern Effect in the Clock Recovery Based on SOA. IEEE Photonics Technology Letters 14(6), 855–857 (2002)

48. Aleksic, S., Ribnicsek, G.: Fast Clock Recovery Methods for Application in All-Optical Networks. In: Conference on Optical Network Design and Modeling (ONDM 2006), Copenhagen, May 2006, pp. 1–5 (2006)

49. Silveira, T.G., Teixeira, A., Tosi Beleffi, G., Forin, D., Monteiro, P., Furukawa, H., Wada, N.: All-Optical Conversion From RZ to NRZ Using Gain-Clamped SOA. IEEE Photon. Tech. Lett. 19(6) (2007)

50. Ono, S., Okabe, R., Futami, F., Watanabe, S.: Novel demultiplexer for ultra high speed pulses using a perfect phase-matched parametric amplifier. In: OFC 2006 (March 2006)

51. Song, X., Yu, F.C., Song, H., Sugiyama, M., Nakano, Y.: All-Optical OTDM DEMUX with Monolithic SOA-MZI Switch by Regrowth-Free Selective Area MOVPE. In: Lasers and Electro-Optics, CLEO/Pacific 2005 (Aug. 2005)

52. Olsson, B.-E., Blumenthal, D.J., et al.: A Simple and Robust 40-Gb/s Wavelength Converter Using Fibre Cross-Phase Modulation and Optical Filtering. IEEE Phot. Technol. Lett. 12(7) (2000)

53. Watanabe, S., et al.: 160 Gb/s Optical 3R-Regenerator in Fibre Transmission experiment. In: Proc. of OFC, PD16-1 (2003)

54. Petropoulos, P., et al.: Rectangular pulse generation based on pulse reshaping using a superstructured fibre Bragg grating. J. Lightwave Technol. 19, 746–752 (2001)

55. Parmigiani, F., et al.: All-optical pulse reshaping and retiming systems incorporating a pulse shaping fibre Bragg grating. J. Lightwave Technol. 24(1), 357–364 (2006)

56. Oxenløwe, L.K., et al.: ECOC, paper We2.3.4 (2006)

57. Park, Y., et al.: Opt. Express 14(26), 12671 (2006)

58. Oxenløwe, L.K., et al.: CLEO-Europe'07, paper CI8-1 (2007)

4 Evolution of Optical Access Networks

P. Kourtessis (chapter editor), C. Almeida, C.-H. Chang, J. Chen,
S. Di Bartolo, P. Fasser, M. Gagnaire, E. Leitgeb, M. Lima, M. Löschnigg,
M. Marciniak, N. Pavlovic, Y. Shachaf (assistant editor), A.L.J. Teixeira,
G.M. Tosi Beleffi, and L. Wosinska

Abstract. This chapter reviews the current developments in access network architectures and protocols to communicate dynamically the emerging broadband services to end-users at low cost. Following a summary of Gigabit Ethernet and Passive Optical Network (PON) standards and deployment issues with reference to Ethernet (EPON) and Gigabit-capable PON (GPON) infrastructures, an original transparent network architecture is presented to allow interoperability of time division multiplexing (TDM) and wavelength division multiplexing (WDM) PONs, by means of coarse routing. To provide flexible connectivity at extended service reach hybrid wireless and free space optic technologies have been investigated to terminate mobile end-users to high bandwidth PON terminals. To demonstrate independent bandwidth management of the constituent sectors of such architectures developed dynamic bandwidth allocation (DBA) algorithms are summarised followed by an original control plane to coordinate the various mandatory access control (MAC) protocols. Finally, to provide reliable service delivery several protection schemes have been analysed.

4.1 Introduction: FTTX Developments

The developments in photonic technologies, optical techniques and fibre-plants deployment achieved a great increase in transport network capacity. This was followed by the seemingly improvements of capacity at the end user side making possible handling of large bandwidth services like symmetric peer to peer, HDTV broadcasting, remote storage, e-services and grid computing [1].

FTTx (fibre to the home/curb/building/premises) is only one of the technologies now explored to spread these broadband services, employing wireless and wired solutions both fibre and copper based [2-5]. FTTx is the most future-proof infrastructure available for providing triple-play services, employing various types of technologies including active Ethernet and standard Passive Optical Networks (PONs). FTTx has become the worldwide network architecture for broadband access deployed by large carriers and small municipalities. Among the different types of PON technologies EPON and the GPON are the mostly deployed standards. EPON was developed and formalised in the IEEE 802.3ah standard [6] to bring Ethernet to residential and business customers in the access network [7] and has been adopted rapidly, primarily due to its ubiquitous and cost-effective tech-

I. Tomkos et al. (Eds.): COST 291 – Towards Digital Optical Networks, LNCS 5412, pp. 97–131, 2009.

nology, allowing for interoperability with a variety of legacy equipment [7, 8]. EPON is capable of providing triple-play services [9] at symmetrical 1.25 Gbit/s data rates using wavelengths at 1490 nm and 1310 nm for downstream and upstream transmissions respectively, with a wavelength at 1550 nm reserved for future extensions or additional services such as analogue video broadcast for a maximum of 32 ONUs at distances of up to 20 km [10, 8, 11]. In upstream, a typical mandatory access control (MAC) protocol is used [12-15], avoiding data collisions in the distribution network. In order to extend the capacity limitation of EPON by means of data rates and splitting ratio, the ITU-T has set standards for the GPON in the G.984 series [16-19].

The GPON is set to accommodate full services and support Gigabit bandwidth for subscribers in the access network. It emerged to remove the bandwidth bottleneck in the first mile [20-22] offering technical advantages over EPON due to higher splitting ratio, data rates and bandwidth efficiency [23]. GPON provides triple-play services at variable bidirectional data rates of up to 2.5 Gbit/s using similar wavelength assignment, allowing network operators to configure transmission rates according to user requirements. Unlike EPON, GPON supports a maximum of 128 ONUs for distances of up to 60 km [16, 23] as well as offering almost double bandwidth efficiency due to less data transmission overheads. To address security issues in downstream, GPON provides an advanced encryption standard [21].

In downstream, signals communicated from the OLT reach the passive optical splitter and broadcasted to all ONUs in the PON. Although all ONUs receive all downstream data, due to broadcasting, secure channels are established to ensure that each ONU only recovers the data intended for itself [24]. In upstream, the splitter combines all ONU packets in a TDM fashion to avoid collisions in the feeder fibre. Consequently, each ONU has to restrict its transmission only to predefined time-slots administrated by a MAC protocol, which can statically or dynamically allocate time-slots among ONUs [24].

The fundamental functionality of MAC protocols is to allocate constant time-slots for each ONU at regular periods regardless of its bandwidth requirement. However, a practical limitation of constant TDM protocols is the demonstrated inefficiency in time-slot assignment since independently of either network ONUs are silent or exhibit low traffic load by means of moderate buffer queuing performance, the same, constant time slots are assigned at each cycle of operation indicating that fractions of the allocated time-slots remain idle and equally important incapable of being transferred to high bandwidth required ONUs. In the latter, random packets will have to be buffered for various polling cycles since their capacity requirement exceeds the relative assigned bandwidth. To overcome these drawbacks TDM-dynamic bandwidth allocation (DBA) protocols have been developed to increase the transmission efficiency and to reduce the packet delay by dynamically allocating upstream time slots according to ONUs' bandwidth demands and overall network capacity [25-28].

To provide dedicated bandwidth per subscriber by allowing virtual point to point links between the OLT and each subscriber where time slot management is not required, wavelength division multiplexing (WDM)-PONs have been extensively researched [29-34] as a viable solution for next-generation optical access, offering also greater security and protocol transparency [29,34]. However, the economic model for a densely penetrated PON deployment in the access has not been justified yet, mainly due to the high cost of components in the ONU and OLT [35]. Consequently, research efforts have been concentrated [31, 33, 36] in providing enhanced PON architectures combining the merits of time and wavelength multiplexing, consequently to serve increased number of customers by assigning more customers on each wavelength, and to share network resources among greater number of PONs, thus increasing cost efficiencies and revenue from invested resources. At the same time they allow minimum modification in the ONUs and outside plant when upgrading to densely penetrated PONs. In a search for another potential cost-effective solution for a flexible high-speed optical access [35], coarse-WDM (CWDM) [37] has been also studied [38-41], employing up to 18 channels, 20 nm spaced, allowing for CWDM devices [42] to utilise at least 7 nm-wide passband windows around each of the coarse channels. Such broad channels relax tolerances for thermal management [43] and optical loss requirements, allowing for reduced network complexity and flexible network power budget management respectively.

4.1.1   FTTX Architectures

European Community has answered to the request for increased bandwidth by financing several projects targeting the delivery of fibre bandwidth all the way to end users [44-54] in the form of specific targeted research projects or the networks of excellence and integrated projects with ATLAS, e-Photon ONe, NOBEL and BONE being the most prominent of the latter. The message that should be retained from this review is that the most common topology for the metro & access network, and in particular for the FTTx sector, is based on a ring interconnecting multiple trees. Although, optical and copper based technologies can coexist, a gradual transition is expected that will potentially lead to the all optical network of the future (AO NoF).

Within the specific targeted research projects arena, HORNET [44, 45] is a given suggestion for next generation metro networks. The main architecture is based on a packet-over-WDM ring network utilising fast-tunable packet transmitters. RINGO [46] presents two versions of WDM ring: the first is based on a unidirectional WDM fibre ring with N nodes and N wavelengths; the second is a folded ring, separating the transmission/reception using two fibres ring solution. WONDER [47, 48] is an evolution of the RINGO architecture for metro applications that limits optical complexity and uses only commercially-available components. The architecture is a bidirectional WDM amplified optical ring.

RINGOSTAR [49] is a hybrid ring-star architecture. This topology not only combines the strengths of both ring and star configurations, but also avoids their drawbacks. Generally, only a subset of ring nodes is directly connected to the star network, resulting in less fibre requirements and node interfaces. The FLAMINGO [50] network is based on a WDM fibre ring with a node called Access Point (AP). The access to the Metropolitan Area Network (MAN) is guaranteed by several access points that connect the MAN with Wide Area Network (WAN). The bandwidth of each channel time-shared with constant length slots that go around the ring. Each AP is capable to transmit and receive at any wavelength. The DAVID [51] network architecture covers the MAN and WAN, each having a distinct operating structure. The network uses a fixed-length packet in a slotted mode. The MAN is based on one or more unidirectional optical physical rings interconnected by a Hub. A ring within the MAN will consist of one or more fibres, each in DWDM, where each wavelength will be used to transport optical packets.

The BORN [52, 53] network connects several edge nodes. The main ring constitutes working and protection fibres, where the latter is able to provide resilience in case of a single link damage. It is also a WDM ring, employing a set of wavelengths for upstream and downstream signals. This architecture clearly identifies a hub node for the metro ring. Transparent ring nodes can read and write dynamically in upstream and downstream bus respectively. BORN network is organised around a hub, where an optical packet add&drop multiplexer can insert or extract the optical packet produced by several nodes.

MATISSE, next generation high-speed Internet backbone networks will be required to support a broad range of emerging applications which may not only require significant bandwidth, but may also have strict quality of service (QoS) requirements and are expected to be highly bursty in nature. For such traffic, the allocation of static fixed-bandwidth circuits may lead to the over-provisioning of bandwidth resources in order to meet QoS requirements. Optical burst switching (OBS) is a promising new technique which attempts to address the problem of efficiently allocating resources for bursty traffic. Design and research issues involved in the development of OBS networks are discussed in the MATISSE Project.

SUCCESS is a next-generation WDM/TDM optical access architecture, providing practical migration path from current TDM-PONs to future WDM optical access networks. The topology is based on a collector ring and several distribution stars connecting the Central Office (CO) with both business and residential users. The SARDANA [54] project targets the performance enhancement of dense FTTH networks, utilising two key performances such as scalability and robustness, since they constitute pillars of such a cost-sensitive segment. Scalability is reached by means of the new adoption of remotely-pumped amplification, a WDM/TDM overlay and cascadable remote nodes. The network is capable of serving up to 4000 users with symmetric several hundred Mbit/s per user, spread along distances up to 100 km. Robustness is achieved by means of passive central-ring protection and new monitoring and electronic compensation strategies over the PON, intelligently supervising and controlling potential impairments when extended to a 10Gbit /s PON.

4.1.2 Current Standard PON Deployment Worldwide

With fibre cabling cost [55] in mind, EPON and GPON architectures have been largely deployed in Asia and North America with Europe moving gradually forward. Throughout Europe there are currently 1 million FTTH subscribers [56] based on GPON networks, although the number of connected homes is approximately 2.6 million [57] with a forecast of 4 million FTTH subscribers by 2010 [58].

On the contrary, South Korea has maintained high broadband service penetration rate with current rapid growth of approximately 7 million subscribers requesting triple-play services in excess of 50 Mbit/s [9] and as a result approximately 12 million subscribers are expected to be served by 2010 [9, 59]. WDM-PON based FTTH has been chosen to be deployed primarily, capable of delivering triple-play services at symmetrical bandwidth of 155 Mbit/s per subscriber [29, 60].

The Japanese broadband market is also entering into total FTTH age. Although initially FTTH was used only for Internet access, at present, network providers employ mainly EPON systems to provide up to 100 Mbit/s per customer focusing on triple-play services as the main drive to full-scale FTTH [61]. The FTTH market in Japan continues to grow rapidly as the number of residential and business customers have exceeded 10 million [56] with a target of serving 30 million by 2010 [23, 61].

In the Hong Kong region, around 23 percent of households are presently connected with FTTH [56]. The first provider, Hong Kong Broadband Network Limited (HKBN), to launch FTTH with 100 Mbit/s and 1 Gbit/s services in 2005, has been operating GPON at 2.5 Gbit/s downstream since the beginning of 2008 and is planning to increase its coverage to 2 million home connected within few years [62].

Finally, in the United States there are currently 10 million [56] FTTH connections with the aim to exceed 18 million households by 2010 [63], out of which nearly 2 million [56] residential and businesses customers are currently receiving services by Verizon as the main player in addition to other suppliers. Recently, GPON has been receiving more attention and has already been in deployment in undeveloped areas.

4.2 Emerging Standards for 100 Gbit Ethernet Access and Beyond

4.2.1 Introduction – Why Higher Speed Ethernet?

Ethernet, being originally a computer networking protocol, nowadays is able to unify long distance, metro and access networking into a single network of the future [64]. The deployment of Fibre-To-The-Home in access observed in Japan, Korea, US and Europe will assure a broad bandwidth for the user at an affordable cost [65]. Computing speed and system throughput doubles approximately every two years. Consequently, fundamental bottlenecks are happening everywhere. Increased number of users together with increased access rates and methods and increased services results in explosion of bandwidth demand. Networking is driven by the aggregation of data from multiple computing platforms. When the number of computing platforms grows fast, it results in a multiplicative effect on networking [66].

Therefore it is necessary to provide a solution for applications that have been demonstrated to need bandwidth beyond the existing capabilities. These include IPTV, downloading/uploading of large files at short time, internet exchanges, high performance computing and video-on-demand delivery. High bandwidth applications, such as video on demand and high performance computing justify the need for a 100 Gb/s Ethernet in metro and access networks. Indeed, even a personal computer will surpass 10 GHz computation speed in few years.

4.2.2 100 Gbit Ethernet Challenges

Ethernet is now widely adopted for communications in local area networks and in metropolitan area networks. The Ethernet is facing the next evolutionary step towards 100 Gbit/s Ethernet, or 100GbE [67, 68]. As Ethernet becomes more prevalent, the issues related to the software, electronics, and optoelectronics need to be addressed. This becomes more evident for 100GbE, since that technology does not simply refer to high bit rate transmission at 100 Gbit/s, but also relates to switching, packet processing, and queuing and traffic management at 100 Gbit/s line rate. This is in parallel with a remarkable progress in transmission as 10 Gb/s and recently 40 Gb/s systems have become commercially deployed standards in optical networking, and multiplying the total aggregate capacity by an use of DWDM technology and transmitting simultaneously several wavelength channels. This has faced problems in view of fibre impairments, one of the most serious ones being fibre Polarisation Mode Dispersion (PMD). In particular, care has to be taken to minimise PMD coefficient when manufacturing the fibres and cables.

The IEEE HSSG (Higher Speed Study Group) objectives are:

- Support full-duplex operation only
- Preserve the 802.3/Ethernet frame format utilizing the 802.3 Media Access Control
- Support a Bit Error Rate (BER) better than or equal to 10-12 at the MAC/PLS (Physical Layer Signalling) service interface
- Support a MAC data rate of 100 Gb/s
- Provide Physical Layer specifications which support 100 Gb/s operation over: At least 40km on SMF (Single Mode Fibre), 100m on OM3 MMF (850nm Laser Optimized Multi-Mode Fibre) and 10m over a copper cable assembly

As an amendment to IEEE 802.3, the proposed project will follow the existing format and structure of IEEE 802.3 MIB definitions providing a protocol independent specification of managed objects (IEEE 802.1F). As was the case in previous IEEE 802.3 amendments, new physical layers specific to either 40 Gb/s or 100 Gb/s operation will be defined.

4.2.3 Transparent Optical Transmission For100 Gbit Ethernet

The technical feasibility of 100 GbE has been already proven, as well as its confidence in reliability. The principle of scaling the IEEE 802.3 MAC to higher speeds has been already established within IEEE 802.3. Systems with an aggregate bandwidth of greater than or equal to 100 Gb/s have been demonstrated and deployed in operational environment. The 100 GbE project will build on the array of Ethernet component and system design experience, and the broad knowledge base of Ethernet network operation. Moreover, the experience gained in the deployment of 10 Gb/s Ethernet might be exploited. For instance, parallel transmission techniques allow reuse of 10 Gb/s technology and testing.

An alternative approach to avoid the development of ultra-fast electronic circuits is to use advanced modulation formats that achieve 100 Gbit/s information rate while allowing lower transmission rates. In such a case, the implementation will require components operating around 50 GHz and since electronic circuitry for 40 Gbit/s is already commercially available, there will be an easier migration to the development of say 50 Gbit/s capable silicon components. Finally, for short reach interfaces there have been a number of implementations that provide 10 or 12 parallel 10 Gbit/s lanes for a total aggregate bit rate of 100 Gbit/s or 120 Gbit/s. Such solutions are being currently under discussion in the IEEE HSSG [69].

The next step in order to increase data rates and speed of the services is the introduction of services based on 100 Gb Ethernet. But 100 Gbit/s transmission is standing on the very beginning and the worldwide level of knowledge and know-how in the field of 100 Gbit/s is still low. A lot of research activities have to be done until the first test links can be prepared for commercial and field exploitation. First of all integrated circuits are necessary which enable transmission equipment, like e.g. transceivers, to provide this high speed data signal with an adapted modulation technique. To make the technology suitable for exploitation basic physical effects must be investigated in order to use them for a future technology or to minimise or overcome them if they contribute impairments. Only then all the processes for the production of necessary components can be controlled with the desired and necessary reliability. Other challenges like the cost reduction of the components, the reduction of the operational expenses of the network operators and the minimisation of the energy consumptions are also a big challenge and subject of research.

The existing 802.3 protocol has to be extended to the operating speed of 40 Gb/s and 100 Gb/s to provide a significant increase in bandwidth while maintaining maximum compatibility with the installed base 802.3 interfaces, previous investment in research and development, and principles of network operation and management.

4.2.4 Future Directions

Optical networks consisting of standard single mode fibres are in principle suitable for transportation of data rates up to 100 Gbit/s and more, are to be widely deployed both in long distance and in metro/access. Physical limitations laid by the fibres themselves require new technologies to overcome these constraints. Noise accumulation, chromatic dispersion, polarisation mode dispersion and nonlinear effects limit data rate and maximum transmission distance. Highly stable 100 Gbit/s Ethernet transmission over different distances through the network would require pushing state of the art in the limits towards optimisation and development of new technologies and components for transmitters and receivers.

A possible solution for 100GbE modulation format can be a pure multi-level amplitude modulation, offering the advantage of lower clock frequency and required signal bandwidth of critical components, e.g. modulators. On the other hand, the robustness of multi-level modulation scheme against such common impairments in the transmission path as optical amplifier noise and fibre dispersion must be carefully analyzed. Bandwidth requirements for computing and networking applications are growing at different rates. These applications have different cost / performance requirements, which necessitates two distinct data rates, 40 Gb/s and 100 Gb/s.

4.3 Interoperability of TDM and WDM PONs

4.3.1 Introduction

The emergence of new bandwidth-intensive applications has ultimately justified the necessity of upgrading the access network infrastructure to provide fat-bandwidth pipelines at subscriber close proximity. PONs offer currently more opportunities to communicate these services than ever before, with potential connection speeds of up to 100 Mbit/s in mind [70]. PON-based access networks envisage the demonstration of scalability to allow gradual deployment of time and wavelength multiplexed architectures in a single-platform without changes in fibre infrastructure and also highly-efficient bandwidth allocation for service provision and upgrade on-demand.

4.3.2 Network Architecture

An access network architecture [39, 42] has been investigated to provide interoperability among dynamic TDM and WDM-PONs through CWDM routing in the OLT. The network architecture in Fig. 1 exhibits a single 4x4 coarse array waveguide grating (AWG) in the OLT to route multiple PONs by means of a single tunable laser (TL1) and receiver (RX1) allowing for coarse-fine grooming to display smooth network upgrade. Proposed coarse AWG devices display 7 nm, 3 dB Gaussian passband windows [42], denoted in Fig. 1 by coarse ITU-T channels λ_1=1530 nm and $\lambda 2$ =1550 nm, set to accommodate up to 16, 0.4 nm-spaced wavelengths to address a total of 16 ONUs per PON. In downstream, TL1 will optimally utilise $\lambda 19$, placed at the centre of the AWG coarse channel λ_1, to broadcast information to all ONUs of PON1. To address a WDM-PON, TL1 will switch on all 16 wavelengths, centred ±3.2 nm around coarse channel λ_2, λ_2^{1-16}, to address jointly all ONUs in PON4 [71]. The established network interoperability allows a smooth migration from single to multi-wavelength optical access to address increasing bandwidth requirements.

Reflective semiconductor optical amplifier (RSOA)-based ONUs are universally employed, avoiding the necessity of wavelength-specific, local optical sources. The use of multiple transceivers in a single OLT to serve all reflective PONs allows for centralised control to distribute ONU capacity among upstream and downstream on demand and concurrently provide each PON with multiple wavelengths for enhanced bandwidth allocation flexibility [71]. The network also exhibits increased scalability since extra TLs can be directly applied at unused AWG ports in the OLT, e.g. TL2 at I/O port 2, to maintain high network performance at increased traffic load.

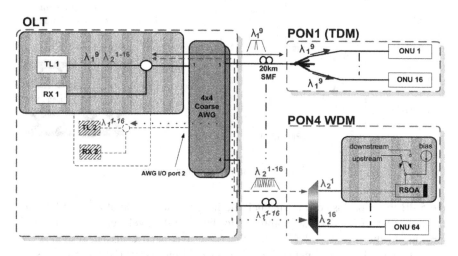

Fig. 4.1. Unlimited-capacity multi-PON access network architecture.

4.3.3 Network Routing Performance

In defining accurate routing of all ONUs comprising a PON, the AWG polarisation-dependent wavelength (PDW) shift [42] and associated polarisation-dependent loss (PDL) have been simulated [39,71] and also experimentally demonstrated. In that sense, Fig. 2 displays the PDW shift of the central channel λ_2=1550 nm, for which the transverse magnetic (TM) response is -1.8 nm shifted [42] with respect to the transverse electric (TE). It can be observed that on top of

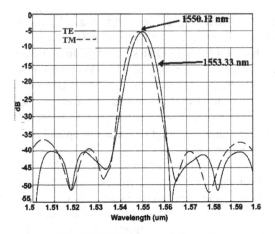

Fig. 4.2. PDW of central channel.

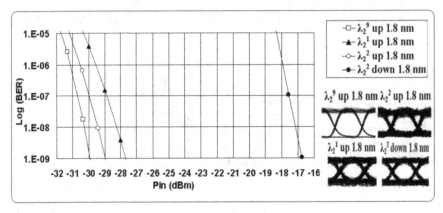

Fig. 4.3. BER down/upstream transmissions.

the 5 dB insertion loss of the channel, the longest wavelength at λ_2^1=1553.33 nm can suffer up to 5.5 dB loss, compared to the central wavelength at λ_2^9=1550.12 nm, due to the increase of PDL and the AWG passband Gaussian response. To evaluate the bidirectional network performance, bit-error-rate (BER) responses, in Fig. 3 [39], have confirmed error-free transmission with measured 10^{-9} BER achieved by 93% of ONUs when 1.8 nm PDW is employed. Further investigations into the highest possible PDW to allow all ONUs achieve a BER of 10^{-9} produced a PDW figure of 0.8 nm. In order to enhance the bandwidth utilisation for each subscriber, a full-duplex transmission has been established [72] using polarisation division multiplexing (PDM) [32]. Consequently, the scheme overcomes time-interleaving in the OLT of the transmission time-slots between CWs and burst-data, allowing for their independent transmission over the two polarisations to avoid multiplexing idle time [72].

4.3.4 Conclusions

Extensive efforts have been carried out to provide enhanced PON architectures combining TDM and WDM to increase user-density and resource-sharing, particularly allowing smooth upgrade to densely penetrated PONs. In that direction, a multi-PON access network architecture has been developed, demonstrating interoperability among dynamic TDM and WDM-PONs through coarse channels of a AWG in a single OLT with coarse-fine grooming features. Simulation results to address a fully-populated reflective WDM-PON have demonstrated error-free routing of 16, 0.4 nm-spaced wavelengths, over a single 7 nm-wide Gaussian passband window of the AWG. This has been subsequently verified experimentally with a commercially-available 2.7 nm-wide coarse-AWG. To achieve full-duplex operation, an orthogonally-multiplexed OLT has been devised, allowing for increased bandwidth utilisation in both downstream and upstream.

4.4 3G Radio Distribution over Fibre

4.4.1 Introduction

Radio waves are nowadays the most popular way to access data and communicate, since they are used in the very front end of every user, as they provide one extremely important capability: mobility without complexity (user logs without any special specific knowledge). This has made communications and data transfer a reality wherever people go. Nowadays it is a surprise when there is no service coverage.

The demand for more services and increased penetration of data and voice, has pushed the operators into several developments and strategies to enable full time, space and, whenever possible, bandwidth coverage to the users by exploring all types of technical solutions that can make the three aforementioned guidelines possible. To achieve the referred target, bandwidth should be delivered and managed in highly dense sporadic places (commercial centres, sport centres, etc) as well as areas with feeble wave propagation (tunnels, galleries, etc). One approach to address these issues is to take the digital data through the fibre and in the end place a complete base station to manage all the required problems. However that is not always possible (accessibility, etc) therefore other strategies emerged, eg. radio over fibre (RoF). UMTS, Wi-fi and WiMAX Radio Distribution over Passive Transparent Optical Networks can be seen as a promising technique to overcome many of the RF spectrum and distribution limitations. However this can only be a reality if it can be shown that it is robust and possible to implement in a cost effective way.

Several studies have been performed within the scientific and corporate communities throughout the world [73]. With the current trends in PON deployments and massive deployment of triple and upper count play (tetra and penta play), will require that new approaches which can be compatible with existing and currently deploying G-PON standards (or in the future with NG-PON) should be considered. These new signals should be directly delivered to the costumer without the need of any expensive equipment at the customer premises or to be distributed to highly inaccessible or difficult to propagate radio waves places in conjunction with existing data and standards. It is still important to demonstrate the propagation capabilities of these signals and their robustness to propagation and other impairments. The basic requirement for any mass deployment strategy is that the emitter/receiver should be cheap and the modulation should not bring increased complexity. A block diagram of this kind of systems is presented in Fig. 4.4.

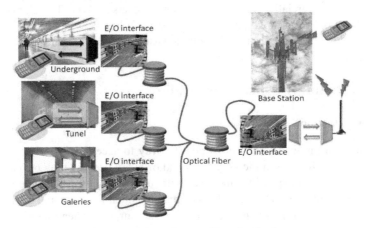

Fig. 4.4. Diagram of a radio over fibre distribution system.

The referred modulation formats UMTS, WiFI and WiMax are different in nature and therefore they will behave naturally differently when propagated.

UMTS:	WLAN:	WIMAX:
– Physical layer mode : WCDMA-3GPP; – Symbol rate: 3.84 Msymb/s; – Modulation: QPSK 45° offset – Filter: root cosine, roll off factor = 0,22 – At receiver EVM<12%.	– Standard : 802.11g; – Physical layer mode: OFDM; – Modulation: 64 QAM; – At the receiver EVM<5.62%.	– Physical layer mode: OFDM; – Duplexing: TDD, downlink; – Frame configuration: QPSK ¾; – At receiver EVM<-18,5dB (≈1.41%).

From the works performed one of the basic conclusion is that cheap lasers can be suitable for the modulation of any the signal formats. This conclusion makes cheap modules a possibility. Due to its nature, even semiconductor amplifiers, which are on of the options for wavelength agnostic ONU's (Optical Network Units), like RSOA's and SOA's (Reflective Semiconductor Optical Amplifiers), can be suitable for this operation making this a viable option for the next generation networks. In any case should be granted that simple antenna site should be granted.

Several demonstrations are being performed showing the viability of this strategy [73]. A set of tests was made in laboratory environment where the EVM (Error Vector Magnitude), which is the metrics used for the characterization of physical transmission in RF propagation was used. A schematic of what can be the up or downstream in a pre-amplified radio distribution system is illustrated in Fig. 4.5. SOAs were used as a booster (downstream) or a pre-amplifier (upstream) since they are the bigger candidates to be integrated and potentially cheap. The usage of SOAs can potentially bring limitations to the system due to its fast response to optical signals and input signal power changes. So, a carrier density change can lead to a amplitude and/or phase shift on the modulated signal resulting in signal distortion.

Fig. 4.6 presents the results obtained when the three different signals are transmitted upstream over 40 Km of fibre. It can be seen that for electrical power

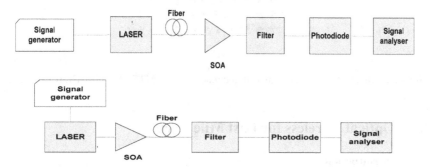

Fig. 4.5. Up and down streams in a centrally preamplified radio distribution.

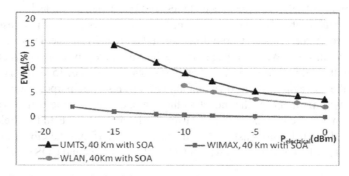

Fig. 4.6. Results obtained for 40 Km of fibre using SOA as pre-amplifier.

higher than -12.5dBm, -6dBm and -15dBm respectively for UMTS, WiFi and WiMAX, any of the tested transmissions will cope with the minimum EVM standards. It can also be concluded that for the same emitter, propagation and receiver conditions the most robust (less requiring in terms of electrical power) signal is the WiMAX.

To better illustrate what was obtained, the constellations at the signal analyzer for the transmission of the signals with RF power of 0dBm are illustrated in Fig. 4.7. Due to its inherent definitions the constellations have different formats and aspects, however it is proven that is viable the usage of low cost devices to extend the reach of radio signals.

The overall conclusions that can be brought out of this work and comments is that the current radio signals are robust enough to be put into fibre and current configuration could be extended with low cost optical interfaces, opening a window of operation for the next generation of optical fibre systems.

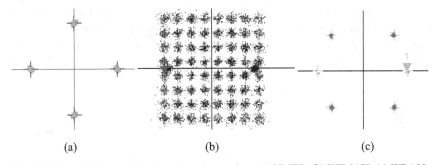

(a)   (b)   (c)

Fig. 4.7. Constellations of the signals at the receiver (a) UMTS, (b) WIMAX, (c) WLAN.

4.5   Optical Wireless for Last Mile Access

4.5.1   Introduction

Today, many people need a high data rate for their internet connection or for other access-services. Therefore Free Space Optics (FSO) is a well suited technology. The high bandwidth of the "Backbone" (fibre network) is also available for the user. The research group for Optical Communication ("OptiKom") at the Department of Communications and Wave Propagation, Technical University Graz conducted the research in the field of Free Space Optics (FSO) over a period of 15 years. The work includes the development of own equipment for research purpose and in co-operation with partners (of the industry) the evaluation of commercially available FSO systems for the climate in Graz (Austria). The main projects in this field have been funded by Telekom Austria AG, InfraServ Gendorf and the Government of Styria and later by co-operations within COST 270, SatNEx and COST 291.

In free space the transmitted light (signal) is reflected, refracted or absorbed by objects, rain, fog, wind or sun. Usually in free space each wavelength can be used. But because of the atmospheric conditions and due to the laser safety regulations the larger ones (i.e. 1550 nm) are better for transmission. The losses due to Mie-scattering in haze or light fog at longer wavelengths (1550 nm) are smaller than at shorter ones (850 nm). Due to atmospheric turbulences the received polarized light has changed its direction slightly, compared to the transmitted beam.

4.5.2 FSO Networks

Optical Wireless is the nomadic broadband solution (high data rates without any cabling) which connects the "Backbone" to the clients (Last-Mile-Access). This technology is an excellent supplement to conventional radio links and fibre optics [74] for short ranges ("Last Mile", max. 1 km). Meshed FSO networks have shown advantages in regard to Ring or Star configurations. Meshed networks combine and sum up the benefits of all available architectures.

In each configuration, the FSO-unit can be connected with a node to satellites, directional radio links, telephone networks, mobile phones or fibre optics. In Fig. 4.8 a connection to a fibre realised with a Point-to-Multipoint architecture is shown. The Optical Multipoint Unit (OZS) is connected with a switch or router to the "Backbone"-network. The clients in the buildings are linked with their FSO units to the OZS. This is a similar solution like the FSO-network at the Department of Communications and Wave Propagation at the TU Graz.

Additional TU Graz has done similar installations in 2003 (in Vienna at the UNO-City and in Graz for a cultural event Graz 2003) and in 2004 the CIMIC exercise "Schutz 04" in Styria (also in combination with Satellite and Wireless LAN [74]. The government of Styria uses such permanent Free Space Optics systems since 2004.

In this part we will show the latest permanent installation, which was started in the summer of 2007 together with the municipality of Dobl, a small suburb of Graz, and the ML11 GmbH, an IT-company. Within this cooperation the elementary school of Dobl was connected to the Internet via FSO. The FSO-units were

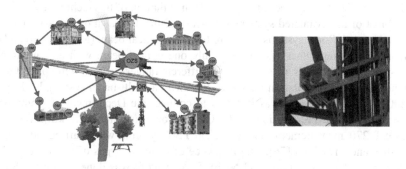

Fig. 4.8. Optical Wireless (meshed architecture).

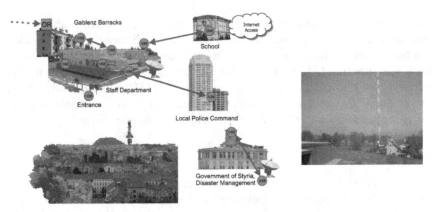

Fig. 4.9. Free Space Optics demonstration in the CIMIC exercise "Schutz 04".

successfully installed and the setup is now used to explore the influence of weather on the FSO-link in the area of Graz.

The installed Optical Wireless system is working with a data rate of 10 Mbps and the school (in a distance of 350 meters from the large antenna) has now a well working Internet connection. This installation is also used for research and investigations on reliability and availability studies well known at TU Graz [75]. One of the FSO-units was installed on an old aerial mast for a radio transmitter (build in the second World War, now used for weather monitoring and as mobile phone base station for all Austrian mobile phone providers).

4.5.3 Propagation Results

Because of the different weather conditions, thunderstorms in summer (often in combination with hail), drizzle, storm and snow in winter, Graz is a very good area for testing the reliability and availability of optical free space links. Since July 2000 such tests have been running on existing optical free space laser systems for Telekom Austria AG and on self-developed optical point-to-point and point-to-multipoint-systems. Models for propagation and prediction of link availability in different climate zones could improve the installation of this technology.

Most of the evaluated systems differ in technical realisation and can hence not be compared. Some systems use more than one optical beam (Multi-beam) and only manual tracking. Other systems use only one optical beam with an "autotracking system". The products are specified for different ranges (from 100 m up to 4000 m). One of our first evaluation tests was done with a multi-beam system, back in July 2000 [75]. This system was installed to connect the Department of Communications and Wave Propagation to the Observatory Lustbühel (2.7 km). Within the COST 270 measurement campaigns on FSO-systems were carried out together with France Telekom [76]. The influence of different types of fog were analysed on various locations (Graz and Nice). Further work was done within the SatNEx

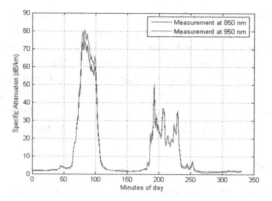

Fig. 4.10. Specific attenuation for 850 & 950 nm at Graz.

project (a Network of Excellence within framework 6). The follow-up SatNEx 2 co-operation with scientists from Slovenia and Italy has already started.

Fig. 4.10 shows the main difference between continental fog and maritime fog. In a typical dense continental fog (occurring in Graz as example) the attenuation is around 100 db/km and the changes of the fog-attenuation are extremely slow, that means the fog is very stable for long times (minutes and hours) and does not change in seconds. But dense fog in maritime regions has much higher attenuation (up to 400 dB/km) and it changes in seconds. In maritime fog the wind on the seaside is moving the air-masses and that results in faster changes of the attenuation. Concentration and size of fog particles is much more different, compare [77]. Measurements in dense maritime fog and continental fog were preformed and compared.

Our results show basically no wavelength dependency for maritime fog conditions. For continental radiation fog conditions (like at Graz), we have found indications for wavelength dependent attenuation using the same measurement

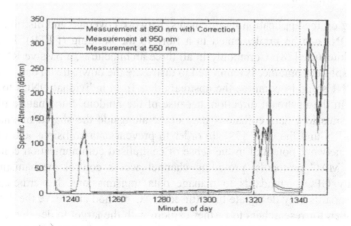

Fig. 4.11. Specific attenuation 550 nm, 850 & 950 nm at Nice.

equipment. This means that different models for attenuation caused by different types of fog can be justified. For designing high performance links and to possibly design an intelligent switchover between the FSO and a redundant RF link, the fluctuations in the time domain are of key importance. If the averaged minute values of visibility are used (as taken by meteorological stations) then according to the models, it would not exceed a certain value; but due to the fluctuations, this value can be exceeded by a percentage as high as 68 %. The attenuation characteristics of snow clearly indicate [76] that it is much less detrimental for the link.

4.5.4   Conclusions

Optical Wireless is an excellent nomadic broadband solution for connecting the backbone to the users. This technology should be understood as supplement to conventional radio links and Fibre Optics. The use of cheaper FSO-systems for short distances makes this technology more interesting for private clients. At the moment the main work in this field is to increase reliability and availability. Those two parameters of the FSO-link are mainly determined by the local atmospheric conditions. Good reliability and availability can be achieved by using Free Space Optics for short distances, with sufficient link budget and optimal network architecture for each FSO application. The combination of FSO and microwave-links is also a possible solution for increasing reliability and availability, because terrestrial FSO is mostly influenced by fog, whereas the microwave propagation is mainly influenced by rain. In that case wireless hybrid (optical / microwave) links are evaluated at the Department of Communications and Wave Propagation [77]. First results show a reliability of 99.9991 % of such a hybrid system.

4.6   Dynamic Bandwidth Allocation Protocols over GPONs

4.6.1   Introduction

Depending on the application requirements, network coverage and distribution of ONUs, PONs can be implemented in a fibre to the building (FTTB), FTTH or FTTC architecture [78]. Commonly in all three architectures, a passive RN in the form of a splitter/combiner is employed to distribute the downstream data from the OLT to ONUs and to combine the upstream data from individual ONUs to a single OLT. In the upstream direction, because of the random, burst nature of each ONU transmission, data collisions may occur alongside the shared medium between the RN and the OLT [79]. In order to prevent data collisions and to fully utilise the network potential in the sense of centralised communication control in the OLT, MAC protocols, providing channel access control mechanisms, are adopted by GPON standards to manage data transmissions. Nevertheless, the GPON standards only designate the utilised MAC method and leave the detail algorithm open for researchers to further explore with the target to develop suitable DBA MAC protocols to demonstrate triple-play transmission by effectively managing diverse services on demand and dynamically allocating channel capacity among network ONUs.

4.6.2 Dynamic Bandwidth Allocation Protocols

To allow centralised, dynamic bandwidth allocation among the architecture's TDM and WDM-PONs, the OLT will assign varying frame time-slots, initially to each PON in order-of-demand and subsequently arrange each PON bandwidth among their ONUs based on their service level and individual bandwidth requirement [78, 79]. In that direction, the dynamic minimum bandwidth (DMB) protocol provides ONUs in a FTTH-based TDM-GPONs with three service levels. As a result service level one for example will acquire the lowest weight which will reflect the amount of time slots it will occupy in one polling cycle. Contrasting the proposed scheme with reported EPON and adapted GPON algorithms, simulation results have demonstrated reduced mean packet delay accomplishing up to a tenfold decrease at high network load [78]. In addition, the demonstrated network performance in packet delay of high service level subscribers is sustained for increased traffic demonstrating network integrity and QoS according to subscriber SLA.

In practical networks, the number of upstream packets is naturally random, increasing during the bandwidth requirement processing period. To further reduce the packet waiting time in the ONUs, the real-time modifications in ONUs' upstream buffers during their bandwidth request-assigning process period are accounted in the advanced DMB (ADMB) protocol to reduce packet waiting time in each ONU [79]. Furthermore, in order to adapt various services in GPONs, the OLT has to receive all report packets from the ONUs before it starts to transmit the grant packets notifying the ONUs about their upstream windows. Accordingly in polling cycle k for example, the upstream channel will remain idle for the period between the last packet arrives at the OLT and the first packet is sent out from the first ONU in cycle k+1. To face the issue, the upstream transmission order of ONUs in each polling cycle is dynamically adjusted according to their time slot occupancy with the longest transmission period ONU assigned to the last upstream time slot to reduce idle intervals between bandwidth request-assigning process period. As shown in Fig. 4.11, performance evaluation results for the ADMB have shown a notable bandwidth efficiency reaching almost ninety-five percent of network capacity with a significant reduction in mean packet delay and packet loss rate to allow effective provisioning of high-bandwidth, time-sensitive multi-media services [79].

By reason of the exponential increase in PON deployment and their envisaged application to terminate widely scattered subscribers effectively to common RNs, reducing considerably the number of COs, longer-spin PON architectures with 100 km reach and split size of 1024 have been attracting substantial attention. To that extend the proposed integrated access/metro networks, could enhance QoS for various multimedia services, such as HDTV and VoD. However, when a protocol devised for standard access GPONs is introduced directly into a 100 km typical long-reach architecture, the overall network throughput performance will be considerably reduced since additional transmission time slots in each polling cycle will remain idle. This is due to an up to 500 µs increase in packet propagation time from 125 µs in 25km-PONs. To overcome this problem, a two-state DMB (TSD) protocol [80] has been proposed to overlap the idle time slots in each packet transmission cycle with a virtual transmission period to significantly increase channel utilisation rates as well as reducing packet delay and loss rate. Fig. 4.12 confirms superior performance of the TSD protocol in terms of achieved network throughput, with network load values increasing up to around 987 Mbit/s, compared to the DMB that stalls at 680 Mbit/s.

In GPON and long-reach PON architectures, the network capacity for each transmission direction is constant and cannot be shared. To increase transmission efficiency by dynamically sharing the network capacity between downstream and upstream transmissions on demand, a multi-wavelength DMB (MDMB) [81] protocol has been developed to jointly manage the downstream and upstream data transfers for loop-back WDM-PONs. Instead of accounting for the upstream and downstream Capacity requirements separately in TDM algorithm, they are integrated in a single parameter and time slots assigned to ONUs according to their service levels and bandwidth demand for both transmission directions. Simulation results have confirmed an increasing of the upstream network efficiency from 94% in DMB to 128% in MDMB at a typical 33 Mbit/s downstream ONU service demand.

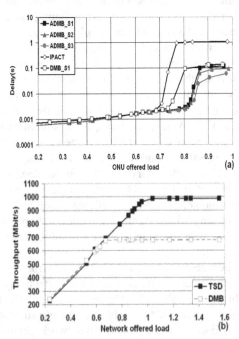

Fig. 4.12. (a) Packet delay IPACT, DMB and ADMB (b) upstream throughput in TSD and DMB

4.6.3 Conclusions

To fully utilise the network potential in the sense of centralised communication control in the OLT, a novel DMB protocol has been developed to that extend to introduce multiple service levels and QoS into a FTTH-based GPON demonstrating a significant improvement in terms of mean packet delay. Depending on the physical-layer architecture, the innovative DMB protocol has successfully been modified to develop the ADMB, TSD and MDMB protocols for GPONs, long-reach PONs and WDM-PONs respectively to efficiently improve the performance of channel throughput and packet delay. It is anticipated that this research objective will assist in the further development of dynamic bandwidth and wavelength allocation MAC protocols for multi-wavelength operation of GPON standards [82].

4.7 Innovative Architecture and Control Plane for Metro-Access Convergence

4.7.1 Motivation for Metro-Access Convergence

Metropolitan networks and access networks differ by their topology, encapsulation technique and bandwidth management scheme with the later relying in most cases on opaque WDM self-healing dual rings. More recently, native optical switched Ethernet applied to meshed topologies is considered as a more cost-effective approach for the metropolitan area than traditional rings. In terms of encapsulation technique, Ethernet frames or GEM (GPON Encapsulation Method) frames are used in PONs. In the metro area, SONET/SDH frames are used on dual rings whereas VLAN-Ethernet frames are used onto switched Ethernet and meshed topologies. At last, in terms of bandwidth management, TDM and packet scheduling techniques are used in PONs such as the MPCP-IPACT protocol for EPONs (the MAC protocol of GPONs remains open to the design of the manufacturers). In the metro area, SONET/SDH dual rings operate with cross-connected bi-directional circuits at fixed data rate and rough granularity. In comparison, VLAN traffic management used in metro-Ethernet is better suited than SONET/SDH to the asymmetry and unpredictability of IP traffic. All these disparities between access networks and metro networks have a strong economical impact for the carriers, both in terms of CAPEX and OPEX.

4.7.2 Unified Metro-Access Networks Criteria

The basic role of an access network is to facilitate traffic concentration is order to reduce the number and the size of access switches. Point-to-multipoint (P2MP) configurations in the access reach this objective. In addition to upstream traffic concentration, a P2MP configuration in the access is very well suited to broadcast services such as video-bouquet distribution or video-conferencing. Due to the low sharing factor of the infrastructure in the last mile, the devices and systems used in the access area must be cheaper and more robust than those used in the metro area. New devices like arrayed waveguide gratings (AWG), erbium doped fibre amplifiers (EDFA), semi-conductor optical amplifiers (SOA), vertical-cavity surface-emitting lasers (VCEL) and optical add-drop multiplexers (OADM) are examples of important innovations in this matter. Two other properties should be ideally satisfied by unified metro-access architectures (UMAN): survivability and transparency. On one hand, traditional automatic protection switching (APS) techniques inherited from SONET/SDH are too costly for the access.

On the other hand, protection is mandatory in the metro area. This statement justifies the use of optical loops in the metro area. At last, the inherent opacity of hybrid optical-copper access systems such as FTTC (Fibre-To-The-Curve) has been widely discussed during the last decade. In spite of important advances in the field of modulation and coding techniques, the provision of data rates of the order of 100 Mbps seems hardly achievable over copper pairs due to attenuation and crosstalk disruptive effects. Let us recall that such data rates are targeted by the major carriers for the year 2010. Today, a consensus promotes the transparent FTTC (Fibre-To-The-Home) configuration. Transparency and flexibility is now achievable in new-generation all-optical WDM dual rings in the metropolitan area thanks to reconfigurable OADMs (ROADM). Optical transparency offers two key advantages in terms of OPEX. First, it facilitates a great flexibility of the infrastructure in terms of capacity, only the opto-electronic modems installed at the ends of the network requiring to be upgraded. Second, compared to electrical devices, optical devices enable to reduce dramatically power consumption. At the scale of a global carrier's network, the economical impact of transparency on power consumption is far from negligible. In brief, unified metro-access networks (UMAN) should satisfy ideally all the previous requirements.

4.7.3 A Few Examples of Unified Metro-Access Networks (UMAN)

Several innovative architectures trying to reduce disparities between the metro are and the access area have been proposed these last ten years. We can mention the SuperPON concept proposed at the end of the years 1990s that consists in an enlarged PON covering the area from the Provider Edge (PE) to the Customer Edges (CE). Typically, the PE corresponds to the point of interconnection between the metropolitan area and the long distance area. A CE corresponds to a residential gateway installed at the customer premises. In a SuperPON, the high splitting ratio with up to 2048 ONU and the geographical range up to 200 kilometers (125 miles) impose optical amplification for both upstream and downstream traffic. Like in a PON, bandwidth allocation in a SuperPON is based on a TDMA MAC protocol relying of a request-permit mechanism. The dynamics of such a request-permit scheme can be discussed since the round-trip-time (RTT) from the OLT to an ONU is ten times larger in a SuperPON than in a PON.

More recently, the Success network developed at Stanford University consists in multiple PONs connected to an all optical ring via low cost AWG routers [83]. The ring is logically divided into two branches managing two contra-directional optical flows. For that purpose, two distinct OLTs made of a pool of tunable transceivers are used at the head-end node. One OLT serves TDM-PONs connected to multiple AWGs inserted along the ring, whereas the other OLT serves WDM-PONs connected to different AWGs also inserted along the ring. The agility of the AWG routers for both upstream and downstream traffic is provided by the tunability of the transceivers installed at the two OLTs. Bandwidth allocation in the Success network relies on an end-to-end MAC protocol operating at the packet or burst level. In reference to the criteria we mentioned in section 4.7.1, the Success network is certainly one of the best solutions proposed in the literature for UMAN.

4.7.4 The Success + Network

The Success hybrid PON scalability in terms of number of connected ONUs can be discussed in the case of WDM-PONs. Indeed, the larger this number, the larger the selectivity of the tunable transceivers to be used at the OLT. For a large number of ONUs per WDM-PON (typically 64 in GPON), the tuning delay of the OLTs and of the ONUs transceivers may become prohibitive for interactive traffic (interactive telephony, gaming etc.). The hybrid TDM-WDM nature of the Success network remains limited since two different modes of operation are adopted for TDM (full-duplex mode) and WDM (half-duplex mode). In addition, the downstream and upstream optical capacity assigned to each PON is fixed a priori. In other terms, the Success approach does not favor a smooth transition from current PONs to UMAN architectures since it ignores the existing standardized MAC protocols used in EPONs and GPONs. This is the reason why we propose an evolution of the Success architecture called "Success+" to which we couple an original control plane that really unifies the TDM and the WDM modes of operation [84, 85].

This control plane can adapt dynamically the upstream and downstream capacity assigned to each PON of the global infrastructure, not at the packet time scale but at the connection time scale. For that purpose, a key innovation is introduced: the use of reflective modulators at each ONU in order to prevent the usage of active laser sources at the CPs. The basic idea of the Success+network consists in feeding dynamically each PON by a given number of continuous waves (CW) generated at the OLTs for its upstream traffic. Such downstream CWs are dynamically routed to the PONs by the AWGs. When they arrive at an ONU, these CWs are modulated externally by users' data thanks to reflective semi-conductor amplifier (RSOA). The modulated signal returns then to the OLT via the same AWGs.

4.7.5   The Success + Network Topology

Unlike the Success network, we do not distinguish in the Success+ network configuration two types of PONs since all the PONs connected to the AWG routers operate in a TDMA mode. As it is illustrated in Fig. 4.13, a single OLT is used at the head-end node, between an AWG and a 40 Gigabit Ethernet switch. Thanks to the agility of the AWGs, it is possible in case of fibre cut to reorganize the distribution of the downstream CWs and of their inherent upstream modulated channels in order to provide APS survivability of the Success+ infrastructure.

Similarly to the TDM-PONs of the Success network, a single pair of optical channels is dedicated to each PON for its full-duplex upstream/downstream transmissions. Upstream traffic is managed by means of an agnostic TDMA MAC protocol, the EPON or BPON/GPON MAC protocols (or any other MAC protocol) being applicable in this context. In Fig. 4.13, the downstream CWs are referenced by means of the $\lambda_{i,j}$ variables whereas once they have been modulated at the level of the reflective modulators are referenced by the $\lambda^*_{i,j}$ variables.

4.7.6   The Success + UMAN Control Plane

This control plane relies on the concepts of "macroscopic" and "microscopic" timescales. The macroscopic timescale corresponds to a relatively long period during which hundreds or thousands of connections are generated at the ONUs. By "Connection", we mean either pure connection-oriented sessions (stream traffic) or connectionless-oriented exchanges (elastic traffic).

In practice, up to W CWs are assigned to a PON in accordance to its upstream traffic requirements. To optimize upstream bandwidth utilization, a first rule has to be determined by which to assign CWs to the P PONs of the network. For a given CWs assignment, we have then to determine a second rule to assign connections generated at the ONUs of each PON to the right CWs. We assimilate this ob-

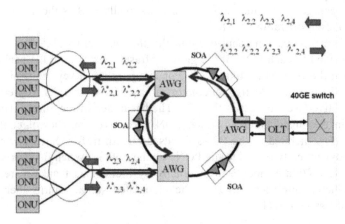

Fig. 4.13. The Success+ UMAN configuration.

jective to a double bin-packing optimization problem [84]. We want to consume globally the least number of CWs as possible and to satisfy the largest amount of upstream traffic as possible. Another strategy consisting in minimizing the number of rejected connections instead of minimizing the volume of rejected traffic is also possible.

The scalability of the global system may be obtained by adding multiple sets of tunable laser diodes and photo-detectors at the two OLTs, these sets being connected to the AWGs on different access ports. The microscopic timescale is considered once the assignment of ω CWs to a PON has been carried out and once the assignment of active connections at the ONUs of this PON to the ω optical channels has also been carried out. The microscopic timescale corresponds then typically to the fair and efficient share of these ω CWs channels to the various packets transmitted during the active connections. In other term, the microscopic timescale operates at the scale of packets' transmission duration. Considering 2.5 Gbps modulation speed and assuming an average IP packet size of 1500 bytes, this timescale is of the order of a few μs.

4.7.7 Conclusion

The current disparity between access and metropolitan networks has a strong impact on the CAPEX and OPEX of the carriers. We have proposed an original UMAN configuration, Success+, which is an evolution of the Success network that really unifies in terms of bandwidth management the access and the metro areas.

4.8   Protection Schemes for PONs

4.8.1 Evolution of PON Protection Schemes

To provide reliable service delivery over PONs several protection schemes have been proposed that could be divided into three evolution phases. In the first phase, the protection architectures were standardised by ITU-T [86] in late 90s. They are based on duplication of the network resources and are referred to as type A, B, C and D depending on which elements of the network, e.g., feeder fibre (FF), optical interfaces at the OLT, PON resources and the FF and distribution fibres (DFs) independently, are duplicated respectively. Type D provides end users with either full or partial protection referred to as Type D1 or D2 respectively.

Among the different types, Type C and Type D1 offer relatively high reliability performance by requiring though duplication of all network resources (and investment cost) to realize the protection function, which may result in capital expenditures (CAPEX) that might be too high for the cost-sensitive access networks. Therefore, in the second phase of the PON protection scheme evolution the effort was put on development of cost-efficient architectures in order to decrease the deployment cost.

Following the trend of minimizing the cost per end user, besides considering CAPEX reduction the third (future) phase of PON protection schemes evolution will probably migrate towards the reduction of operational expenditures (OPEX). Meanwhile, OPEX is related to both protection architecture and maintenance strategy. Therefore, each PON protection architecture should be evaluated by a comprehensive reliability analysis along with both CAPEX and OPEX study.

4.8.2   Recent PON Protection Architectures

In this section we review some recent PON protection schemes where the effort was put on the cost-efficient solutions. These recent architectures correspond to the second phase of the evolution in PON protection development. In the considered schemes the neighbouring ONUs protect each other utilizing the interconnection fibres (IFs) between adjacent ONUs in TDM PON, WDM PON and hybrid WDM/TDM PON, respectively. Consequently, the cost for burying redundant DFs can be avoided.

Fig. 4.14 shows a 1:1 dedicated link protection architecture proposed in [87] for TDM-PONs. Two geographically disjoint fibres provide dedicated protection against the FF cut between OLT and RN. Furthermore, every two adjacent ONUs form a pair to realize dedicated protection of DFs. This is a protection scheme based on the IFs between neighbouring ONUs in TDM PON. Each ONU contains an optical switch (OS) to initiate recovery from the DF failure. If the fibre break occurs between ONU and the RN (see ONU Pair N/2 in Fig. 4.14), ONU_{N-1} will detect the loss of the downstream signal power and the OS will be triggered from port 1 to port 2 (for the protection state) shown in the figure. Thus, the corresponding IF between port 2 of the OS in ONU_{N-1} and the Nth output of the RN works for both upstream and downstream traffic associated with ONU_{N-1}. ONUN is protected in a similar way through the IF connecting port 2 of the OS in ONUN with the N-1st output at the RN.

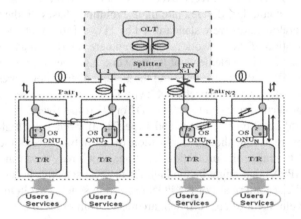

Fig. 4.14. TDM PON with 1:1 link protection in [87].

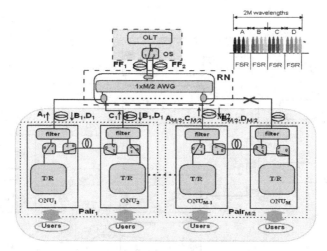

Fig. 4.15. WDM PON with protection in [88].

The protection architecture for WDM PONs shown in Fig. 4.15 [87] is also based on interconnection between the neighbouring ONUs. The RN consists of an AWG and couplers/splitters. One 1 x M/2 AWG (M = 8) and M/2 1 x 2 couplers/splitters are used to route the wavelength channels between the OLT and the ONUs. Ai, Bi, Ci and Di denote the wavelengths from different wavebands (A, B, C and D) each of which covers one whole free-spectral range (FSR) passing through the i-th of the AWG.

Wavebands A and B are referred to as the blue bands while wavebands C and D the red bands (see Fig. 4.15). The wavelengths in the wavebands A and C are for upstream, and those in wavebands B and D for downstream. Each odd ONU (ONU_{2i-1}, $i > 0$) receives the traffic carried by wavelength Bi, while wavelength Di is removed by the Red/Blue filter. Similarly, each even ONU (ONU_{2i}) receives the traffic at wavelength Di while wavelength B_i is removed by Red/Blue filter. The OLT and the RN are connected by a single working FF_1 and a (geographically disjoint) single protection FF_2. Automatic protection switching is performed at the OLT. The ONUs can detect DF failures and control the OS status to perform neighboring protection.

4.8.3 Hybrid WDM/TDM PON

In this section we describe one of two hybrid WDM/TDM PON architectures with protection based on interconnection between the neighbouring ONUs, on the basis of increased applicability and reduction in network resources. As opposed to the approach [89], based on the TDM PON protection [87], Fig. 4.16 displays a protection scheme whereby the OLT hosts M TDM PONs using M wavelengths, and each TDM PON supports N ONUs. Utilizing the cyclic property of M x M AWG, only the 1st, 2nd, (M/2)+1st, (M/2)+2nd ports are used on the input side to create

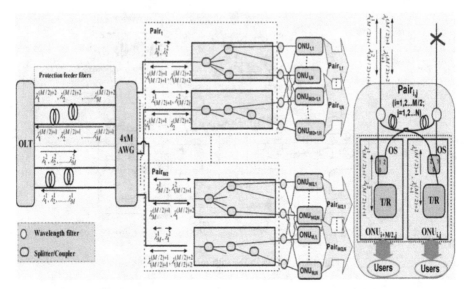

Fig. 4.16. Hybrid WDM/TDM PON with protection in [90].

4 x M AWG in the RN. These four selected ports connected to the OLT by the FFs can be used for upstream/downstream and working/protection paths, respectively.

Since the same wavelength can be reused for one TDM PON's upstream and another TDM PON's downstream, the hybrid WDM/TDM PONs with M TDM PONs only need M wavelengths which can save 50% wavelengths compared with protection scheme in [88]. We define $ONU_{i,j}$ as the j-th ONU in the i-th TDM PON. The i-th and (M/2)+i-th TDM PONs are the geographical neighbours and form $Pair_{i,j}$ for the j-th ONUs of these two neighbouring TDM-PONs. In contrast to [89] the scheme presented here has one more advantages, i.e., it can be applied to any arbitrary tree (not only star) topology of each TDM PON.

4.8.4   Reliability Performance Evaluation

In this section we compare availability of a connection between OLT and ONU for unprotected scheme, Type A-D standard protection schemes [86] and protection schemes proposed in [87-90]. Two typical PON deployment alternatives referred to as collective and dispersive cases are considered in order to make the reliability results more general. The collective case corresponds to areas with relative dense population of users while the dispersive case is applied to sparsely populated areas. Consequently, for the collective case we assumed 19.5 km long FF, 0.5 km long DFs and 0.2 km long IFs while for the dispersive case it is assumed that FF is 15 km long, DFs are 5 km long and IFs are 2 km long.

Table 4.1 shows the reliability data [89, 92] of components and devices used in PON. Utilizing the model in [91] the reliability performance of the considered

PON architectures in terms of asymptotic connection unavailability and mean downtime per year (MDT) is evaluated. The results are presented in Table 4.2. It can be seen that Type C, D_1 and schemes [87-90] can offer very high connection availability (higher than 99.999%) in the both dispersive and collective cases. Furthermore, connection unavailability and MDT in Type A, B and D_2 are much better in the collective case than in the dispersive case. It is because in the dispersive case DFs are much longer than in the collective case and the unprotected DFs in Type A, B, and D_2 can significantly deteriorate the connection availability.

Consequently, in the collective case, connection availability close to or higher than 99.999% can be achieved in Type B and D_2. Thus, the results reveal that in order to achieve high connection availability in dispersive case, all fibre links (FFs and DFs) should be protected while for the collective case it can be sufficient to protect only the shared parts (FFs) of PON.

Table 4.1. Reliability Data [89, 92]

Component	Failure rate [FIT]*	Mean Time To Repair (MTTR) [hours]	Value
OLT	256	2	5.12E-07
ONU	256	6	1.54E-06
Optical Switch	200	6	1.20E-06
Wavelength Filter	50	6	3.00E-07
AWG	200	6	1.20E-06
1:2 Splitter	50	6	3.00E-07
1:N Splitter	120	6	7.20E-07
2:N Splitter	170	6	1.02E-06
Fibre (/km)	570/km	24	**1.37E-05**

*1 FIT = 1 failure / 10E09 hours

Table 4.2. Connection availability.

Scheme	Asymptotic Connection Unavailability		MDT (min/year)	
	collective	dispersive	collective	dispersive
Unprotected	2.76E-04	2.76E-04	145.26	145.26
Type A	1.12E-05	7.27E-05	5.88	38.22
Type B	9.47E-06	7.10E-05	4.98	37.32
Type C	7.64E-08	7.64E-08	0.04	0.04
Type D_1	7.17E-08	4.74E-08	0.04	0.02
Type D_2	9.17E-06	7.07E-05	4.82	37.16
Scheme [87]	5.54E-06	5.52E-06	2.91	2.90
Scheme [88]	6.32E-06	6.30E-06	3.32	3.31
Scheme [89]	6.74E-06	6.72E-06	3.54	3.53
Scheme [90]	5.12E-06	5.10E-06	2.69	2.68

References

1. Kazovsky, L.G., Shaw, W.-T., Gutierrez, D., Cheng, N., Wong, S.-W.: Next Generation Optical Access Networks. JLT 25(11), 3428–3443 (2007)
2. ITU-T Recommendation G.922, Gigabit-capable passive optical networks (GPON): transmission convergence layer specification (2003)
3. ITU-T Recommendation J.112/122, Transmission systems for interactive cable television services/Second-generation transmission systems for interactive cable television services – IP cable modems (1998/2002)
4. IEEE Recommendation 802.11, LAN/MAN Wireless LANS (2007)
5. IEEE Recommendation 802.16, LAN/MAN Broadband Wireless LANS (2001)
6. Ghani, N., Shami, A., Assi, C., Raja, M.Y.A.: Quality of service in Ethernet passive optical networks. In: 2004 IEEE/Sarnoff Symposium on Advances in Wired and Wireless Communication, pp. 26–27 (2004)
7. Kramer, G., Pesavento, G.: Ethernet Passive Optical Network (EPON): Building a Next-Generation Optical Access Network. IEEE Communications Magazine, 66–73 (2002)
8. Effenberger, F., Clearly, D., Haran, O., Kramer, G., Ruo Ding, L., Oron, M., Pfeiffer, T.: An introduction to PON technologies [Topics in Optical Communications]. IEEE Communications Magazine 45, S17–S25 (2007)
9. Lee, C.-H., Lee, S.-M., Choi, K.-M., Moon, J.-H., Mun, S.-G., Jeong, K.-T., Kim, J.H., Kim, B.: WDM-PON experiences in Korea (Invited). OSA Journal of Optical Networking 6, 451–464 (2007)
10. Gutierrez, D., Kim, K.S., Rotolo, S., An, F.-T., Kazovsky, L.G.: FTTH Standards, Deployments and Research Issues (Invited). Presented at Joint International Conference on Information Sciences (JCIS), Salt Lake City, UT, USA (2005)
11. Ethernet in the First Mile Task Force. IEEE 802.3ah (2004)
12. Xie, J., Jiang, S., Jiang, Y.: A dynamic bandwidth allocation scheme for differentiated services in EPONs. IEEE Communications Magazine 42, S32–S39 (2004)
13. Clarke, F., Sarkar, S., Mukherjee, B.: Simultaneous and interleaved polling: an upstream protocol for WDM-PON. Presented at Optical Fibre Communication Conference and the National Fibre Optic Engineers Conference (OFC/NFOEC) 2006, Anaheim, California, USA (2006)
14. Zhang, N., Xu, M., Liao, R., Yoshiuchi, H., Ji, Y., Saren, G.: A Service-Classified and QoS-Guaranteed Triple Play Mode in FTTH Network. Presented at 1st International Conference on Communications and Networking, ChinaCom 2006, Beijing, China (2006)
15. Mynbaev, D.K.: Analysis of quality of service provisioning in passive optical networks. Presented at 2nd International Conference on Broadband Networks, Boston, Massachusetts, USA (2005)
16. ITU-T Recommendation G.984.2, Gigabit-capable passive optical networks (GPON): Physical media dependent (PMD) layer specification (2003)
17. ITU-T Recommendation G.984.3, Gigabit-capable passive optical networks (GPON): transmission convergence layer specification (2003)
18. ITU-T Recommendation G.984.1, Gigabit-capable Passive Optical Networks (GPON): General characteristics (2003)
19. ITU-T Recommendation G.984.3, Gigabit-capable passive optical networks (GPON): transmission convergence layer specification (2003)

20. FlexLight Networks and BroadLight, Comparing Gigabit PON Technologies ITU-T G.984 GPON vs. IEEE 802.3ah EPON
21. Li, A.: Comparing the techniques and products of EPON and GPON (2004)
22. Qiu, X.-Z., Ossieur, P., Bauwelinck, J., Yi, Y., Verhulst, D., Vandewege, J., De Vos, B., Solina, P.: Development of GPON Upstream Physical-Media-Dependent Prototypes. Journal of Lightwave Technology 22, 2498–2508 (2004)
23. Chanclou, P., Gosselin, S., Palacios, J.F., Álvarez, V.L., Zouganeli, E.: Overview of the Optical Broadband Access Evolution: A Joint Article by Operators in the IST Network of Excellence e-Photon/ONe. IEEE Communications Magazine 44, 29–35 (2006)
24. Cauvin, A., Tofanelli, A., Lorentzen, J., Brannan, J., Templin, A., Park, T., Saito, K.: Common technical specifications of the G-PON system among major worldwide access carriers. IEEE Communications Magazine 44, 34–40 (2006)
25. Ma, M., Zhu, Y., Cheng, T.H.: A bandwidth guaranteed polling MAC protocol for Ethernet passive optical networks. In: INFOCOM 2003, 22nd Annual Joint Conference of the IEEE Computer and Communications Societies, vol. 1, pp. 22–31. IEEE, Los Alamitos (2003)
26. Naser, H., Mouftah, H.T.: A fast class-of-service oriented packet scheduling scheme for EPON access networks. IEEE Communications Letters 10, 396–398 (2006)
27. Miyoshi, H., Inoue, T., Yamashita, K.: QoS-aware dynamic bandwidth allocation scheme in Gigabit-Ethernet passive optical networks. Presented at IEEE International Conference on Communications (ICC), Paris, France (2004)
28. Kramer, G., Mukherjee, B.: Supporting differentiated classes of service in Ethernet passive optical networks. Optical Networking 1, 280–298 (2002)
29. Park, S.-J., Lee, C.-H., Jeong, K.-T., Park, H.-J., Ahn, J.-G., Song, K.-H.: Fibre-to-the-Home Services Based on Wavelength-Division-Multiplexing Passive Optical Network (Invited). IEEE/OSA Journal of Lightwave Technology 22, 2582–2591 (2004)
30. Kani, J.-I., Teshima, M., Akimoto, K., Takachio, N., Suzuki, H., Iwatsuki, K.: A WDM-Based Optical Access Network For Wide-Area Gigabit Access Services. IEEE Optical Communications 41, S43–S48 (2003)
31. Shin, D.J., Jung, D.K., Shin, H.S., Kwon, J.W., Hwang, S., Oh, Y., Shim, C.: Hybrid WDM/TDM-PON With Wavelength-Selection-Free Transmitters. IEEE/OSA Journal of Lightwave Technology 23, 187–195 (2005)
32. Tsalamanis, I., Rochat, E., Walker, S.D.: Experimental demonstration of cascaded AWG access network featuring bi-directional transmission and polarization multiplexing. OSA Optics Express 12, 764–769 (2004)
33. Hsueh, Y.-L., Rogge, M.S., Shaw, W.-T., Kazovsky, L.G., Yamamoto, S.: SUCCESS-DWA: A Highly Scalable and Cost-Effective Optical Access Network. IEEE Optical Communications 42, S24–S30 (2004)
34. Aldridge, J.: The best of both worlds. Lightwave Europe, pp. 18–19 (2002)
35. Langer, K.-D., Grubor, J., Habel, K.: Promising Evolution Paths for Passive Optical Access Networks. Presented at International Conference on Transparent Optical Networks (ICTON 2004), Wroclaw, Poland (2004)
36. Bock, C., Prat, J., Walker, S.D.: Hybrid WDM/TDM PON using the AWG FSR and featuring centralized light generation and dynamic bandwidth allocation. IEEE/OSA Journal of Lightwave Technology 23, 3981–3988 (2005)
37. ITU-T Recommendation: G.694.2, Spectral Grids for WDM Applications: CWDM wavelength grid (2003)

38. An, F.-T., Gutierrez, D., Kim, K.S., Lee, J.W., Kazovsky, L.G.: SUCCESS-HPON: A Next-Generation Optical Access Architecture for Smooth Migration from TDM-PON to WDM-PON. IEEE Communications Magazine 43, S40–S47 (2005)
39. Shachaf, Y., Chang, C.-H., Kourtessis, P., Senior, J.M.: Multi-PON access network using a coarse AWG for smooth migration from TDM to WDM PON. OSA Optics Express 15, 7840–7844 (2007)
40. Wellen, J., Smets, R., Hellenthal, W., Lepley, J., Tsalamanis, I., Walker, S., Ng'oma, A., Koonen, G.-J.R., Habel, K., Langer, K.D.: Towards High speed Access Technologies: results from MUSE. Presented at SPIE Broadband Access Communication Technologies, Boston, USA (2006)
41. Langer, K.-D., Habel, K., Raub, F., Seimetz, M.: CWDM access network and prospects for introduction of full-duplex wavelength channels. Presented at Conference on Networks & Optical Communications (NOC 2005), UCL (2005)
42. Jiang, J., Callender, C.L., Blanchetière, C., Noad, J.P., Chen, S., Ballato, J., Dennis, J., Smith, W.: Arrayed Waveguide Gratings Based on Perfluorocyclobutane Polymers for CWDM Applications. IEEE Photonics Technology Letters 18, 370–372 (2006)
43. Davey, R., Kani, J., Bourgart, F., McCammon, K.: Options for future optical access networks. IEEE Communications Magazine 44, 50–56 (2006)
44. White, I.M., Rogge, M.S., Shrikhande, K., Kazovsky, L.G.: A Summary of the HORNET Project: A Next-Generation Metropolitan Area Network. IEEE Journal on Selected Areas in Communications 21(9), 1478–1494 (2003)
45. White, I.M., Hu, E.S., Hsueh, Y., Shrikhande, K., Rogge, M.S., Kazovsky, L.G.: Demonstration and System Analysis of the HORNET Architecture. Journal of Lightwave Technology, vol. 21
46. Carena, A., De Feo, V., Finochietto, J.M., Gaudino, R., Neri, F., Piglione, C., Poggiolini, P.: RingO: An Experimental WDM Optical Packet Network for Metro Applications. IEEE Journal on Selected Areas in Communications
47. Antonino, A., Birke, R., De Feo, V., Finocchietto, J.M., Gaudino, R., La Porta, A., Neri, F., Petracca, M.: The WONDER Testbed: Architecture and Experimental Demonstration. Dip. di Elettronica, Politecnico di Torino, Torino, Italy
48. Kliazovich, Granelli, Woesner: Bidirectional Optical Ring Network Having Enhanced Load Balancing and Protection. DIT - University of Trento, Italy, CREATE-NET, Italy
49. Herzog, Maier, Wolisz: RINGOSTAR: An Evolutionary AWG-Based WDM Upgrade of Optical Ring Networks. Journal of Lightwave Technology 23(4) (2005)
50. Dey, Van Bochove, Koonen, Geuzebroek, Salvador: FLAMINGO: A Packet-switched IP-over-WDM Alloptical MAN
51. Dittmann, Develder, Chiaroni, Neri, Callegati, Koerber, Stavdas, Renaud, Rafe, J. Solé-Pareta, Cerroni, Leligou, Dembeck, Mortensen, Pickavet, Le Sauze, Mahony, Berde, Eilenberger: The European IST Project DAVID: a Viable Approach towards Optical Packet Switching. JSAC, Special Issue on High-Performance Optical/Electronic Switches/routers for High-Speed Internet
52. Le Sauze, Dotaro, Ciavaglia, Dupas, Chiaroni, Ge, Sridhar, Dembeck, Koerber, Wolde: Optical Packet Switched Metro Networks. Presented at 28th European Conference and Exhibition on Optical Communication (ECOC), Copenhagen, Denmark (2002)
53. Le Sauze, Dupas, Dotaro, Ciavaglia, Nizam, Ge, Dembeck: A Novel, low cost optical packet metropolitan ring architecture
54. Lazaro, Prat, Tosi Beleffi, Teixeira, Tomkos, Soila, Koratzinos: Scalable Extended Reach PON. In: IEEE OFC2008, San Diego USA, Invited (2008)

55. Lakic, B., Hajduczenia, M.: On optimized Passive Optical Network (PON) deployment. In: 2nd International Conference on Access Networks &Workshops, Ottawa (August 2007)

56. FTTH Council Europe General Presentation. FTTH Council Europe (Jan. 2008), http://www.ftthcouncil.eu/documents/presentation/2007-11-29_FTTH_CE_General_presentation_V.2.1.pdf

57. Chanclou, P., Gosselin, S., Palacios, J.F., Álvarez, V.L., Zouganeli, E.: Overview of the Optical Broadband Access Evolution: A Joint Article by Operators in the IST Network of Excellence e-Photon/ONe. IEEE Communications Magazine 44, 29–35 (2006)

58. FTTH situation in Europe (2007), http://www.idate.fr/pages/download.php?id=371&rub=news_telech&nom=PR_IDATE_FTTH_CONF_2007.pdf

59. 2004 Broadband IT Korea Information white paper (2004), http://www.mic.go.kr/index.jsp

60. Hyun Deok, K., Seung-Goo, K., Chang-Hee, L.: A low-cost WDM source with an ASE injected Fabry-Perot semiconductor laser. IEEE Photonics Technology Letters 12, 1067–1069 (2000)

61. Shinohara, H.: FTTH experiences in Japan (Invited). OSA Journal of Optical Networking 6, 616–623 (2007)

62. Alcatel-Lucent and Hong Kong Broadband Network introduce first carrier-class GPON in Hong Kong. Press Releases, Alcatel Lucent (2007)

63. Effenberger, F.J., McCammon, K., O'Byrne, V.: Passive optical network deployment in North America (Invited). OSA Journal of Optical Networking 6, 808–818 (2007)

64. Marciniak, M.: Future Networks – beyond Next Generation Networking. In: 10th Anniversary International Conference on Transparent Optical Networks, Conference Proceedings, Athens, Greece, June 22-26, 2008, vol. 1, pp. 25–28 (2008)

65. Cochrane, P.: Fibre-to-the-home (FTTH) Costs Are Now In! Proceedings of the IEEE 96(2), 195–197 (2008)

66. IEEE 802.3 Higher Speed Study Group tutorial: An Overview: The Next Generation of Ethernet. IEEE 802 Plenary, Atlanta, GA, November 12 (2007)

67. McDonough, J.: Moving Standards to 100 GbE and Beyond. IEEE Applications & Practise, Online Magazine 45(Suppl. 3), 6–9 (2007)

68. Marciniak, M.: 100 Gb Ethernet over Fibre Networks– Reality and Challenges. ICTON - 'Mediterranean Winter' 2007, Sousse, Tunisia, December 6-8 (2007)

69. Muller, S., Bechtolsheim, A., Hendel, A.: HSSG Speeds and Feeds Reality Check (January 2007), http://www.ieee802.org/3/hssg/public/jan07/muller_01_0107.pdf

70. Fournier, P.-F.: From FTTH pilot to pre-rollout in France. Presented at CAI Cheuvreux, France (2007)

71. Shachaf, Y., Kourtessis, P., Senior, J.M.: An interoperable access network based on CWDM-routed PONs. Presented at 33rd European Conference and Exhibition on Optical Communication (ECOC), Berlin, Germany (2007)

72. Shachaf, Y., Kourtessis, P., Senior, J.M.: A Full-duplex Access Network Based on CWDM-Routed PONs. Presented at Optical Fibre Communication and the National Fibre Optic Engineers Conference (OFC/NFOEC 2008), San Diego, USA (2008)

73. Koonen, A., et al.: Perspectives of Radio over Fibre Technologies. OFC/NFOEC 08, San Diego, USA, pp OThP3 (2008)

74. Leitgeb, E., Gebhart, M., Birnbacher, U., Schrotter, P., Merdonig, A., Truppe, A.: Hybrid wireless networks for civil-military-cooperation (CIMIC) and disaster management. In: Proceedings and Presentation at SPIE's European Symposium on Optics and Photonics for Defence and Security, London, 28th October 2004, vol. 5614, pp. 139–150 (2004)

75. Leitgeb, E., Gebhart, M., Birnbacher, U.: Optical Networks, Last Mile Access and Applications. In: Free-Space Laser Communications: Principles and Advances, Springer, Heidelberg (2008)

76. Flecker, B., Chlestil, C., Leitgeb, E., Sheikh Muhammad, S., Gebhart, M.: Results of Attenuation Measurements for Optical Wireless Channels under Dense Fog Conditions. Presented at the SPIE Optics and Photonics Symposium, San Diego, USA (August 2006)

77. Leitgeb, E., Sheikh Muhammad, S., Flecker, B., Chlestil, C., Gebhart, M., Javornik, T.: The influence of dense fog on Optical Wireless systems, analysed by measurements in Graz for improving the link-reliability. In: Proceedings and Invited Presentation at the IEEE-conference ICTON 2006, Nottingham, UK, 18-22 June (2006)

78. Chang, C.-H., Kourtessis, P., Senior, J.M.: GPON service level agreement based dynamic bandwidth assignment protocol. Journal of Electronics Letters 42, 1173–1174 (2006)

79. Chang, C.-H., Kourtessis, P., Senior, J.M.: Dynamic Bandwidth assignment for Multiservice access in GPON. Presented at 12th European Conference on Networks and Optical Communications (NOC), Stockholm, Sweden (2007)

80. Chang, C.H., Alvarez, N.M., Kourtessis, P., Senior, J.M.: Dynamic Bandwidth assignment for Multi-service access in long-reach GPON. Presented at 33rd European Conference and Exhibition on Optical Communication (ECOC), Berlin, Germany (2007)

81. Kourtessis, P., Shachaf, Y., Chang, C.-H., Senior, J.M.: PON Topologies for Dynamic Optical Access Networks. Presented at 10th International Conference on Transparent Optical Networks (ICTON), Athens, Greece (2008)

82. ITU-T Recommendation G.984.5, Enhancement band for gigabit capable optical access networks (2007)

83. An, F.-T., Soo Kim, K., Gutierrez, D., Yam, S., Hu, E., Shikhande, K., Kazovsky, L.G.: SUCCESS: a next-genertion hybrid WDM/TDM optical access network architecture. IEEE Journal of Ligthwave Technology 22(11), 2557–2569 (2004)

84. Gagnaire, M., Koubaa, M.: A new control plane for Next-Generation WDM-PON access systems. In: Proceedings of the second IEEE Accessnets conference, Ottawa (August 2007)

85. Gagnaire, M.: Transparent WDM Metro-Access Networks. International Journal of Communication Networks and Distributed Systems 2(2-3), 281–301 (2009)

86. ITU-T recommendation G983.1, Broadband optical access systems based on passive optical network (PON) (1998)

87. Chen, J., Chen, B., He, S.: Self-protection Scheme against Failures of Distributed Fibre Links in an Ethernet Passive Optical Network. OSA JON 5, 662–666 (2006)

88. Chan, T.J., Chan, C.K., Chen, L.K., Tong, F.: A self-protected architecture for wavelength division multiplexed passive optical networks. IEEE Photon. Technol. Lett. 15, 1660–1662 (2003)

89. Chen, J., Wosinska, L.: Protection Schemes in PON Compatible with Smooth Migration from TDM-PON to Hybrid WDM/TDM PON. OSA JON 6, 514–526 (2007)

90. Chen, J., Wosinska, L., He, S.: High Utilization of Wavelengths and Simple Interconnection Between Users in a Protection Scheme for Passive Optical Networks. IEEE Photon. Technol. Lett. 20, 389–391 (2008)
91. Wosinska, L., et al.: Scalability limitations of Optical networks due to reliability Constrains. In: Proc. NFOEC'01 (2001)
92. COST 270 reliability data base

5   Novel Switch Architectures

W. Kabaciński (chapter editor), J. Chen, G. Danilewicz, J. Kleban,
M. Spyropoulou, I. Tomkos, E. Varvarigos, K. Vlachos, S. Węclewski,
L. Wosinska, and K. Yiannopoulos

Abstract. In this paper we discussed different switch architectures. We focus mainly on optical buffering. We investigate an all-optical buffer architecture comprising of cascaded stages of quantum-dot semiconductor optical amplifier-based tunable wavelength converters, at 160 Gb/s. We also propose the optical buffer with multi-wavelength converters based on quantum-dot semiconductor optical amplifiers. We present multistage switching fabrics with optical buffers, where optical buffers are based on fibre delay lines and are located in the first stage. Finally, we describe a photonic asynchronous packet switch and show that the employment of a few optical buffer stages to complement the electronic ones significantly improves the switch performance. We also propose two asynchronous optical packet switching node architectures, where an efficient contention resolution is based on controllable optical buffers and tunable wavelength converters TWCs.

5.1   Introduction

Optical fibres introduced in transmission systems offer a huge transmission bandwidth unavailable for copper cables. Transmission bit rates of 2.5, 10 and 40 Gb/s are now available and soon rates of 160 Gb/s will be available commercially. Electronic switching cannot be used at such high rates, so incoming signals have to be not only converted from optical to electrical form but also have to be demultiplexed to lower bit rates. To omit this inconvenient and expensive signal conversion and demultiplexing, switching systems based on the optical technology have been elaborated in research laboratories and industry. Optical switching, called also photonic switching, enables optical signals to be switched directly from inputs to outputs without conversion to electronic form. To construct optical switching elements, different technologies are being used. They exploit various optical effects in materials. In general, these technologies can be grouped into two main categories [1]: guided lightwave based switches and free-space switches. Each category can be further divided into different classes, depending on the physical phenomena used to switch lightwaves between inputs and outputs. We can distinguish: electro-optic switches, acousto-optic switches, thermo-optic switches, MEMS switches, liquid-crystal switches, SOAs based switches. Detailed description of different technologies and examples of optical switching elements can be found in [1]. One of popular electro-optic switches is the titanium diffused lithium-

I. Tomkos et al. (Eds.): COST 291 – Towards Digital Optical Networks, LNCS 5412, pp. 133–160, 2009.

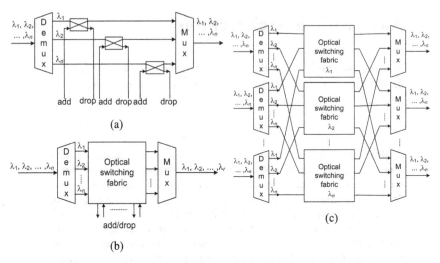

Fig. 5.1. (a) OADM with switching elements (b) OADM with the switching fabric (c) OXC with separate switching fabrics for each wavelength.

niobate (Ti:LiNbO₃) directional coupler. It has a capacity of 2×2 and can be in one of two states: cross or bar. In MEMS technology switches may have different capacities (on-off switch, 2×2, $1 \times N$, $N \times N$) and different moving parts (mirrors, prisms, lenses, fibres).

Functions performed by an optical switching note are closely related to the transfer mode used in the optical transport network. At present, most optical nodes provide circuit switching either on the fibre level or wavelength level. These nodes are Optical Add/Drop Multiplexers (OADMs) and Optical Cross-Connects (OXCs). Different architectures of OADMs and OXCs were proposed in literature and implemented in practice [2]. Some of them use optical switching fabric to switch fibres or wavelengths. Examples of such architectures are shown in Fig. 5.1. OXCs and OADMs mostly use MEMS switches. This technology is relatively cheap and mature, however the switching time is rather slow, and cannot be used in Optical Packet Switches (OPS). Much faster switching elements are needed in OPS. However, fast optical switches are at the moment very expensive and not mature. The compromise between the circuit switching and packet switching is the Optical Burst Switching (OBS). OPS/OBS switches are considered for use in core routers [3, 4, 5].

In packet (burst) switching one of the important functions of the switch it to solve the output contention problem. This contention appears when two or more packets are to be directed to the same output at the same time. The output contention can be solved in the space, wavelength, or time domains. In the space domain, one packet is directed to the desired output, while other competing packets are directed to other outputs. This approach is also called the deflection routing. Such deflected packets can be directed back to their original destination in one of the next nodes. When wavelength multiplexing is used in the output ports, contention may be solved in wavelength domain by sending congested packets to the same output but on different wavelength. This requires the wavelength conversion capability in the switching node. Finally, when one of competing packets is sent to its original output, other congested packets are directed to the memory and delayed for some time.

The latter issue – buffering is an important functionality in optical burst-switched networks. It allows the temporal storing of data bursts or packets to resolve contention for the switch outputs. In case of all-optical buffering, we are facing a lack of RAM memory, wide available in electronics. Thus, electronic buffering has been extensively utilized in currently installed optical networks at a great cost and complexity and is limited by the electronic processing speeds and the relative slow O/E and E/O conversion times [6]. On the other hand, programmable fibre delay lines have been extensively used to form feed-forward [7] or re-circulating schemes, employing in addition wavelength conversion to enhance buffering capabilities [8, 9, 10]. In particularly, feedback loops theoretically provide infinite storage time, but they suffer from noise accumulation and OSNR degradation. In contrast, feed-forward delay line buffers allow for short buffering times but recent studies indicate that statistically multiplexed optical networks will require only minimal buffering, provided some traffic engineering is performed [11]. Other storing technology such "slow light" are even more immature and up to date.

The paper gives some latest results in Optical Packet/Burst Switch architectures, with the special attention put on optical buffering. In Section 2 we demonstrate the applicability of QD-SOAs in an optical buffer architecture that can support ultra-high speed optical packet switching. Packet buffering is implemented in a multi-stage Time-Slot-Interchanger (TSI) that consists of quantum-dot semiconductor optical amplifiers (QD-SOAs) based wavelength converters exploiting the cross-gain modulation (XGM) effect and feed-forward delay lines. QD-SOAs constitute candidate technology for this purpose due to their advantageous properties such as strong nonlinearities, high gain and ultra-fast carrier dynamics which lies on the subpicosecond scale [12, 13] are heralded as the main technology able to support signal processing applications at high bit rates that reach up to 160 Gb/s.

A significant reduction in the number of active and passive components needed, can be achieved when using switches with multi-wavelength switching capabilities. QD-SOAs have been reported as a candidate technology to provide multi-wavelength operation at high bit rates [14]. Other technology candidate is acousto-optic tunable filters, [15], which are slower, with less flexibility in processing a multi-wavelength spectrum, but it is a proven technology. The switch architecture with QD-SOAs performing a multi-wavelength conversion is the subject of Section 3.

In Section 4 we describe an architecture of the single-stage shared-FDL switch and FDL assignment algorithms. Next we show how to build a multistage optical switch using the single-stage shared-FDL switch. The architecture of the three-stage Clos-network is considered in our research as a potential solution to overcome the limited scalability of single-stage switches. Two FDLs assignment algorithms for the three-stage optical Clos-network are presented: sequential FDL assignment algorithm for Clos-network switches (SEFAC) and multicell FDL assignment for Clos-network switches (MUFAC). Furthermore, the results of simulation experiments are shown.

In Section 5 we describe a photonic asynchronous packet switch and show that the employment of a few optical buffer stages to complement the electronic ones significantly improves the switch performance. Furthermore we propose two asynchronous optical packet switching node architectures, where an efficient contention resolution is based on controllable optical buffers and tunable wavelength converters TWCs. First, we describe the new types of variable optical memory and evaluate the benefits of implementing them in the optical packet switching node. Furthermore, two structures of the switching node with dedicated and shared hybrid buffer respectively are presented along with two switching node architectures with contention resolution based on both buffering and wavelength conversion. We show that providing a few shared optical buffers significantly boosts the performance improvement obtained by TWCs.

5.2   Application of Quantum-Dot SOAs for the Realization of All-Optical Buffer Architectures up to 160 Gb/s

The buffer architecture comprises cascaded programmable delay stages, each consisting of two Tunable Wavelength Converters (TWCs) and two delay line banks. Each TWC provides w separate wavelengths at its output, and each wavelength is routed to the respective branch of the delay line bank by means of a wavelength demultiplexer. Adjacent TWCs and stages are connected by wavelength multiplexers. The system architecture has been designed based on [16] modified such as to utilize the maximum number of available wavelengths and it is presented in detail in [17].

Full wavelength utilization is of practical importance when considering QD-SOAs, since the devices exhibit a limited wavelength range due to the fact that wavelengths that do not reside in the same homogenous gain peak do not interact. Moreover, the channel spacing is many hundreds of GHz in systems that employ QD-SOA based TWCs, because of the ultra high rates that the devices are required to operate at (160 Gbps and beyond). The combined effects of the limited homogenous bandwidth and the increased wavelength spacing in high data rates mean that the total number of available wavelengths is drastically reduced in QD-SOA based TWCs when compared to the number of available wavelengths in conventional SOA and fibre based TWCs.

In buffer architectures, however, the number of available wavelengths determines the buffer storage capacity, and more wavelengths enable the design of buffers with increased buffering capabilities. On the contrary, if the number of available wavelengths is limited or if the available wavelengths are not optimally used, increased storage capacities will be realized at the expense of the buffer size. That is, more buffering stages are required in a multistage architecture, leading to increased system cost, as well as severe optical signal degradation due to the cascade of TWCs. Since QD-SOA based TWCs inherently lack a wide wavelength range, the buffer designer should focus on fully exploiting the number of wavelengths that are available by the QD-SOA. Moreover, the designer should pursue an architecture that minimizes the number of required stages for a given number of available wavelengths.

A well known multi-stage architecture that ensures minimization of the number of buffer stages is the Benes network. The principle of operation of the Benes network in buffering architectures is identical to its operation in space switches; still, Benes buffering involves the time-domain interchange of the packet positions, whereas Benes switching involves the spatial interchange of the packets. Under this scheme, if each Benes buffering stage is capable of interchanging n packets in total, then the i-th cascade stage must be able to interchange n packets that are spaced by ni packet durations (slots). In space switching terms this corresponds to having a Benes network that is formed from $n \times n$ switches: each switch has the capability of spatially interchanging n packets, while switches in stage i of the network interchange packets that are spaced ni positions apart.

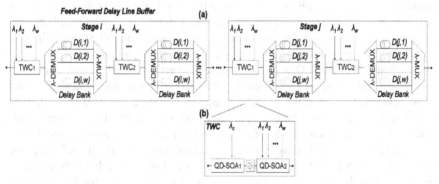

Fig. 5.2. (a) The structure of each delay stage. (b) The QDSOA tunable wavelength converter (TWC) setup. λ-MUX/DEMUX are the wavelength multiplexers and demultiplexers, respectively [17].

Following the above, the challenge is to design Benes packet interchanging stages that fully utilize the number of available wavelengths w provided by the QD-SOA based TWC SOAs. It has been shown in [17] that the maximum number of packets that can be interchanged per stage are $n = w - 1$. Even though the analysis is beyond the scope of the current chapter, it is easy to show that it is not possible to utilize all available wavelengths by considering two packets and two wavelengths ($n, w = 2$). Clearly, there is no possible means of interchanging two successive packets in a Benes stage with two wavelengths, since the stage will provide delays that equal '0' or '1' packet durations only. Should the first packet be delayed by '1' packet duration, both packets will arrive simultaneously at the output of the delay line, therefore packet collision and consequent loss will occur. If the first packet is not delayed at all then interchanging cannot be performed. However, the two packets can be interchanged if three wavelengths are used, since an additional delay of '2' packet durations is available. By assigning first packet a '2' packet delay and the second packet a '0' delay, packet positions are interchanged at the output of the delay lines. Additionally, having a single TWC followed by w delay lines does not suffice to interchange all n packets. This is so, since interchanging packets 1 and n requires that packet 1 is delayed by $2n - 1$. This delay can only be achieved by cascading a second TWC with delay lines inside the Benes stage and the maximum delay in such case is $2(w - 1)$ or $2n$. Therefore the construction of the Benes stages results in a cascade of two TWCs plus delay lines per stage.

In agreement with the theoretical analysis, the main elements of TWC subsystem are two QD-SOAs in serial configuration as illustrated in Fig. 5.3(a). Wavelength conversion is performed in two steps based on the cross-gain modulation (XGM) effect between a data signal (pump) and a continuous wave signal (probe) in each QD-SOA, respectively. The incoming data signal modulates the carrier density and consequently the gain of the QD-SOA, resulting in the respective modulation of the cw probe signal which at the output of the QD-SOA has inverse polarity. Wavelength conversion from λ_m to λ_n is facilitated by intermediate conversion from λ_m to λ_c in QD-SOA1 and from λ_c to λ_n in QD-SOA2, where λ_c represents the center of the inhomogeneously broadened gain profile of the device. On one hand, the second QD-SOA is used to convert the signal polarity back to its initial state and on the other, to account for the case that the input data signal is not to be converted (in case this is defined by the routing algorithm) hence the cw wavelength needs to be different from the pump wavelength at each intermediate conversion. Furthermore, each QD-SOA is followed by a tunable optical filter aiming to cut off the unnecessary pump signal and keep only the modulated probe signal which will be fed into the next QD-SOA and play the role of the modulating pump. Finally, the saturable absorber is used to compensate for the extinction ratio degradation of the converted signal which is attributed to the XGM effect in SOA-based devices [18]. As a result, the absorber will suppress the level of the spaces of the converted signal thus improving the extinction ratio along the cascaded stages of the buffer.

The buffer architecture is designed such as to exploit the maximum number of the available wavelengths which are symmetrically located around the center of the inhomogeneously broadened gain spectrum of the QD-SOA (Fig. 5.3(b)). However, it is necessary that all wavelengths are constrained within the spectral bandwidth of a single-dot group gain, known as the homogeneous broadening, for effective XGM to occur. In addition, the wavelength separation between adjacent wavelengths should be such that it prevents spectral overlap of the transmitted signals at 160 Gb/s.

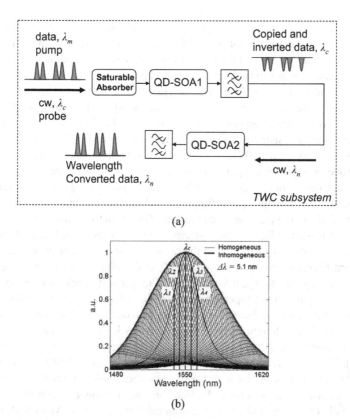

(a)

(b)

Fig. 5.3. (a) Configuration setup of TWC. (b) Available wavelengths for wavelength conversion lying within the single dot-group homogeneous broadening.

The performance of the TWC subsystem cascade is based on extensive numerical analysis which simulates the carrier dynamics of the QD-SOA device. The implemented model is used to solve a set of rate equations each of which corresponds to the changes of the carrier density at each energy state of the quantum-dots: the ground state, the excited state, the continuum state and the wetting layer. The rate equations are presented in [19] The homogeneous single-dot group bandwidth is considered equal to 16 meV (~ 31 nm at 1550 nm communication window) at room temperature. The QD-SOA parameters have been obtained from [19], [20]. The available wavelengths $\lambda_{m,n}$ ($m, n = 1, 2, 3, 4$) are spaced at 5.1 nm around the central wavelength λ_c which is used for intermediate wavelength conversion. Even if QD-SOAs have broad gain spectrum which is attributed to the size variation of the dots, only the spectrum under a homogeneous broadening can be exploited due to the fact that, XGM-based wavelength conversion should be feasible from an input available wavelength to an output available wavelength. In addition, for the simulation study that will follow, each tunable filter has been considered as a simple passive element of 2dB loss. Additional losses of 6dB have been considered for the MUX and DEMUX at the input and at the output of the TWC respectively, hence 8 dB losses have been considered for each TWC subsystem, in total.

The steady state response of the saturable absorber is simulated based on a simple transfer function which is introduced by the loss parameter $\alpha(t,P)$ that tracks the signal envelope according to $\alpha(t,P) = \alpha_0/(1+P(t)/P_{sat})$ [21], where P_{sat} is the saturation power and α_0 is the steady state loss being equal to -0.1 dB in this case. Extensive simulations have indicated an optimum value of +20 dBm for the P_{sat} parameter.

The input data consists of 32% duty cycle RZ-Gaussian pulses modulated by a 2^7-1 PRBS bit pattern at 160 Gb/s. The average input pulse power is 27 dBm and the cw probe power levels input to QD-SOA1 and QD-SOA2 are -15 dBm and 0 dBm, respectively. These values have been determined based on an optimization study in terms of the output extinction ratio and relative Q-factor ratio of the converted signal along the cascade of converters. The Q-factor ratio is directly related to the actual Bit-Error-Rate of the system, only when the signal degradation follows Gaussian statistics. Although in the present work this is not the case, this figure of merit function can still be used to reflect the efficiency of the converter illustrating its regenerative capabilities. In the present work and in order to highlight the regenerative properties of the TWC subsystem, the input data signal has 13 dB extinction ratio but suffers from amplitude jitter at the marks leading to input Q-factor 7. The length of SOA-based devices is an important parameter for their fast response and it has been previously studied and reported in the literature [22], [23]. Fig. 5.4 illustrates the gain recovery time of the QD-SOA as a function of the QD-SOA length at the impulse of very short pulses (500 fs) at 100 GHz repetition frequency. It is noteworthy that for lengths longer than 6 mm the gain response falls under 1ps indicating the applicability of QD-SOAs to high bit rate processing functionalities. It should be noted that experimental results have been published with up to 25 mm QD-SOA device [24]. The length of the QD-SOA device in the current simulation study is considered 10mm and the current density that drives the QD-SOAs is 36 kA/cm^2 in order to make sure that the upper layers are full of carriers.

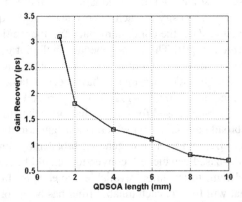

Fig. 5.4. QDSOA gain recovery time as a function of the device length.

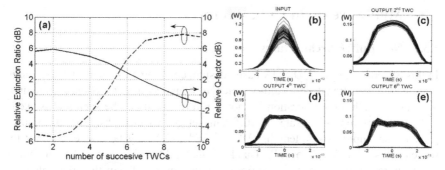

Fig. 5.5. (a) Relative extinction ratio and Q-factor ratio as a function of the number of successive TWCs (average values over 10 iterations). (b-e) eye diagrams of the input data signal and the converted data signals after the first, second and third stage of the buffer (worst case scenario out of 10 iterations).

Fig. 5.5(a) illustrates the relative extinction ratio and the Q-factor ratio of the converted signal with respect to the input values, after each TWC. The results depicted here are the averaged over 10 iterations. It is clear that, there are regions where both the extinction ratio and the Q-factor ratio show significant improvement. However, there is an inverse relationship between them as the number of successive converters increases imposing a trade-off. In particular, the extinction ratio gradually improves along the cascade reaching 8 dB at the output of the eighth TWC. On the contrary, the Q-factor ratio reaches 6 dB at the output of the first TWC, back tends to drop back to its input value at the output of the ninth TWC.

The worst case scenario out of ten iterations illustrating wavelength conversion from λ_1 to λ_4 and so on and so forth has been considered. Fig. 5.5(b-e), illustrate the eye diagrams of the input signal and the converted output signals after the first, second and third stage (second, fourth and sixth TWC). It is clear that, the level of spaces is suppressed owing to the saturable absorber. In addition, power overshooting at the leading edge of the pulse is observed which is attributed to self-phase modulation effects in the QD-SOA [21]. At the output of the third stage of the buffer architecture, the extinction ratio and Q-factor improvement is 5dB and 3dB, respectively, illustrating the regenerative performance of the TWC subsystem. Finally, according to the equation which relates the number of cascaded stages with the number of packets served, we reach the conclusion that, under the aforementioned conditions, for 3 cascaded stages 9 input packets can be served.

5.3   Multiwavelength Optical Buffers

5.3.1   New Buffer Architectures

In this Section we use a QD-SOA based design, combing with an array waveguide grating router to transform multi-wavelength conversion to functional switching operations. Such a configuration is shown in Fig 5.6. It can be seen that packets from different inputs operate on different set of wavelengths. For example a packet from input 1(wavelengths: λ_1^I, λ_1^{II}, λ_1^{III}, λ_1^{IV}) may be converted only to wavelengths λ_1^I, λ_1^{II}, λ_1^{III}, or λ_1^{IV}. Thanks to that, all the signals may be converted and then transmitted in one shared fibre simultaneously. One of the other interesting functions of QD-SOAs and SOAs in general is that they can copy one signal to more than one wavelengths, when multiple probe signals are used [25]. This feature can be used for multi-casting purposes on packet per packet based within such optical switches.

In [26] we proposed 2 architectures for implementing optical buffers. Both use multi-wavelength selective elements like QD-SOAs as multi-wavelength converters and fixed-length delay lines that are combined to form both an output queuing and a parallel buffer switch design. The output queuing buffer design requires less active devices (QD-SOA) when implementing large buffers, but the parallel buffer design becomes more profitable, when the number of wavelength channels that can be simultaneously processed by the wavelength selective switches (QD-SOAs) increases.

The "parallel buffer" architecture is depicted in Fig 5.7. We have assumed a 4×4 switch, where an input stage (not shown in the figure) converts all input signals to four different ($\lambda_1 \dots \lambda_4$) wavelengths. The main buffer stage of the size equals to 15 time slots consists of QD-SOAs followed by cyclic arrayed waveguide gratings (AWG) with banks of fibre delay lines (FDLs) at theirs outputs. To this end, for each input wavelength a set of four output wavelengths exist, equal to the AWG ports. Thus, a total of 16 internal wavelengths are needed (see Fig. 5.6). The first and the second QD-SOAs on the packet route simultaneously convert all packets from all the four wavelengths to four different internal wavelengths in order to access the desirable delay in the following banks of FDLs. In the sequence, a third QD-SOAs acting as the output switch fabric forwards the data signals to the desired output link.

The packets from the different inputs operates on different set of wavelengths (Fig. 5.6), so the packet contention within the buffer doesn't occur. However, the contention problem arises when two or more packets from the same input (go to different outputs) arrive to the same QD-SOA simultaneously. In the "*parallel buffer*" architecture, it doesn't take place in the buffering stage but could appear in the output switching stage (the third QD-SOA on the packet route). To overcome such a packet contention, a simple scheduling algorithm given in the next section is proposed.

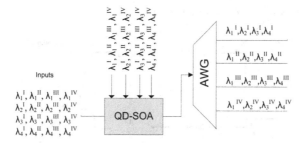

Fig. 5.6. QD-SOA acts as a multi-wavelength converter.

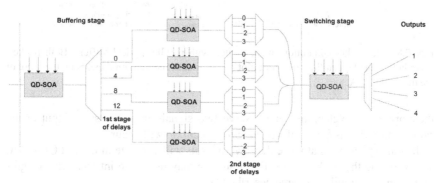

Fig. 5.7. The first and the second QD-SOA on the packet route simultaneously convert all packets from all the four wavelengths to four different internal wavelengths in order to access the desirable delay in the following banks of FDLs. In the sequence, a third QD-SOAs acting as the output switch fabric forwards the data signals to the desired outgoing link.

5.3.2 Scheduling Algorithms

In this section a simple scheduling algorithm according to "parallel buffer" scheme is presented. It is based on the algorithm [27]. To visualize a possible packet contention, let us consider the example shown in Fig. 5.8. We assume that packets are of fixed size and their duration is one time slot. All packets share the same fibre delay lines so, it is not visible in the figures from which input and to which output come packets. The route of the specific packet is indicated by bolding the proper delay lines. Each delay line owns also a corresponded "number" mark which stands for its time slot delay. We assume that, some of the delay lines has been yet occupied thus, the starting point of our consideration is a xth timeslot.

In such a time slot, only one packet – $P_{1,2}$ (from the first input to the second output) appears at the input of the switch. The next timeslot when the second output is free is $x + 8$, thus the packet delay is set to 8 timeslots. It goes then through the third branch and enters the first delay line. Six time slots later, packet $P_{1,4}$ enters the switch. In $(x + 8)^{th}$ time slot both packets enter the switching stage. The QD-SOA at the switch output that is commissioned to convert the signals back to

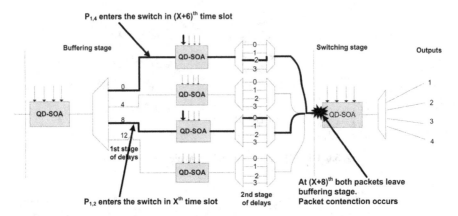

Fig. 5.8. The packet contention in the switch with parallel optical buffers. Both packets leave buffering stage at the same time slot and the packet contention appears on the input of the third QD-SOA.

their original wavelengths cannot serve two signals from the same input at the same time and thus a case of *wavelength contention* will occur.

It is also possible that more than two packets will compete in the last QD-SOA. To overcome the packet contention issue we may use more internal wavelengths or a simple scheduling scheme described below.

Every input is represented with a single- dimension matrix (vector). The length of each input vector equals a buffer depth + 1 (0th timeslot means that packet will be sent immediately, without entering the delay line). Every cell within it, is a boolean value and corresponds to relevant timeslot. If the input is sending the packet out in X^{th} time slot, the Xth cell value in the vector must be set to 1.

In addition, every output owns a similar vector that indicates whether packets will be sent through this output in an appropriate time slot or not. A scheduling algorithm should prevent packet contention. In proposed structure, the scheduling algorithm must to choose the first time slot when both input and output are free. This time slot is considered as a buffering time and then packets go to relevant delay lines. Contention can be resolved with a simple OR logic function. In particular, when a packet enters the Xth input, trying to reach Y^{th} output, the values corresponded to the same time slot in the Xth input vector and Y^{th} output vector will be considered and OR function will be executed. We assume a 15 timeslot buffer, so that the number of cells in each input (or output) vector is 16 (0-15 timeslots). The algorithm runs until the result of the OR function takes zero, which means in such a time slot both input and output are free and it is possible to send the packet without contention. Next, the values in the Xth input vector and Y^{th} output vector corresponded to the chosen time slot have to be set to one. When the algorithm has been executed $(N+1)$-times, where N is the length of the buffer and the result of OR function is still one, the packet will be lost. In the next time slot all cells are shifted left.

5.3.3 Performance Evaluation

In the simulation experiments we compared few performance metrics of the *parallel buffer* switch architecture and the output queued (OQ) switch. We considered the Bernoulli arrival model, where packets arrive at each input in the slot-by-slot manner. Under the Bernoulli process, the probability that there is a packet arriving in each time slot is identical and independent of any other slot. The probability that a packet may arrive in a time slot is referred to as the load of the input. The experiments have been carried out for a wide range of traffic loads: from 0.05 to 0.9 with the step 0.05. Time proceeded the exact simulation was 10000 cycles and exact simulation lasts 90000 time slots.

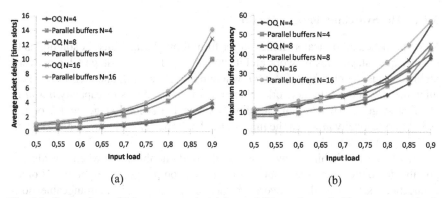

(a)   (b)

Fig. 5.9. Comparison of (a) average packet delay and (b) maximum buffer occupancy for *b* = ∞, versus input load for different switch sizes.

We have evaluated the following performance metrics: the maximum occupancy of the buffer and a mean packet delay in case of an infinite buffer size (buffer size is denoted by *b*) as well as the packet loss probability (PLP) and mean packet delay when buffer size becomes 15, 31 and 63. The "parallel buffer architecture" suffers from a higher mean packet delay and maximum buffer occupancy than OQ switch model. It is worth noting that in case of the parallel buffer architecture with *b* = ∞; (see Fig. 5.9) and for loads lower than or equal to 0.7, the mean packet delay was less than 3 time slots, which can be considered to be satisfactory. For workloads less than 0.85, the mean delay does not exceed 8 time slots. Maximum buffer occupancy for N=16, in case of output buffered switch was found to be 57 packets.

In the "parallel buffer" architecture it is possible to implement 15 time slots buffer using a limited number of QD-SOAs. This is crucial to optical switching systems because the optical signal suffers from OSNR degradation. In the proposed architecture, the total number of QD-SOAs is five but the optical signal has to go only through two devices. This case applies when assuming a 4×4 switch and thus the input signal consists of $w=4$ different wavelengths. Alternatively, the proposed switch architecture could be cascaded to employ more stages [26].

In the case of the switch with a large number of inputs/outputs, processing time of this algorithm may be too slow because every input has to match a free timeslot to send a packet, one after another (not simultaneously). Thus, considering large switches, it is also possible to use other scheduling algorithms [28] to solve a packet contention in "parallel buffer" architecture.

5.4   Multi-Stage Optical Switches with Optical Recirculation Buffers

5.4.1   The Switching Fabric Architecture

Many proposed all optical switches suffer from limited scalability. A solution is to employ multistage Clos network architectures [29]. The three-stage Clos-network architecture is denoted by $C(m, n, k)$, where parameters m, n, and k entirely determine the structure of the network. There are k input switches of capacity $n \times m$ in the first stage, m switches of capacity $k \times k$ in the second stage, and k output switches of capacity $m \times n$ in the third stage. The capacity of this switching system is $N \times N$, where $N = nk$.

A multistage switching network is called strictly nonblocking when it is always possible to connect any idle input port to any idle output port irrespective of other connections set up in the network. A switching network is rearrangeable nonblocking if it is possible to connect any idle input port to any idle output port, but some of the existing connections have to be reconfigured to do so. A switching network is called internally blocking when it is not able to guarantee connection between an idle input and an idle output. The three-stage Clos-network switching fabric is strictly nonblocking if $m \geq 2n-1$ and rearrangeable nonblocking if $m \geq n$.

In a packet switching concept, like optical packet switching (OPS), time-sliced optical burst switching (TSOBS), and optical cell switching (OCS) the switch adopts a slotted mode of operation. In one time slot packets form the input side are switched to the appropriate outputs, therefore in the next time slot the switch may be completely reconfigured. In this case it is sufficient to have a rearrangeable nonblocking switch, because in each time slot we can choose other second-stage switch [30]. In all-optical switching schemes time is divided into fixed size slots. In each time slot we can send one optical packet called cell. While a cell is being routed in a packet switching system, it can face a contention problem resulting from two or more cells competing for a single resource. In terms of place, where the contention points occur it is possible to categorize the contention states into output port contention and internal contention (internal blocking) [31]. In general, there are three ways to avoid collisions i.e. optical buffering, optical wavelength conversion and deflection routing [32], [33]. Using optical buffering strategy would make the structure of an optical switch strictly close to that of a traditional electronic packet switch, therefore it is extensively investigated. Since the optical RAM buffer is not available yet, the only practical optical buffers today are fibre delay lines (FDLs), which are fixed-length fibres. A packet which has entered the fibre must emerge from the other end after a fixed amount of time and cannot be removed before that time. The implementation of large buffers requires a large number of fibre delay lines of different length and causes a high hardware cost. The fibre loop memory may be also used for buffering (delaying) cells when contention occurs. The contending packet is pushed into the loop on the available wavelength after converting it through tunable wavelength converter. The storage time is equal to the number of recirculations.

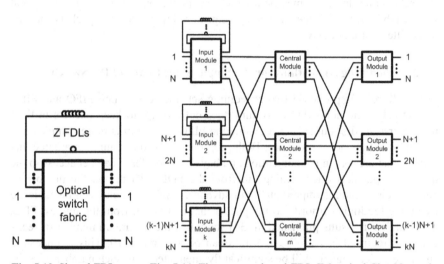

Fig. 5.10. Shared-FDL switch.

Fig. 5.11. Three-stage shared-FDL-IM optical Clos-Network switch.

The architecture of the single-stage shared-FDL switch was first proposed by Karol in [34] (Fig. 5.10). The switch called Shared-Memory Optical Packet (SMOP) switch uses feedback FDLs to resolve packet contentions and can achieve good performance with proposed control algorithms.

The switch has N input ports, N output ports and Z feedback FDLs of appropriately-selected length, that are shared by all input ports. The optical switching fabric used within the SMOP switch is memoryless, rearrangeable non-blocking e.g. an optical crossbar switch fabric and has a $(N+Z) \times (N+Z)$ dimension. The lengths of the delay lines could be: $d_1, d_2, d_3, ..., d_m$ packet duration (slots). A total of $B = (d_1 + d_2 + d_3 + ... + d_m)$ packets can be stored in the recirculation fibres. This reduces to $B = m(m+1)/2$ when $d_1 = 1$, $d_2 = 2$, $d_3 = 3,...,$ and $d_m = m$. Each FDL delays cells by a fixed number of time slots, and in general, any two FDLs may have the same or different delay values. Delay lines of length greater than one packet duration reduce the number of recirculation loops needed. As a result the number of amplifiers may be reduced and less noise is added to the signal. Another limitation is that each FDL requires one switch port, thereby increasing the overall switch cost with increased number of FDLs. In the case of two or more cells collision only one of them gets access to the output port and the others are routed to the available FDLs according to how much delay each packet needs. A packet that is buffered in a loop of length d_i will exit the delay line after d_i time slots. The scheduling algorithm selects packets to be sent to outputs, and also assigns the remaining packets to the recirculation loops. The required delay of k time slots for the particular packet may be achieved by buffering it in a recirculation delay line of length k (if possible) or by combinations of delay-line lengths that sum to k. The scheduling decisions may be revised each time a packet returns to the optical switch fabric from the feedback loops e.g. new higher priority packets may be transmitted without delay.

5.4.2   Scheduling Algorithms for the Single-Stage Shared FDL Switch

In [34] Karol has proposed two scheduling schemes called non-FIFO and FIFO respectively. The non-FIFO algorithm does not attempt to keep packets in their proper first-in, first-out sequence, while the FIFO algorithm is more complex and maintains FIFO packet sequence. Both proposed algorithms cannot guarantee that packets that are lost in the contention and have to be buffered can get access to the desired output port after coming out from the FDLs. Therefore, the number of packet recirculation is unpredictable in advance. Minimum delay can be achieved by giving the higher priority to the packet that comes out from the longest FDL. The simulation results published in [34] show that the maximum number of recirculations required in Karol's algorithm can be as high as 10. This is undesired since the optical signals will be significantly attenuated with such number of recirculations.

Three FDL assignment algorithms for the single-stage shared-FDL optical switch, namely sequential FDL assignment (SEFA), multicell FDL assignment (MUFA), and parallel iterative FDL assignment (PIFA), were proposed by S. Y. Liew et al. in [35]. These algorithms alleviate the recirculation problem by the implementation of FDLs and output port reservation. The algorithms can make the output port matching for current time slot and the FDLs assignment for the entire journey of a delayed packet so that it can be scheduled to match with the desired output port in the future time slot. The number of packet recirculation here depends on the maximum number of FDLs. If the FDLs are unavailable and the packet fails to be scheduled it will be discarded before entering the switch to avoid any FDL resources occupation.

The SEFA algorithm searches FDL routes for cells in a cell-by-cell basis, so the cells which arrive in the same time slot are scheduled one after another. The decisions are taken on the basis of a configuration table which is maintained by the shared FDL switch. The configuration table is used to making all possible FDL routes and indicating the switching schedule of the switch. It can be formulated into a slot transition diagram that includes all possible FDL routes for cells. The MUFA algorithm uses sequential search to assign FDLs for multiple cells simultaneously. The algorithm also maintains the configuration table as in SEFA algorithm, but the slot transition diagram is modified to guarantee that the FDL routes with fewer delay operations are searched and assigned for cells earlier. The PIFA algorithm uses a distributed method to assign FDL routes for multiple cells simultaneously. The same modified slot transitions diagram as for the MUFA algorithm is used. The simulation results concerning the SEFA, MUFA and PIFA algorithms are presented in [35]. To avoid packets out-of-order problem X. Wang at al. have proposed and evaluated modified MUFA algorithm called SMUFA (Sequence MUFA) [36].

5.4.3 Scheduling Algorithms for the Three-Stage Shared FDL Optical Clos-Network Switch

The three-stage optical Clos-network switch (OCNS) is a potential solution to overcome the limited scalability of single-stage switches. In general, the FDLs can be placed at the input modules (IMs), central modules (CMs) and/or at the output modules (OMs). Different FDL location influencing scheduling complexity and performance. The three-stage optical Clos-network switch with FDLs placed at IMs is shown in Fig. 5.11. In this kind of the optical Clos-network cells may be delayed only at the first stage, while the second and third stages are used only for switching.

Two assignment algorithms for the OCNS were proposed by S. Jiang et al. in [37] namely sequential FDL assignment algorithm for Clos-network switches (SEFAC) and multicell FDL assignment for Clos-network switches (MUFAC). The former assigns FDL routes and determines central-module routes in the Clos-network at a cell-by-cell basis, while the latter assigns FDL routes for multiple cells simultaneously and then assigns central-module routes in a heuristic manner.

In the SEFAC algorithm each input module maintains its own slot transition diagram, while the whole system maintains a configuration table to check the availability of all outputs in each time slot. For each input port it is necessary to find the earliest time slot that satisfies the following three conditions: (1) the destined output port is idle in the time slot; (2) on the corresponding input module there is the FDL route which can delay the cell to that time slot; (3) a connecting path between the input module and required output module is available at the time slot.

If there is possible to find the time slot that fulfill these three conditions, the SEFAC algorithm assigns the FDL route, departure time, and randomly selects available central-module of the three-stage optical Clos-network for set up the connecting path to the desired output port. For each input port the searching process is performed sequentially and in addition the round robin mechanism is employed for selection of an input module with the highest priority for the searching process.

The MUFAC algorithm is a modification of the MUFA algorithm and it can assign FDL routes and departure times for multiple cells simultaneously in a distributed manner. It has to perform three steps: (1) assignment of FDLs routes in the IMs; (2) scheduling cell departure times according to the output port availability; (3) assignment of connecting path between input modules and output modules for multiple cells in the same time slot.

The heuristic algorithm proposed by M. Karol in [40] is used in MUFAC scheme for set up connecting paths between input modules and output modules in three-stage optical Clos-network switch. The optimized algorithms provided guaranteed routes for all matches were proposed in [38], [39].

The MUFAC scheme is based on transition diagrams maintained by each IM, where each level-k node keeps information about available FDLs of that IM for time slot t. In addition, each OM keeps information about availability of their output-ports, and each of the IMs and OMs keeps the corresponding connecting path availabilities. The FDL assign process uses the transition diagram to find one or more delay lines to delay cell to required time slot n.

In the next step the MUFAC algorithm attempts to pair up each IM to a particular OM using Karol's matching algorithm. After this step there are k IM-OM pairs and only between these matched modules it is possible to send cells. The algorithm has to go through four phases namely request, grant, accept and update to select cells to be sent to output ports. The request phase is performed independently in each IM. In this phase the parent node sends the unfulfilled requests to its child nodes, and each child node collects information about availability of the output port from the paired OM for the corresponding time slot. This information is used in the grant phase to grant unfulfilled requests with the available output ports. In the next step grant decisions are sent back to the parent node. The parent node collects the central routes availabilities from the corresponding IM and the paired OM. The accept decisions made by the parent node are based on the following criteria: (1) unfulfilled input port requests; (2) availability of FDL on that IM for the corresponding time slots; (3) availability of connecting paths from that IM to the paired OM in the corresponding time slots; (4) in the case of several grants, the parent node accepts the grant with the earliest departure time. At the end of the accept phase a parent node passes the accept decision to its child nodes for updating. In the update phase, the parent node updates central route and FDL availabilities, while the child nodes update output port availabilities on paired output module. All phases mentioned above can be executed in a distributed manner.

5.4.4 Simulation Experiments

The performance of SEFAC and MUFAC algorithms was evaluated using computer simulation. The three-stage optical Clos-network switch of size 1024×1024 with FDLs placed at IMs was considered in simulation experiments (the same architecture as in [37]). The investigated switching network consists of 32 IMs, 32 CMs, and 32 OMs of capacity 32×32. Each IMs employs 32 FDLs, and there are 5, 5, 5, 5, 4, 4, and 4 FDLs with delay values 1, 2, 4, 8, 16, 32, and 64 cell times respectively. We have carried out the simulation experiments also for doubled number of FDLs (64) at each IM. The delay operation for each cell was limited to two in SEFA as well as in MUFA algorithms. The Bernoulli arrival model and uniform traffic distribution pattern are considered in simulation experiments. Two performance measures were evaluated: the average cell delay in time slots and cell loss rate. The simulation results are shown in Fig. 5.12 and Fig. 5.13.

We can see that both FDL assignment algorithms proposed for the three-stage optical Clos networks switch with FDLs placed at IMs can achieve ~10^{-8} loss rate at 0.86 input load (Fig. 5.12). The results obtained for both algorithms are comparable for the case with 32 FDLs at each IM. It is possible to observe that the SEFAC algorithm performs better at a load below 0.94, while the MUFAC algorithm performs better at a load above 0.94. The results are a little bit different when 64 FDLs are employed and the better results gives the SEFAC algorithm. For the large number of FDLs the SEFAC algorithm uses the FDLs more efficiently than the MUFAC algorithm.

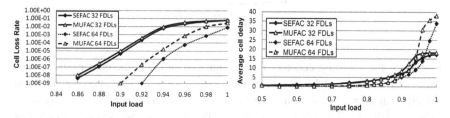

Fig. 5.12. Cell loss rate for SEFAC and MUFAC; 32 and 64 FDLs at each IM.

Fig. 5.13. Average cell delay for SEFAC and MUFAC, 32 and 64 FDLs at each IM.

The average cell delay for both algorithms is very low for wide range of input load and do not exceed 20 for very high input load – close to 1 in the case with 32 FDLs at each IM (Fig. 5.13). We observed that for input load greater than 0.88 the SEFAC algorithm gives slightly lower average cell delay than the MUFAC algorithm. For 64 FDLs at each IM and input load greater than 0.9 the average cell delay grows faster for MUFAC than for SEFAC, but do not exceed 40 and 30 cells respectively. This result is obvious due to lower cell loss rate for the case with 64 FDLs at each IM. Note that under heavy input traffic load the cell delay factor depends also on availabilities of connecting paths between IM and OM. The heuristic algorithms used in simulation experiments become also a recourse limitation.

Taking into account the time complexity it is necessary to emphasize that MUFAC is more feasible than SEFAC. The MUFAC algorithm can handle multiple packets at the same time and is scalable, while SEFAC has scalability limitation mostly due to time complexity, which is in linear proportion to the switch size. The time complexity of SEFAC as well as MUFAC was evaluated in [37].

5.5   Optical Asynchronous Packet Switch Architectures

5.5.1   All-Optical Buffer Technologies

To implement an optical memory using current technologies, we believe that two methods are practically relevant, namely the one based on electromagnetically induced transparency EIT and one based on coupled cavity waveguides.

5.5.1.1   Electromagnetically Induced Transparency

When light interacts with a three atomic energy levels it is possible to observe electromagnetically induced transparency. This effect allows for using one optical field to control the absorption and dispersion of another optical field. More precisely, by varying the intensity of the controlling field it is possible to drastically modify the group velocity of a signal beam. The group velocity is given by:

$$v_g = \frac{c}{1 + \frac{g^2 N}{\Omega_c^2}} \qquad (5.1)$$

where g is the coupling constant for the signal beam, N is the density of the medium, and Ω_c is the Rabi frequency for the control field. Thus, very small group velocity can be achieved by decreasing the control field intensity. Also, by adiabatically reducing the control field to zero the effective group velocity becomes zero, accompanied by storage of the light pulse as a material excitation. This has been observed in cold atomic gases, hot gases, and in doped crystals. The allowed storage time is determined by the coherence of the involved atomic levels, i.e. how well one can keep a quantum mechanical superposition of the atomic states. This coherence time is determined by the particular choice of material and the environment of the medium. States encoded in cold atomic gasses are the most robust allowing storage times up to 1 ms. For doped crystals where the environment is more uncontrollable, storage times up to 100 μs has been observed at cryogenic temperatures.

For practical use of electromagnetically induced transparency (EIT) in telecommunication systems it is not favourable to use alkali metals, such as rubidium or sodium where the effect has been demonstrated. The reason for this is that optical transitions for these atoms are far from wavelengths suitable for optical communication (1.3-1.5 μm). Atoms are also not subject to engineering, so it is difficult to engineer optical components based on atomic transitions, so one either adapt other components to suit the used atom, or one is lucky enough to find an atom that has the desired transitions. These two inflexibilities seem to imply that atoms will remain highly impractical to use as the active medium at the telecommunication wavelengths. The doped crystals are also unsuitable for practical applications because of the low crystal temperatures involved. For a successful implementation of a three-level system required by EIT one needs an optically active material with absorption in this interval with good coherence properties. There are plenty of semiconductor materials and structures with suitable optical transitions in this wavelength interval. These are frequently used to build lasers, modulators etc. Unfortunately, all these materials and structures have exceptionally poor coherence properties compared to the alkali atoms, so any application relying on a semiconductor implementation would have to operate at timescales faster than the decoherence time. For atoms, the coherence times are around seconds at room temperature, which is significantly longer than for any semiconductor material where the coherence rarely exceeds seconds, even at low temperatures. Error correction could in principle be used to increase the coherence time of semiconductor systems, but this remains experimentally untested.

For an atomic implementation of EIT, there is one atom that is usable at telecom wavelengths. The lutetium atom (175Lu) has a ground state consisting of hyperfine levels (F=2, 3, 4, 5) with electronic configuration [Xe]5d16s2 (2D3/2). The main transition of this atom occurs at the wavelength 1.337 μm to the state [Xe]6s26p1 (2P3/2). This transition fulfils all the requirements needed for an implementation of EIT. These transitions are within the O-band (1.260-1.360 μm) used for some telecommunication systems. The allowed storage time is determined by the coherence of the involved atomic levels, i.e. how well can one keep a quantum mechanical superposition of the atomic states. Expressed in another way, the storage time is roughly equal to the time the two (e g atomic) states storing the optical pulse can retain their relative phases. The performance of EIT based optical signal processing schemes, such as an optical delay line or buffer, depends on the coherence between the atomic levels.

The most realistic semiconductor system would be quantum dot arrays. These would have the advantage that the coupling constant for the optical interaction is much larger than for the atoms. In addition the coherence properties are among the best for the semiconductor systems. Also, the quantum dots can be integrated with other opto-electronic components using standard semiconductor technology.

5.5.1.2 Coupled Cavity Waveguides

Another promising candidate for an optical memory is to use a coupled cavity waveguide where the light may couple into auxiliary cavities next to the waveguide. This approach has the advantage that the cavities can be microfabricated and engineered for specific applications. To obtain storage times much longer than 50 ns will be a challenge using existing technologies and at the same time keep the device small. Therefore, also the cavity scheme is suitable only for high-speed systems that require short storage times. In the cavity storage scheme, it is necessary to control the coupling strength between the different cavities to switch the device from a guiding mode to a storing mode. Such switching can be performed either using electro-optic elements, or mechanical/thermal positioning of the cavities. This will unfortunately increase the device complexity.

5.5.2 Node Architectures

5.5.2.1 Contention Resolution Based on Hybrid Buffers

The considered optical packet switching node has a capacity of $n \times N$ wavelength channels, where N is a number of input and output fibres and n is a number of wavelength channels multiplexed on each fibre. We show two architectures of the optical asynchronous packet switching node with hybrid buffers (see Fig. 5.14 and Fig. 5.15) [41]. The first one, referred to as Structure I, is based on a dedicated output buffer while the second one, referred to as Structure II, is based on a shared buffer.

Structure I consists of $N \times 2N$ optical switch matrices, $N \times N$ buffer blocks, and $n \times N$ 2-to-1 optical switches. The buffer block consists of one pipeline buffer per output and wavelength. After passing the switching matrix, packets can either be terminated to the addressed output link or, if the output port is occupied, the packets are routed to the dedicated buffer. As soon as the corresponding output link is available packets from the buffer are sent directly to the output link. The optical 2×1 switches are to select a packet at the specific wavelength from either the switch matrix or from the buffer. This architecture is very simple and the pipeline is easy to control because no virtual queuing algorithms are needed.

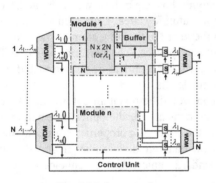

Fig. 5.14. Structure I.

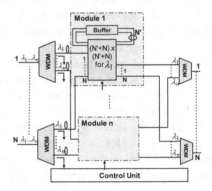

Fig. 5.15. Structure II.

The second architecture (Structure II) is based on $(N' + N) \times (N' + N)$ optical switch matrices and $N' \times N'$ buffer blocks. This structure allows for sharing the buffer resources. The buffer positions are not dedicated to the specific output links and can be shared by several channels. A number of parallel buffer positions (N') is offered per wavelength and is shared by all output ports (see Fig. 5.15). Structure II is more flexible, but it requires a more sophisticated control system. Shared, parallel memory requires advanced virtual queue management that has to be done by the control unit. It requires more complicated algorithms and more computing time.

For both node architectures the switch fabric can be divided into n identical modules. Each module serves one wavelength channel. The traffic load for a wavelength channel is assumed to be uniformly distributed between the outputs.

Further, we assumed that the traffic load at all inputs is identical. These assumptions make evaluation of one module representative for the entire switch.

5.5.2.2 Contention Resolution Based on Optical Buffers and Tunable Wavelength Converters

Fig. 5.16 and Fig. 5.17 [42] schematically illustrates the switch architectures A1 and A2. The switch fabric of A1 can be divided into n identical wavelength modules. Each module serves one wavelength channel and consists of a buffer block with b optical buffers and $(m + b) \times (2m+b)$ strictly non-blocking optical switch matrix. The wavelength conversion part of each module consists of m TWCs which are connected to all m output ports. Arriving packets are delivered to the addressed output fibre at the same wavelength if available. Otherwise, if another wavelength at the addressed output fibre is available, packets are sent to the TWC, converted to the available wavelength and delivered to the addressed output fibre. Since in A1 m TWCs are provided in each module (i.e., the same number as the input/output fibres) there will always be TWCs available if needed. However, if all the wavelengths at the addressed output fibre are occupied, packets are sent to the all-optical buffer block and wait until the appropriate output channel is available. If the time of packets stored in the buffer is longer than the maximum storage time of the optical buffer, packets are expired.

The switch fabric of A2 also can be divided into n identical modules. However, each module of A2 does not serve the specific wavelength, since the recirculation wavelength converters are attached to each switch matrix as shown in Fig. 5.17. Therefore, in A2 packets that have been converted to another wavelength pass the switching matrix twice. In the next section we evaluate how both the number of buffers b and the number of TWCs c can improve contention resolution at the node.

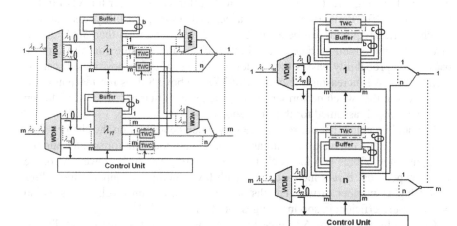

Fig. 5.16. OPS architecture A1 based on optical buffers and tunable wavelength converters

Fig. 5.17. OPS architecture A2 based on optical buffers and tunable wavelength converters

5.5.3   Performance Evaluation

Switch performance has been evaluated through a computer simulation tool that allowed simulations of ATM (Asynchronous Transfer Mode) and IP traffic. For Structures I and II we study the benefits of implementing optical memory in the optical packet switching node in addition to electronic ones. In [43] it is shown that Structure I is less suited for this kind of evaluation due to the limited flexibility of the scheduling algorithm that can be applied. We proposed to add a few all-optical buffer positions to the electrical buffer block and evaluated the switching node based on the shared hybrid buffer.

We studied the admission algorithm giving a priority for the service that requires optical signal transparency in order to minimize the time a transparency packet has to spend in the buffer. We assumed that the load of this class of packets is 20% of the total load. The transparency packets were scheduled to be transmitted before the packets that could wait in the electrical buffer. In order to avoid non-priority packets waiting for ever, the transparency packets did not get the "total" priority. We implemented the following scheduling algorithm. If, at arrival of a transparency packets which need to be buffered, there are already packets in the electrical buffer waiting to be transmitted at the same output up to seven optical packets will be transmitted prior to the electrical and then packets will be transmitted in the order of arrival.

We have shown that the packet loss probability can be significantly improved by implementing a few optical buffer positions. At the same time a new optical memory technology allows for building an asynchronous optical packet switch with optical buffers. Furthermore, we evaluate architectures A1 and A2. The simulation results are presented in Fig. 5.18. It is shown that improvement achieved by TWCs is the higher, the larger optical buffer size is. Thus, for ATM traffic we can observe a boosting effect of caused by optical buffers. It is clearer for ATM traffic than for IP traffic. Furthermore, for IP traffic, packet loss probability less than 1%, which is often required, can be achieved only at low load (less than 30%) due to insufficient maximum storage time for IP traffic pattern while for ATM traffic packet loss probability passes the 1% level with a few buffer positions. It can be seen that for A2 increasing the number of TWCs above 8 does not make any improvement in the switch performance. It is due to small number of wavelengths on each fibre ($n = 4$) assumed for the simulations.

It should be noted that our results are very much dependent on the choice of parameters m, n and the maximum storage time. It is obvious that for smaller number of input/output fibres (m) the packet loss probability would be lower. Also, we would expect that if n/m increased (i.e. with lower number of input/output fibres and/or higher number of wavelengths per fibre) to a certain level, TWCs would become more efficient in solving congestion than optical buffers.

Our results reveal that A2 outperforms A1 with regard to number of TWCs, which is related to the cost of the switching node. However, one should take into account that fabrication of switching matrices for multiwavelenth operation can be more difficult and expensive than for one selected wavelength.

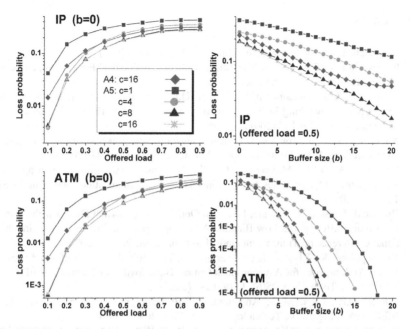

Fig. 5.18. Evaluation of A1 and A2. Packet loss probabilities for IP and ATM traffic patterns.

Finally, it should be mentioned that the two main advantages of our architectures for implementing in an optical packet switching network are: the asynchronous operation and relatively small optical buffer and number of TWCs needed to obtain low packet loss probability.

5.6   Conclusions

The new architectures of optical packet switches and optical buffers were studied in this Chapter. We have demonstrated the applicability of QD-SOAs in the realization of a 160 Gb/s line rate buffer architecture. Physical layer simulation results have shown regenerative performance in terms of extinction ratio and Q-factor improvement along the cascade of converters up to the third stage of the buffer. We have also considered applicability of QD-SOAs as multi-wavelength converters for constructing parallel optical buffers with high buffer depth and low number of wavelength converters. We have also discussed the switching fabric architecture of greater capacity, constructed from optical switches arranged in stages and with recirculation buffering in the first stage. We have presented several scheduling algorithms for such architecture and shown by simulation that performance of these algorithm are sufficient for using them in practice. Finally, we have discussed two switch architectures with optical buffering, which enables the asynchronous operation. These architectures have been compared in the number of TWCs needed and the packet loss probability.

References

1. El-Bawab, T.S.: Optical Switching. Springer, Heidelberg (2006)
2. Ramaswami, R., Sivarajan, K.N.: Optical Networks. A Practical Perspective, 2nd edn. Morgan Kaufmann, San Francisco (2002)
3. O'Mahony, M.J., Simeonidou, D., Hunter, D.K., Tzanakaki, A.: The application of optical packet switching in future communication networks. IEEE Communications Magazine 39(3), 128–135 (2001)
4. Xu, L., Perros, H.G., Rouskas, G.: Techniques for optical packet switching and optical burst switching. IEEE Communications Magazine 39(1), 136–142 (2001)
5. Yao, S., Yoo, S.J.B., Mukherjee, B.: All-optical packet switching for metropolitan area networks: Opportunities and challenges. IEEE Communications Magazine 39(3), 142–148 (2001)
6. Bjørnstad, S., Stol, N., Hjelme, D.R.: An Optical Packet Switch Design with Shared Electronic Buffering and Low Bit Rate Add/Drop Inputs. In: Proceedings of International Conference on Transparent Optical Networks, pp. 69–72 (2002)
7. Yiannopolous, K., Vlachos, K., Varvarigos, E.: The Multiple Input Buffer and Shared Buffer Architectures for Asynchronous Optical Burst Switching Network. IEEE/OSA J. Lightwave Technology 25(6), 1379–1389 (2007)
8. Hunter, D.K., Chia, M.C., Andonovic, I.: Buffering in optical packet switches. IEEE/OSA J. Lightwave Technology 16(12), 2081–2094 (1998)
9. Chia, M.C., Hunter, D.K., Andonovic, I., Ball, P., Wright, I., Ferguson, S.P., Gulid, K.M., O'Mahony, M.J.: Packet Loss and Delay Performance of Feedback and Feed-Forward Arrayed-Waveguide Gratings-Based Optical Packet Switches With WDM Inputs–Outputs. IEEE/OSA J. Lightwave Technology 19(9), 1241–1254 (2001)
10. Gauger, C.M.: Dimensioning of FDL buffers for optical burst switching nodes. In: 6th IFIP Working Conference on Optical Network Design and Modeling (ONDM 2002), Torino (2002)
11. Enachescu, M., Ganjali, Y., Goel, A., McKeown, N., Roughgarden, T.: Routers with very small buffers. In: Proceedings of IEEE INFOCOM, pp. 1–11 (2006)
12. Berg, T., Bischoff, S., Magnusdottir, I., Mork, J.: Ultrafast gain recovery and modulation limitations in self-assembled quantum-dot devices. IEEE Photon. Technol. Lett. 13, 541–543 (2000)
13. Uskov, A.V., et al.: On high speed cross-gain modulation without pattern effects in quantum-dot semiconductor optical amplifiers. Optics Communications 227, 363–369 (2003)
14. Akiyama, T., Kuwatsuka, H., Simoyama, T., Nakata, Y., Mukai, K., Sugawara, M., Wada, O., Ishikawa, H.: Application of spectral-hole burning in the inhomogeneously broadened gain of self-assembled quantum dots to a multi wavelength-channel nonlinear optical device. IEEE Photon. Technol. Lett. 12(10), 1301–1303 (2000)
15. Chikama, T., Onaka, H., Kuroyanagi, S.: Photonic Networking Using Optical Add Drop Multiplexers and Optical Cross-Connects. FUJITSU Sci. Tech. J. 35(1), 46–55 (1999)
16. Varvarigos, E.: The "packing" and the "scheduling packet" switch architectures for almost all-optical lossless networks. IEEE/OSA J. Lightwave Technology 16(10), 1757–1767 (1998)

17. Spyropoulou, M., et al.: 160 Gb/s simulation of a quantum-dot semiconductor optical amplifier based optical buffer. In: 11th IFIP Working Conference on Optical Network Design and Modeling (ONDM), pp. 107–116 (2007)
18. Lee, H., Yoon, H., Kim, Y., Jeong, J.: Theoretical Study of Frequency Chirping and Extinction Ratio of Wavelength-Converted Optical Signals by XGM and XPM using SOA's. IEEE J. of Quantum Electronics 35, 1213–1219 (1999)
19. Sugawara, M., et al.: Quantum-dot semiconductor optical amplifiers for high-bit-rate signal processing up to 160 Gb/s and a new scheme of 3R regenerators. Meas. Sci. Technol. 13, 1683–1691 (2002)
20. Sugawara, M., et al.: Effect of homogeneous broadening of optical gain on lasing spectra in self-assembled $In_xGa_{1-x}As/GaAs$ quantum dot lasers. Physical Review B 61, 7595–7603 (2000)
21. Massoubre, D.: High Speed Switching Contrast Quantum Well Saturable Absorber for 160Gb/s Operation. In: Proceedings of Lasers & Electro-Optics (CLEO), pp. 1593–1595 (2005)
22. Girardin, F., Guekos, G.: Gain Recovery of Bulk Semiconductor optical Amplifiers. IEEE Photon. Technol. Lett. 10, 784–786 (1998)
23. Yazaki, T., et al.: Device length dependency of cross gain modulation and gain recovery time in semiconductor optical amplifier. In: Proceedings of the Conference on Indium Phosphide and related materials, pp. 510–512 (2003)
24. Akiyama, T., et al.: Pattern-effect free amplification and cross-gain modulation achieved by ultra-fast gain nonlinearity in quantum-dot semiconductor optical amplifiers. Phys. Stat. Sol (b) 238, 301–304 (2003)
25. Pleumeekers, J.L., Leuthold, J., Kauer, M., Bernasconi, P., Burrus, C.A., Cappuzzo, M., Chen, E., Gomez, L., Laskowski, E.: All-optical wavelength conversion and broadcasting to eight separate channels by a single semiconductor optical amplifier delay interferometer. In: Proceedings of Optical Fibre Conference, pp. 596–597 (2002)
26. Vlachos, K., Kabaciński, W., Węclewski, S.: New Architectures for Optical Packet Switching using QD-SOAs for Multi-Wavelength Buffering. In: International Workshop on High Performance Switching and Routing, Shanghai (2008)
27. De Zhong, W., Tucker, R.S.: Wavelength Routing-Based Photonic Packet Buffers and Their Applications in Photonic Switching Systems. IEEE/OSA J. Lightwave Technology 16(10), 1737–1745 (1998)
28. Pavon-Mariño, P., Garcia-Haro, J., Jajszczyk, A.: Parallel Desynchronized Block Matching: A Feasible Scheduling Algorithm for the Input-Buffered Wavelength-Routed Switch. Computer Networks 51(15), 4270–4283 (2007)
29. Clos, C.: A Study of Non-Blocking Switching Networks. Bell Sys. Tech. Jour., 406–424 (1953)
30. Cheyns, J., Develder, C., Van Breusegem, V., Colle, D., De Turck, F., Lagasse, P., Pickavet, M., Demeester, P.: Clos Lives On in Optical Packet Switching. IEEE Communication Magazine 42(2), 114–121 (2004)
31. Chao, H.J., Lam, C.H., Oki, E.: Broadband Packet Switching Technologies: A Practical Guide to ATM Switches and IP Routers. Wiley, New York (2001)
32. Yao, S., Mukherjee, B., Dixit, S.: Advances in Photonic Packet Switching: An Overview. IEEE Communication Magazine 38(2), 84–94 (2000)
33. Yang, J., Li, J., Zeng, Q., Ye, T., Zhu, G.: A Novel Optical Packet Switch: Control Algorithm and Performance. Optical Transmission, Switching, and Subsystems II. Proceedings of SPIE 5625, 913–922 (2005)

34. Karol, M.J.: Shared-Memory optical packet (ATM) switch. In: Proceedings of SPIE, Multigigabit Fibre Communications Systems, vol. 2024, pp. 212–222 (1993)
35. Liew, S.Y., Hu, G., Chao, H.J.: Scheduling Algorithms for Shared Fibre-Delay-Line Optical Packet Switches – Part I: The Single-Stage Case. IEEE/OSA J. Lightwave Technology 23(4), 1586–1600 (2005)
36. Wang, X., Jiang, X., Horiguchi, S.: Maintaining Packet Order in Reservation-based Shared-Memory Optical Packet Switch. In: 22nd International Conference on Advanced Information Networking and Applications, pp. 912–917 (2008)
37. Jiang, S., Hu, G., Liew, S.Y.: Scheduling Algorithms for Shared Fibre-Delay-Line Optical Packet Switches – Part II: The Three-Stage Clos-Network Case. IEEE/OSA J. Lightwave Technology 23(4), 1601–1609 (2005)
38. Hwang, F.K.: Control Algorithms for Rearrangeable Clos Networks. IEEE Transactions on Communications COM-31(8), 952–954 (1983)
39. Jajszczyk, A.: A Simple Algorithm for the Control of Rearrangeable Switching Networks. IEEE Transactions on Communications COM-33(2), 169–171 (1985)
40. Karol, M.J.: I, C-L.: Performance Analysis of a Growable Architecture for Broadband Packet (ATM) Switching. In: Proceedings of the Global Telecommunications Conference GLOBECOM '89, pp. 1173–1180 (1989)
41. Wosinska, L., Haralson, J., Thylen, L., Öberg, J., Hessmo, B.: Benefit of Implementing Novel Optical Buffers in an Asynchronous Photonic Packet Switch. In: Proceedings of European Conference on Optical Communication ECOC'04, Stockholm, Sweden (2004)
42. Chen, J., Wosinska, L., Thylén, L., He, S.: Novel Architectures of Asynchronous Optical Packet Switch. In: Proc. of European Conference on Optical Communication ECOC'07, Berlin, Germany (2007)
43. Wosinska, L., Karlsson, G.: A photonic packet switch for high capacity optical networks. In: Proceedings Batonal Fibre Optic Engineers Conference, NFOEC'02, Dallas, TX (2002)

Future Outlook (Part I)

K. Ennser (part editor)

The increase of traffic demands large bandwidth and high performance next generation network (NGN). The key features that drive the NGN are flexibility, reconfigurability, scalability and cost effectiveness. In part I relevant research topics were presented and discussed to address these requirements.

The optical signal processing allows manipulating the signal in the optical domain avoiding to convert to electronic and back to optical domain. This permits to increase the speed of processing so far limited by electronics and reduce significantly the cost. Chapter 3 proposes several techniques to handle optically the signal.

Another way to increase the network capacity is to develop novel transponders interfaces by using multilevel coding and optically processing the data at the transmitter side by means of packet compression and similar techniques to optically time division multiplexing to increase the transmission speed.

Although general tendency to move to optical domain in the signal processing, it has been found out that an electronic equalization technique can be an efficient solution to compensate linear and nonlinear transmission impairments. Research progress reported in chapter 2 demonstrates the benefits.

A bottleneck in the network evolution is the development of access haul. The bandwidth hungry services demand more capacity at the access network. A way to solve this is to bring optical fiber closer to the users. A hybrid solution could also be beneficial mixing wireless and wired network. One of the dominant requirements is the cost effectiveness. Chapter 4 discusses these issues and includes relevant aspects as network management and protocols.

The network evolutes towards a meshed architecture type were the resources can be used within higher flexibility and dynamically. A key architecture element is the optical switch and optical buffer. The content resolution can be addressed by use of optical buffers and wavelength converters as reported in chapter 5. This allows higher flexibility and agile network.

In summary there are several stimulating topics to research and develop for future generation network and this book contributes towards them.

Part II

Introduction (Part II)

M. Köhn (part editor)

When COST action 291 started in 2004, communication networks started to employ wavelength division multiplexing (WDM) to interconnect discrete network locations and provided thus high transmission capacities. Nevertheless, it was foreseeable that the application of conventional networking technologies for high capacity networks will not be the most cost efficient solution. This was mainly driven by the dramatic growth of required transmission and switching capacity as well as the agility of the network to react rapidly on changes in the traffic patterns.

In the background of this, the two working groups WG 2 and WG 3 focused on *novel network architectures* and a *unified control plane, network resilience and service security*, respectively. WG 2 focused on the evolution of network scenarios including the study of novel network architectures. It also studied different node architectures and technologies in terms of network performance and functionality. WG 3 addressed network survivability and security issues in such networks, covering topics such as protection and restoration, its impact on routing and wavelength assignment algorithms, fault isolation, disaster recovery, etc.

Both working groups focused their studies on three network architectures, i.e., wavelength routed networks, optical burst switching and optical packet switching.

In the first architecture, end-to-end lightpaths are provisioned depending on the chosen protection scheme and QoS requirements such that overall performance metrics are fulfilled. Also multi-layer scenarios have been considered where customer connections are provisioned on an electrical layer. The second and third architecture rely on the packet switching principle and thus can efficiently support applications with highly bursty traffic. End-to-end QoS schemes, contention resolution and scheduling schemes determine the overall network performance and network scalability in terms of throughput and cost efficiency.

This part summarizes the work of the two working groups. It introduces the issues that have been studied, presents the most important results in detail and gives hints for further reading. The remainder is structured as follows:

Chapter 6 is devoted to transparent wavelength routed networks based on DWDM. In such networks, the signal propagates through the network without O/E conversion and is distorted. To overcome this signal degradation different approaches have been investigated in the past, e.g. based on regenerators. In this chapter, a novel approach is investigated. The authors extend the routing and wavelength assignment process (RWA) such that it takes the distortions into consideration. They present a dynamic network planning tool residing in the core network nodes that incorporates real-time assessments of optical layer performance into impairment aware RWA algorithms. Furthermore, they show the integration into a unified control plane.

In Chapter 7, most important aspects of optical burst switching (OBS) and optical packet switching (OPS) are presented which have been investigated and discussed in COST Action 291. These all-optical network architectures rely on the packet switching principle and only convert the signal at the network edge from/to the electrical domain. While OPS is more advanced to be realized due to its technological requirements, OBS seems to be a compromise.

Beyond a general introduction into the different flavors of OBS and OPS, the chapter presents work on burstification algorithms, QoS provisioning and routing. Furthermore, cross-layer issues are tackled, i.e. the impact of routing and scheduling in the optical layer on the overall system performance including higher layer protocols.

Multi-layer Traffic Engineering is addressed in Chapter 8. Multi-layer networks are a very attractive solution to cope with the increasing dynamics and capacities in today's core networks. In such electro/optical multi-layer networks, client layer connections are groomed to wavelength channels and transported using end-to-end lightpaths. Also, intermediate grooming can yield to a more efficient utilization of network resources. In contrast to many other IP-over-WDM network architectures, a clear and efficient evolutionary path exists to upgrade today's networks.

This chapter presents different aspects of Multi-layer Traffic Engineering that have been investigated in COST action 291. These range from integrated routing schemes to mechanisms that improve the overall performance and increase the fairness among different users. Also, the impact of fundamental traffic and network characteristics on the performance are reviewed.

Chapter 9 is dedicated to network resilience in future optical networks. This topic is of deep concern to network operators due to on the one hand the significant loss of revenue in case of network failures, on the other hand the significant capital expenditures required for legacy resilience concepts. The chapter introduces a common terminology and resilience techniques. Based on this, the authors discuss network reliability from different perspectives ranging from models for calculation of availability and recovery times to concepts for differentiated resilience. Furthermore, security against malicious signals induced by attackers or component faults is discussed. Finally, the authors extend the scope to multi-layer networks and present approaches for recovery in IP-over-OTN networks.

An application of optical networks is investigated in Chapter 10. Storage area networks (SANs) are a promising technology to efficiently manage the ever-increasing amount of business data. The authors study unidirectional WDM ring networks operated in a slotted mode with one or more cross section links and investigate them with respect to traffic models, scheduling aspects or the MAC protocol.

6 Cross-Layer Optimization Issues for Realizing Transparent Mesh Optical Networks

S. Azodolmolky (chapter editor), T. Cinkler, D. Klonidis, Z. Szilard, and I. Tomkos

Abstract. In transparent optical networks as the signal propagates through a transparent network it experiences the impact of a variety of quality degrading phenomena that are introduced by different types of signal distortions. In this chapter we present a new optical networking paradigm for mesh topology, in which the impact of physical layer impairments are considered in lightpath routing (routing and wavelength assignment) process and also is integrated in the control plane and network planning and operational tools. We also show that by taking into account both, the physical impairments characterized by the Q-factor, as we propose and the features of the electrical layer will have a strong impact onto the impairments constraints based routing The number of regenerator ports in optical nodes is another constraint which has a strong impact onto the routing. We show the impact of these ports to a routing that is based on impairment constraints where the electronic layer can support traffic grooming.

6.1 An Impairment Aware Networking Approach for Transparent Mesh Optical Networks

6.1.1 Introduction

Increasing traffic volume due to the introduction of emerging broadband services and bandwidth demanding applications with different QoS requirements are driving carriers to search for a cost-effective core optical networking architecture that are tailored to the new Internet traffic characteristics. The optical network evolution and migration should aim at improved cost economics, reduced operations efforts, scalability and adaptation to the future services and application requirements. The main drivers for this migration are: a) requirement for high bandwidth and end-to-end QoS-guaranteed connectivity and b) on demand (dynamic) technology-independent service provisioning.

The network infrastructure of existing core networks is currently undergoing a transformation [1]. All-optical core WDM networks using reconfigurable optical add/drop multiplexers (ROADMs) and tunable lasers appear to be on the road toward widespread deployment and could evolve to all-optical mesh networks based on optical cross connects (OXCs) in the future. In order to realize the vision of transparency, while offering efficient resource utilization and strict quality of ser-

I. Tomkos et al. (Eds.): COST 291 – Towards Digital Optical Networks, LNCS 5412, pp. 167–188, 2009.

vice guarantees based on certain service level agreements the core network should efficiently provide high capacity, fast and flexible provisioning of links, high-reliability, and intelligent control and management functionalities. To accommodate all the above requirements in a cost effective manner, an optical core network must planned in a way that can be easily managed and upgraded in terms of transparency and reconfigurability.

6.1.2   Transparent Optical Network Challenges

Optical transparency has an impact on network design, either by adapting the size of WDM transparent domains in order to neglect physical impact on quality of transmission, or by introducing physical considerations in the network planning process (e.g. extra rules for WDM systems, or performance monitoring). Thus, measurement databases and physical impairments aware algorithms are required. Similarly, new management and control plane functionalities must be defined for dynamic connection management and fault monitoring.

The realization of fully automated and dynamic transparent core optical networks is a difficult task although highly desirable due to the expected cost & performance benefits. This goal has not been yet achieved in commercial exploitation due to: a) limited system reach and overall transparent optical network performance and b) difficulties related to the fault localization and isolation in transparent optical networks.

In transparent optical networks as the signal propagates through a transparent network it experiences the impact of a variety of quality degrading phenomena that are introduced by different types of signal distortions. These impairments accumulate along the path and limit the system reach and the overall network performance. There are distortions of almost "deterministic" type related only to the pulse stream of a single channel, such as Group Velocity Distortion (GVD) or the optical filtering introduced by the multiplexer/demultiplexer elements at the OXCs. The other category includes degradations with disruptive nature such as Amplified Spontaneous Emission (ASE) noise, WDM nonlinearities (four-wave-mixing and cross-phase-modulation) and finally crosstalk.

In a transparent optical network, failures also propagate transparently and therefore cannot be easily localized and isolated. The huge amount of information transported in optical networks, makes rapid fault localization and isolation a crucial requirement for providing guaranteed quality of service and bounded unavailability times. The identification and location of failures in transparent optical networks is complex due to three factors: a) fault propagation, b) lack of digital information and c) large processing time. A single failure may trigger a large number of alarms, which results in redundancy and/or false alarms for some failures. Supervisory information located in the overhead and/or payload of the data transported can only be processed at the source or destination of an optical connection, where the O/E conversions could take place. Transparency also limits the amount of performance parameters available in the core nodes, which are fully analog. The selection of performance parameters to cover the maximum range of faults while assuring cost effectiveness and maintaining transparency, the placement of monitoring equipment to reduce the number of redundant alarms and to lower the capital expenses, and the design of fast localization algorithms are among challenges of fault localization in transparent optical networks.

6.1.3 Proposed Approach

The most commonly adopted approach to overcome the mentioned issues of optically layer transparency is the utilization of optoelectronic regenerators on per channel basis on all (opaque architecture) or selected (managed reach) optical nodes. The other approach is the exploitation of impairment management techniques that may be implemented in-line or at the optical transponder interfaces. However in addition to physical layer impairment management techniques, a third approach could be considered, in which special Routing and Wavelength Assignment (RWA) algorithms are used for lightpath routing that take into account the physical characteristics of the lightpaths. We categorize this class of algorithms as Impairment Aware RWA (IA-RWA) algorithms. The proposed approach, as depicted in Fig. 6.1, is that intelligence in core optical networks should not be limited to the functionalities that are positioned in the management and control plane of the network, but should be extended to the data plane (optical layer).

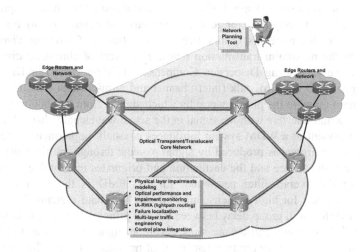

Fig. 6.1. The proposed approach

The key innovation of this approach is the development of a dynamic network planning tool residing in the core network nodes that incorporates real-time assessments of optical layer performance into IA-RWA algorithms and is integrated into a unified control plane. The proposed solution serves as the key enabler for automated network reconfiguration capability, which also provides network resiliency and QoS-guaranteed connectivity. In order to realize it, several building blocks should be considered in an orchestrated fashion. In the following sections we briefly present these building blocks.

6.1.3.1 Physical Layer Modelling and Monitoring

In order to realize the IA-RWA algorithms, covered later in this section, physical impairment should be carefully identified and modelled. Physical layer impairments may be classified as linear and non-linear. Linear impairments are independent of the signal power and affect each of the optical channels (wavelengths) individually.

The important linear impairments that should be modelled and monitored are Amplified Spontaneous Emission Noise (ASE), Chromatic Dispersion (CD), Crosstalk, Filter Concatenation, and Polarization Mode Dispersion (PMD).ASE noise is the principal source of noise in doped fibre amplifiers. In these amplifiers the initial spontaneous emission is amplified in the same manner as the signals, which degrade the optical to signal noise ratio (OSNR). This noise limits the reach and capacity of WDM all-optical networks. Chromatic dispersion is the impairment due to which different spectral components of a pulse (frequencies of light) travel at different velocities. Chromatic dispersion arises for two reasons. The first is that the dependency of refractive index of the fibre to the optical wavelength (material dispersion) and the other is due to the waveguide dispersion. Waveguide dispersion occurs when the speed of a wave in a waveguide (such as an optical fibre) depends on its frequency for geometric reasons, independent of any frequency dependence of the materials from which it is constructed. Chromatic dispersion also limits the maximum transmission reach. The effect of chromatic dispersion can me minimized using Dispersion Compensation Fibres (DCF) or Dispersion Shifted Fibres (DSF). Crosstalk (inter-channel and intra-channel crosstalk) is the general term given to the phenomenon by which signals from neighbouring wavelengths leak and interfere with the signal in the actual wavelength channel. Almost every component in a WDM system introduces crosstalk impairment. Filter concatenation impairment is produced by signal passage through multiple WDM filters between the source and the destination and originates mainly due to the narrowing of the overall filter pass-band. Finally, PMD is the most important polarization effect for high capacity, high bit rate long haul systems. PMD gives rise to the differential group delay between the two principle states of polarization.

The effects of nonlinearities are more severe at higher bit rates and at higher transmitted powers. There are two categories of nonlinear effects. The first arises due to the interaction of light waves with photons (molecular vibrations) in the silica medium. The two main effects in this category are Stimulated Brillouin Scattering (SBS) and stimulated Raman scattering (SRS). The second set of nonlinear effects arises due to the dependence of the refractive index on the intensity of the applied electric field, which in turn is proportional to the square of the field amplitude. The most important nonlinear effects in this category are self-phase modulation (SPM) and four-wave mixing (FWM). References [2,3] provide good overall starting points.

In our previous work [4] the benefits of 2R regeneration on the blocking probability was investigated. In the following figures, the corresponding performance is evaluated as a function of the dispersion-management scheme and for different values of the nonlinear parameter γ, considering only the effect of the random perturbations and having ignored the deterministic eye closure. The efficiency of the ICBR is compared to the SP, where the link lengths are considered as cost parameters. In Fig. 6.2(a), the blocking percentage is demonstrated as a function of the γ-parameter considering the optimum threshold value, L=30 (relative position of the threshold of the nonlinear element with respect to their full-width at half-maximum 'FWHM' level), and two different pairs of pre- and inline residual dispersion values. It is clear that the ICBR scheme outperforms Shortest Path (SP) for lower values where the suppression of the amplitude/jitter distortion is more pronounced. in Fig. 6.2(b) for both ICBR or shortest path (SP) routing schemes where the benefits of the ICBR comparing to the conventional shortest path (SP) routing algorithm are evident almost over the whole range of input powers.

In addition to analytical and simulation techniques for modelling the physical impairment, optical impairment and performance monitoring techniques are required to realize the impairment aware lightpath routing (i.e. IA-RWA) mechanism. The monitoring could be implemented at the impairment level (Optical Impairment Monitoring – OIM) or at the aggregate level where the overall performance is monitored (Optical Performance Monitoring – OPM) [6]. In the former approach, every tunable network element should report its status in terms of e.g. input/output power, noise figure, and dispersion. Then the effect of physical

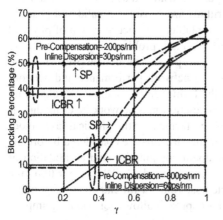

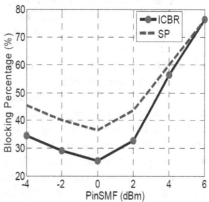

a) Blocking percentage as a function of the γ-parameter for different routings

b) Blocking percentage as a function of the input power in the SMF spans for ICBR and SP

Fig. 6.2. Performance evaluation results

degradation is identified using appropriate analytical models. In the latter approach, optical monitoring systems are located at each node for assessing the overall impact of each degradation. An effective OIM/OPM strategy, which evaluate the status of the link, would support the network Control Plane in performing lightpath establishment or rerouting functions. The most important link performance parameters can be summarized as: a) Residual dispersion, b) Total EDFA input/output powers, c) Channel optical power & wavelength, d) OSNR (Optical Signal-to-Noise Ratio) and e) Q-factor – as an estimator of the overall system performance.

The development of a physical layer modelling and monitoring scheme will provide the intelligence to the proposed approach to: a) implement failure localization methods of single and multiple failures in transparent optical networks b) implement novel impairment aware lightpath routing (i.e. IA-RWA) schemes that will consider all key physical impairments and their interplay and c) construct and control complex network topologies while efficiently maintaining a high QoS and the fulfilment of service level agreements.

6.1.3.2 Impairment Aware Lightpath Routing

In optical networks, the wavelength of the path should be also determined. The resulting problem is considered as RWA problem in literature. If wavelength conversion is allowed in the network, a lightpath can exit an intermediate node on a different wavelength. If no wavelength conversion is allowed then the wavelength continuity constraint is imposed to the generic RWA problem. This constraint implies that a lightpath should occupy only a specific single wavelength, throughout its path from the source to the destination node [7].

In most RWA proposals the optical layer is considered as a perfect medial and therefore all outcomes of the RWA algorithms are considered valid and possible even though the performance may be unacceptable. The incorporation of physical impairments in transparent optical network planning problems has received more attention from research communities. These proposals can be classified into two main categories: a) effects of impairments on network performance and network design with impairment consideration. In the former category the RWA algorithm is treated in two steps: first a lightpath computation in a network layer module is provided, and then lightpath verification is performed by the physical layer module. In the other category the physical layer impairments are considered before the network layer module proceeds to the lightpath computation and a validation of the signal quality requirements follows. Our proposed IA-RWA algorithms the cost of a link will be a vector (not a single cost value) with entries corresponding to individual impairments. This conceptual approach allows for handling impairments differently and more efficiently.

The impairment aware RWA proposals can be also classified as static and dynamic depending on whether or not the impairments and overall network conditions are assumed to be time dependent. Physical impairments may vary during time (i.e. dynamic network conditions) and thus change the actual physical topology characteristics. We refer to this situation as the "dynamic network condition". Network traffic (i.e. request for lightpath establishment) can also be static or dynamic. The great majority of the proposed RWA algorithms in the literature only consider static traffic (permanent lightpaths demands) and network conditions (time invariant impairments). IA-RWA algorithms in our approach try to address other possible scenarios as indicated in Table 1. In Case 3, we consider the more realistic situation, in which dynamic traffic demands may induce a different behavior from certain devices like amplifiers or OXCs.

Table 6.1. Network and traffic conditions captured by IA-RWA algorithms in DICONET

	Static traffic conditions	Dynamic traffic conditions
Static network conditions	N/A	Case 2
Dynamic network conditions	Case 1	Case 3

6.1.3.3 Failure Localization

The peculiar behaviour of all-optical components and architectures bring forth a new set of challenges for network reliability and resilience. Failure management is one of the crucial functions and a prerequisite for protection and restoration schemes. An important implication of using all-optical components in communication systems is that available methods used to manage and monitor the health of the network may no longer be appropriate. All-optical components are not by design able to comprehend signal modulation and coding, therefore intermediate switching nodes are unable to regenerate data, making segment-by-segment testing of communication links more challenging. As a direct consequence, failure detection and localization using existing integrity test methods is made very difficult.

In the proposed framework an algorithm that solves the multiple failure location problem in transparent optical networks is proposed where the failures are more deleterious and affect longer distances. The proposed solution also covers the non-ideal scenario, where lost and/or false alarms may exist. Although the problem of locating multiple faults has been shown to be NP-complete, even in the ideal scenario where no lost or false alarms exist, the proposed algorithm keeps most of its complexity in a pre-computational phase. Hence, the algorithm only deals with traversing a binary tree when alarms are issued. This algorithm locates the failures based on received alarms and the failure propagation properties, which differ with the type of failure and the kind of device that are in the network. Another algorithm has been proposed to correlate multiple security failures locally at any node and to discover their tracks through the network. The algorithm is distributed and relies on a reliable management system since its overall success depends upon correct message passing and processing at the local nodes. To identify the source and nature of detected performance degradation, the algorithm requires up-to-date connection and monitoring information of any established lightpath, on the input and output side of each node in the network. This algorithm mainly runs a generic localization procedure, which will be initiated at the downstream node that first detects serious performance degradation at an arbitrary lightpath on its output side. Once the origins of the detected failures have been localized, the network management system can then make accurate decisions to achieve finer grained recovery switching actions. In this scope, proposed approach aims at developing efficient and innovative failure localization algorithms based on the information received by the network management system to enable physical layer aware protection/restoration schemes. In addition novel attack detection and localization methods will be investigated and deployed by the control plane to provide advance security and reliability in future optical networks

6.1.3.4 Network Planning Tool

The key goal is the development of a dynamic network planning tool residing in the core network nodes that incorporates real-time measurements of optical layer performance into IA-RWA algorithms and is integrated into a unified control plane. As depicted in Fig. 6.3, this tool will integrate advanced physical layer

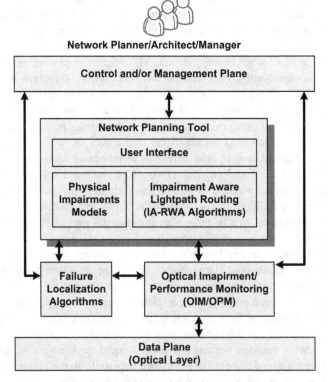

Fig. 6.3. Network planning and operation tool

models with novel impairment aware lightpath routing algorithms. It will serve as an integrated framework that considers both physical layer parameters and networking aspects (cross-layer) and will optimize automated connection provisioning in transparent optical networks.

The network planning tool has two operational modes: a) off-line mode and b) on-line (or real-time) mode. In off-line mode a full map of network traffic and network conditions will be fed to the tool in order to produce the planning outcomes. These results can be disseminated to the network management system, controlled by an operator. For on-line use of the network planning tool an online traffic engineering solution is required utilizing an interface between the control plane and the management plane so that network situation could be evaluated in real time and its results could be periodically disseminated into the network. In on-line mode, this dynamic network planning tool can be used to support optimum network operation and engineering under dynamically changing traffic and physical network conditions.

6.1.3.5 Control Plane Extensions

In order to realize an impairment aware control plane (impairment aware light path routing, topology and resource discovery, path computation, and signalling), existing protocols should be properly extended. The extended control plane will in turn address traffic engineering, resiliency, and QoS issues and will support automated and rapid optical layer reconfiguration. The GMPLS protocol suite [6] has gained significant momentum as a candidate for unified control plane [7]. There are some proposals to address the integration of physical layer impairments into the GMPL control plane.

One direction deals with enhancement of GMPLS signalling (e.g. Resource Reservation Protocol with Traffic Engineering extensions 'RSVP-TE') and management (e.g. Link Management Protocol 'LMP') protocols. In this approach, lightpaths from source to the destination are dynamically computed using current routing protocols (e.g. OSPF-TE), without considering the optical layer impairments. Only upon lightpath establishment, the enhanced reservation protocol computes the amount of impairments and based on the results the lightpath setup request can be either accepted or rejected. Following this approach a local database in each node (e.g. OXCs or ROADM) is required to store the physical parameters that characterize the node and its connected links. In order to setup a lightpath, the source node generates an extended version of the RSVP PATH message, which includes the physical information of the transmitting interface and corresponding link. Each node along the path updates this message by adding its own local values. Admission control at the intermediate or the destination node compares the accumulated values with thresholds and decide to accept or reject the lightpath setup request.

In the second approach, physical layer information are inserted into the some Interior Gateway Routing Protocol (IGRP) (e.g. OSPF-TE). The source node of the lightpath interacts with the Traffic Engineering Database (TED). Network-wide information, which are stored in TED serves as the input information for IA-RWA algorithms in order to optimal lightpath taking into account the physical layer information. Physical layer information are carried on the TE link state advertisements (TE-LSA), in order to provide an updated and accurate inputs to the IA-RWA algorithms. By using the appropriate extensions to the OSPF-TE, physical layer information are flooded to the entire network. As a result of this flooding mechanism, the local TED database of nodes will be updated accordingly.

In order to address the scalability requirements while maintaining TE support, Path Computation Element (PCE) architecture is also considered. The PCE can reside within or external to a network node, in order to provide optimal lightpath and interact with control plane for the establishment of the proposed path. The PCE could represent a local Autonomous Domain (AD) that acts as a protocol listener to the intra-domain routing protocols (e.g. OSPF-TE). Using the information of the global topology stored in the TED the PCE constructs a reduced topology of the network, based on which the IA-RWA algorithms proceed to the path computation taking into account the physical layer parameters.

The main control plane aspects that are addressed by this approach relate to: a) Multilayer network control and b) Routing and signalling-related mechanisms and physical network characteristics information dissemination.

6.2   Mutual Impact of Physical Impairments and Traffic Grooming Capable Nodes with Limited Number of O/E/O

6.2.1 Motivation

The tremendous growth in broadband communication services, brought for the phenomenal expansion of the internet, has triggered an unprecedented demand for bandwidth in telecommunication networks. Wavelength division multiplexing (WDM) has been introduced to increase the transmission capacity of existing optical links. Multi-wavelength technology appeared as the solution for the bandwidth hungry applications. WDM has been introduced to increase the transmission capacity of existing optical links. It has been soon recognized that the switching decision can be made according to the incoming wavelength without any processing of the data stream. In single hop WDM based All Optical Networks (AON) a wavelength is assigned to a connection in such a way that each connection wavelength is handled in the optical domain without any electrical conversion during the transmission [10, 11]. Routing and Wavelength Assignment (RWA) takes a central role in the control and management of an optical network. Many excellent papers deal with design, configuration and optimization of WDM networks. See e.g. [12-14]. The majority of these RWA algorithms assume that once the path and wavelengths have been identified, connection establishment is feasible. This is true when we consider that in each node the signal is regenerated but may not be true in transparent networks, where the signal quality degrades as it is transmitted through optical fiber and nodes. Impairment constraint-based routing (ICBR) may be used in transparent networks as a tool for performance engineering with the goal of choosing feasible paths while obtaining the optimal routes regarding the RWA problem. Many excellent papers have been written about constraint based routing which obeys physical effects [13-16].

There is no doubt, that the near future info-communications will be based on optical networks. In general for networks of practical size, the number of available wavelengths is lower by a few orders of magnitude than the number of connections to be established. The only solution here is to join some of the connections to fit into the available wavelength-links. This is referred to as traffic grooming. The main idea of our optimization was that in optical layer we do not make signal regeneration. We assume that in the optical layer, there is no signal regeneration, and the noise and signal distortion accumulate along a lightpath. Actually, re-amplification, re-shaping, and re-timing, which are collectively known as 3R regeneration, are necessary to overcome these impairments. Although, 3R optical regeneration has been demonstrated in laboratories, only electrical 3R regeneration is economically viable in current networks.

We have already mentioned that in the electric layer it is possible to do traffic grooming. If we investigate the physical limitations in the optical domain, and take them into consideration, we will have to include new optical-electronic-optical conversion just to ensure the quality prescriptions. These new optical-electronic-optical conversions will have influence onto the RWA process. The idea described below was presented in [17] – see Fig. 5.2. We assume that the numbers of opto-electro-opto conversion points are limited in one optical node. This leads to a limited number of grooming capabilities in nodes. Using this kind of approach it is possible to determine the number of conversion point in the nodes.

6.2.2 Modelling the Physical Layer Impairments

The signal quality of a connection is characterized by Bit Error Ratio (BER). Experimental characterization of such systems is not easy since the direct measurement of BER takes considerable time. Another way of estimating the BER is to degrade the system performance by moving the receiver decision threshold value, as proposed in [18]. This technique has the additional advantage of giving an easy way of estimating the signal quality (Q) of the system, which can be more easily modelled than the BER. [19] explains well and gives a definition to it. The Q-factor is the signal-to-noise ratio of the decision circuit in voltage or current units, and can be expressed by:

$$Q = \frac{\langle I_1 \rangle - \langle I_0 \rangle}{\sigma_1 + \sigma_0} \qquad (6.1)$$

where: $I_{1,0}$, are the mean values of the marks/spaces voltages or currents, and $\sigma_{1,0}$ are the standard deviations.

In our model we consider a chain of amplifiers and optical cross-connects (OXC). The calculation of the Q is based on [20] where fully transparent optical cross connection architecture is presented. In this study the OXC architecture is based on wavelength selective architecture, as can be seen in Fig. 6.4.

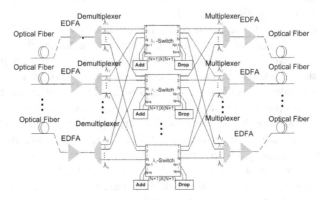

Fig. 6.4. Architecture of a switch

The switching is done for each wavelength by an (N+k)x(N+k) switch that is included between the demultiplexer and the multiplexer. Where k is the number of add-drop ports for one wavelength as it can be seen in Fig. 6.5.

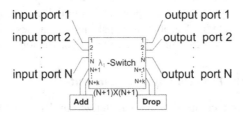

Fig. 6.5. Architecture of a switch

In this approach the noise, power and distribution for ones and zeros are calculated recursively. Assuming we know the ASE and crosstalk parameters at node m-1 and the parameters of node m, then we can calculate the ASE and crosstalk parameters at node m. In this approach the crosstalk is introduced only in the OXC nodes and ASE is introduced by the erbium-doped fibre amplifiers (EDFA), which the signal passes through. We assume that in every 80 km there is an inline amplifier.

The impact of PMD onto the signal quality can be calculated based on [21], where the PMD-induced degradation is assessed by an eye-opening penalty (EOP) along the lines. This EOP is subsequently translated to a Q-factor penalty [22].

6.2.3   The Routing Model

The routing algorithm is a highly complex algorithm which can handle optical nodes, electrical nodes and optical nodes with electrical regenerations. We consider two layer architecture, an electrical layer and an optical layer. The electrical layer supports some features such as traffic grooming and λ-conversion. The routing is realized by a shortest path algorithm. Each link and node has its own cost. In this way we can choose the lowest cost path by implementing Dijkstra's algorithm. This algorithm can route demands dynamically. The input of the optimization is the network topology and the demands. The output of the algorithm is the set of optimal routes and statistical data on the blocking in the network. The routing parameters contain information about the blocking ratio and the reason why the route has been blocked. A route can be blocked due to the RWA problem, or because of the physical impairments. A route is blocked due to RWA problem if there is not enough resource to route the demand between the source and destination node. This happens when all the wavelengths are used or in case of grooming there is not enough free capacity to groom the demand We consider a route blocked due to physical impairments if Q value of the route is lower than 3.5 which is still acceptable if using coherent detection schemes.

6.2.3.1   Routing Algorithm

The setup of the algorithm can be split in two main parts. The first one is the routing part and the second one is the calculation of physical impairments (CPI), which can be switched on or off, (Fig. 6.6). The communications between these two parts are as follow: The routing algorithm chose an optimal route, between the source and destination node and if the CPI is switched on, it sends the description of the route to CPI. The description of the route contains the lengths of the optical fibres between the nodes. The CPI calculates the signal quality and if it is adequate it sends a message back to the routing part, that the connection can be established. If the signal quality is not adequate the CPI determines the maximum reachable node (MRN) along the path and sends this information back to the routing model. The routing model establishes the connection between the source and the MRN, then chooses another route between the MRN and the destination node. If the MRN is the source node e.g. there is no possible connection due to the physical layer, the route is blocked.

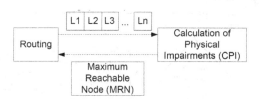

Fig. 6.6. Set-up of the algorithm

To perform the effects of grooming onto the ICBR we made four simulation types.

- The first one when there is no grooming in the RWA and the physical effects are negligible
- The second one when there is grooming and the physical effects are negligible
- The third one when there is no grooming in the RWA and we take into consideration the physical effects
- The fourth one when there is grooming and the physical effects are taken into consideration

As it was mentioned before the routing is done by a shortest path algorithm, when each link has its own cost. By using different cost values for the links of the network we can optimize an RWA oriented, or a physical impairments oriented routing. For this purpose we use four metrics.

- The first one where the cost of each link is the same. Will be referred as hop routing.
- The second one when the cost of each link is equal to the length of the link. Will be referred as length routing.

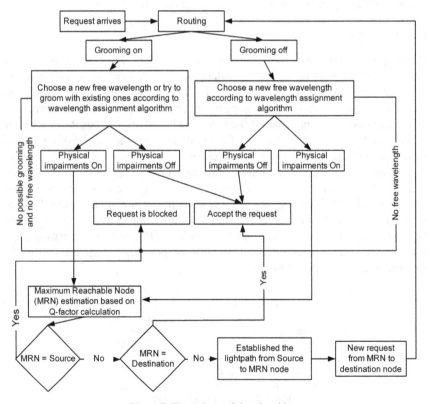

Fig. 6.7. Flow chart of the algorithm

- The third one when the cost of the link is equal to the $1/Q$ where the Q is the Q-factor of the link
- The fourth one when the cost of the link is equal to the $1/Q2$ where the Q is the Q-factor of the link

In the case of the third and of the fourth metrics we calculate the Q-factor of each link as a point-to-point connection between the two end nodes of the link. The Q-factor based routings are not obviously the best routings for the point of view of the physical layer. This is due to the nonlinear behaviour of the Q-factor. If we have two lightpaths, each lightpath has its representative Q, for example Q1 and Q2. Consider a route which contains these two lightpaths in chain. The overall Q can not be calculated from these two Q-factors, if we take both the PMD and ASE effects into calculation. The only assumption which we can make, is that, if Q1 and Q2 have a high values than the overall Q will be high as well. The exact flow of the algorithm can be seen in Fig. 6.7.

6.2.3.2   Network and Traffic Generation

The used network scenario is one of the COST266 basic topology [22]. Each link contains 24 wavelengths. The used bit rate is 10Gbit/s. The generation of the demands is based on the traffic matrices for year 2006 of the COST266 European Reference Network. More than 9000 demands were generated and routed in each simulation. The arrival of the demands occurred according to a Poisson process with the intensity of 0.005.

6.2.4   Simulation Results

We compared the four metrics used for representing the cost values of the links, in Fig. 6.8, 6.9. In Fig. 6.8 the calculation of the physical impairments was switched off and the grooming capability was switched on, and in Fig. 6.9 both modules were switched on. In the X axis the scale of the network can be seen. The meaning of it is that we changed the used network link lengths by multiplying the original lengths with the scale parameter. This resulted in increase of impairments. It was also use infinite number of O/E/O ports to ensure the grooming capability of the nodes. On the Y axis the blocking ratio is plotted. As shown in Fig. 6.8 the best metric from the point of view of the blocking ratio is the hop-metric followed by $1/Q$ and $1/Q^2$ metrics while length metric yields the worst results. We expected that in case when the physical impairments are switched off the scale of the network has no influence onto the blocking ratio. This is true when the grooming is switched off. In case of grooming there are several routing decisions which have the same ratio so it is done randomly. These random decisions lead to the non-deterministic behaviour.

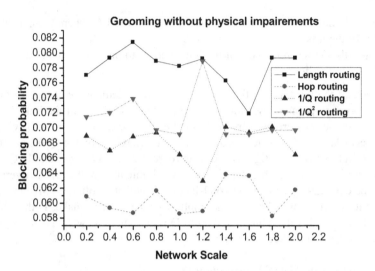

Fig. 6.8. Blocking ratio dependency from the scale of the network in case of grooming without physical effects

In Fig. 6.9 where the physical effects are taken into consideration the differences between the four metrics decrease. To understand this behaviour we investigated the blocking ratio dependency on to the physical effects (Fig. 6.10). In the X axis the network scale and in the Y axis the blocking ratio due to physical effects is plotted. This blocking ratio contains only the blockings due to physical effects without rerouting. This means that the routing module chooses an optimal lightpath and the CPI module calculates its Q-factor. If the Q is lower than 3,5 then the request is blocked. As depicted in Fig. 6.10 the characteristics of the curves are what we expected. In case of low network scales, where the lengths of the links are very small, where the physical effects have no influence, the blocking ratio is very low. While increasing the link lengths, we increase the influence of the physical effects, the blocking ratio is increasing. We compared the four metrics from the point of view of blocking ratio due to physical impairments i.e. grooming and rerouting capabilities were not used at all. Length routing has the best performance while hop routing has the worst. Between these two are the Q-based routings. Of course it is possible to find a metric which is the function of Q, f(Q), that gives better results than the length based metric, however this is not the scope of this paper.

Returning to Fig. 6.9 the blocking ratio subsidence between the four metrics is due to the constraints on the physical effects. In the aspect of physical effects the best metric is the length followed by the $1/Q^2$, and the $1/Q$ while the worst is the hop metric. From the point of view of RWA the order of these four metrics is reverse. Taking into account both the physical effects and the RWA problem, as we did, will leads to the behaviour.

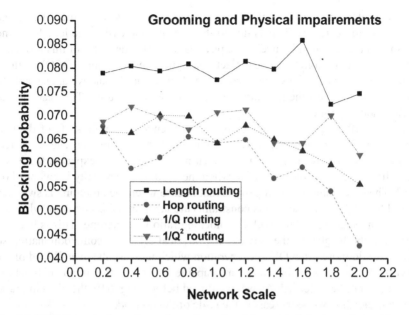

Fig. 6.9. Blocking ratio dependency from the scale of the network in case of grooming and physical effects.

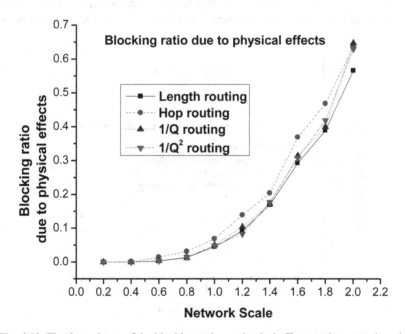

Fig. 6.10. The dependence of the blocking ratio on physical effects as the network scales

The other interesting property is that while increasing the scale of the network the blocking ratio decreases. This is due to the fact that increasing the lengths of the network increases the influence of the physical effects. The effect of this influence is that we have to do more optical-electrical-optical regenerations (OEO). If there are more points where the signal goes to the electrical layer, and we are capable to groom in these nodes, the network will be more optimally used. This leads to decreased blocking ratio.

In Fig. 6.10 we plotted the blocking ratio dependency on the scale of the network for the four routing scenarios using the length routing metric. The characteristics of the curves were the same for each metric. As it was expected there is a huge difference in the blocking ratio when the grooming capability is switched on or off. The other interesting property is that in case when the grooming is switched off and the physical impairments constraints are taken into consideration while increasing the scale of the network the blocking ratio is increasing. This is because increasing the lengths of the network the physical effects become dominating so we have to do more often OEO regeneration which increase the overall load of the network. In case when the nodes are capable to groom this trend of blocking growth can not be observed. As we mentioned before (Fig. 6.9), the blocking ratio is even decreasing while increasing the scale of the network.

Another interesting problem is if we assume that the numbers of O/E/O conversion points are limited in one optical node. This leads to a limited number of grooming capabilities in nodes. Using this kind of approach it is possible to determine the number of conversion point in the nodes. To analyze the influence of

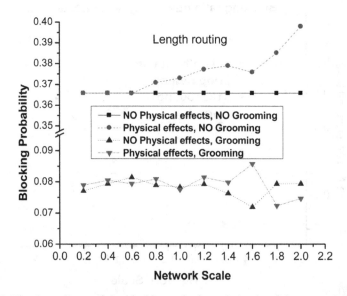

Fig. 6.11. The dependency of the blocking ratio from the scale of the network from four routing scenarios using length routing metric

the number of O/E/O (opto-electro-optical) ports onto the traffic routing we set different number of ports in the optical nodes. The chosen values were 10, 20, 40, 80, and 1000 to simulate the infinite number of O/E/O ports. This number of ports per node denoted by k is the number of add-drop links multiplied by the number of input ports as can be seen in Fig. 6.10 and 6.11. For example if we have 16 wavelengths and 5 input fibres in an OXC to have non-blocking switching we need 5 pairs of add-drop links (k = 5) for all 16 switching matrices (one per wavelength). In overall this means that we need 5x16= 80 O/E/O ports to have a non-blocking switching architecture. Of course most of the traffic does not need to be regenerated or dropped therefore in real networks the number of O/E/O ports should be lower than this. The more O/E/O ports we have the more likely we can use traffic grooming i.e. the network can be better utilized.

In Fig. 6.12 the Y axes shows the blocking ratio as the scale of the network grows (X axes). The high blocking ratio of over 10% is due to the high network usage of 80 %. As it was expected in case of scale larger than one the blocking ratio is increasing due to the physical effects. The results obtained for scale 2,718 show an extreme situation where nearly all point-to-point links are blocked and the original topology is split in multiple sub–topologies ("islands"). The curves corresponding to port numbers of 1000 and 80 are the same since even in case of 80 ports they are not all utilized for this network scenario. Decreasing the number of ports the blocking ratio increases. From the point of view of the blocking ratio, 40 and even only 20 ports are still acceptable in real network scenarios.

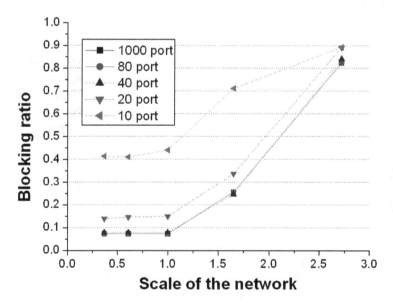

Fig. 6.12. Blocking ratio dependency on the scale of the network for different number of O/E/O ports

In Fig. 6.13 the average number of wavelength conversion is plotted versus the network scale for different O/E/O ports. The average number of wavelength conversions describes the average number of O/E/O conversions in intermediate nodes. This is a good parameter to illustrate the grooming capabilities of the network since if a demand is groomed the average number of wavelength conversions is increasing by one. The curves for 1000 and 80 ports are the same for the reason mentioned before. As it is to be seen in case of a scale less than one where the point-to-point links do not block due to physical effects the average wavelength conversion is increasing. This effect has been investigated previously, and it is because of the physical limitation of the demands. There is another aspect of decreasing the number of O/E/O ports. On the one hand while decreasing the number of ports there will not be enough ports to make the signal regeneration due to physical layer impairments. This leads to the increasing blocking ratio. On the other hand due to decreasing the number of ports the average number of physical links between electronic regeneration points will increase, as shown before.

The average hop count is the number of physical links the signal passes without O/E/O conversion. As it can be seen while decreasing the number of ports the average number of hops is decreasing in case of scale factors less than one where the point-to-point links are not blocking. In the case of scales higher than one even the point-to-point links start to block the demands due to increased physical impairments. In this case in nearly every node electronic regeneration must be done which leads to an increased blocking ratio, particularly in case of lower number of O/E/O ports.

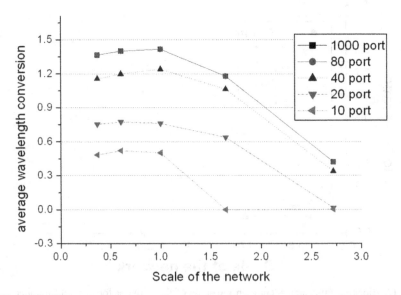

Fig. 6.13. The average number of wavelength conversions as the scale of the network grows for different numbers of O/E/O ports

6.3   Conclusion

Transparent optical networks are the next evolution step of managed reach (translucent) optical networks. Both of them have been recognized as the evolution of static WDM networks. In order to provide high-speed and QoS guaranteed connectivity, with high reliability, considering the realistic optical layer, the proposed approach in this article is a disruptive and novel solution for optical networking. Two main challenges of transparent networks, which are: a) limited system reach and overall network performance due to physical impairments and b) challenges related to failure localization and isolation, presented in this article as the main motivations behind the proposed approach. Intelligence in the core optical networks should not be limited only to certain functionalities of control and management planes, but should also be extended to the physical layer. Following this vision, the main physical impairment and the essential role of optical impairment and performance monitoring discussed and the consequently the impairment aware lightpath routing (IA-RWA) algorithms and failure localization algorithms complemented with an impairment aware control plane presented in this article.

In this chapter we have also shown how the physical impairments and the limited number of O/E/O ports can be taken into account while routing the demands in a grooming capable two layer network. We have turned attention to the mutual impact of grooming and physical impairments, i.e. of using the electronic time- and space-switching capable layer for signal regeneration and better resource utilization jointly. We have shown, that having too few O/E/O ports leads to significant performance deterioration, while having more O/E/O ports than a certain number does not change the performance at all, since they will not be used at all.

References

1. Wagner, R.: Evolution of Optical Networking. In: LEOS 2000 Proceedings, Puerto Rico, TuC1 (2000)
2. Ramaswami, R., Sirvarajan, K.N.: Optical Networks — A Practical Perspective, 2nd edn. Morgan Kaufmann, San Francisco (2001)
3. Agrawal, G.P.: Nonlinear Fiber Optics, 3rd edn. Academic Press, London (2001)
4. Markidis, G., Sygletos, S., Tzanakaki, A., Tomkos, I.: mpairment Constraint Based Routing in Ultralong-Haul Optical Networks with 2R Regeneration. IEEE Photonic Technology Letters 19(6) (2007)
5. Tomkos, I., Sygletos, S., Tzanakaki, A., Markidis, G.: Impairment Constraint Based Routing in Mesh Optical Networks. In: Proceedings of IEEE/OSA OFC2007, Anaheim, USA, OWR1 (2007)
6. Kilper, D.C., Bach, R., Blumenthal, D.J., Einstein, D., Landolsi, T., Ostar, L., Preiss, M., Willner, A.E.: Optical Performance Monitoring. Journal of Lightwave Technology 22(1) (2004)
7. Zang, H., Jue, J.P., Mukherjee, B.: A Review of Routing and Wavelength Assignment Approaches for Wavelength-Routed Optical WDM Networks. Optical Networks Magazine 1 (2000)

8. Green, P.E.: Optical Networking Update. IEEE Journal on Selected Areas in Communications 14(5), 764–779 (1996)
9. Chlamtac, I., Ganz, A., Karmi, G.: Lightpath Communications: An Approach to High Bandwidth Optical WANs. IEEE Transactions on Communications 40(7), 1171–1182 (1985)
10. Poor, H.: An Introduction to Signal Detection and Estimation. Springer, New York (1985)
11. Wauters, N., Demister, P.: Design of the Optical Path Layer in Multiwavelength Cross-Connected Networks. IEEE Journal on Selected Areas in Communications 14(5), 881–892 (1996)
12. Ramaswami, R., Sivarajan, K.N.: Routing and Wavelength Assignment in All-Optical Networks. IEEE Transaction on Networking 3(5), 489–500 (1995)
13. Banerjee, D., Mukherjee, B.: A practical Aproach for Routing and Wavelength Assignment in Large Wavelength-Routed Optical Networks. IEEE Journal on Selected Areas in Communications 14(5), 903–908 (1996)
14. Ali, M., Elie-Dit-Cosaque, D., Tancevski, L.: Enhancements to Multi-Protocol Lambda Switching (MPlS) to Accommodate Transmission Impairments. In: GLOBECOM '01, vol. 1, pp. 70–75 (2001)
15. Ramamurthy, B., Datta, D., Feng, H., Heritage, J.P., Mukherjee, B.: Impact of Transmission Impairments on the Teletraffic Performance of Wavelength-Routed Optical Networks. IEEE/OSA J. Lightwave Tech. 17(10), 1713–1723 (1999)
16. Tomkos, I., et al.: Performance Engineering of Metropolitan Area Optical Networks through Impairment Constraint Routing. OptiComm (2004)
17. Zsigmond, S., Németh, G.Á., Cinkler, T.: Mutual impact of physical impairments and grooming in multilayer networks. In: Tomkos, I., Neri, F., Solé Pareta, J., Masip Bruin, X., Sánchez Lopez, S. (eds.) ONDM 2007. LNCS, vol. 4534, pp. 38–47. Springer, Heidelberg (2007)
18. Bergano, N.S., Kerfoot, F.W., Davidson, C.R.: Margin measurements in optical amplifier systems. IEEE Photon. Technol. Lett. 5, 304–306 (1993)
19. Agrawal, G.P.: Fiber-Optic Communication Systems. Wiley, New York (1997)
20. Ramamurthy, B., Datta, D., Feng, H., Heritage, J.P., Mukherjee, B.: Impact of Transmission Impairments on the Teletraffic Performance of Wavelength-Routed Optical Networks. IEEE/OSA J. Lightwave Tech. 17(10), 1713–1723 (1999)
21. Chen, C.J.: System impairment due to polarization mode dispersion. In: Proc. Optical Fiber Conference and Exhibit (OFC), paper WE2-1, pp. 77–79 (1999)
22. Kissing, J., Gravemann, T., Voges, E.: Analytical probability density function for the Q factor due to pmd and noise. IEEE Photon. Technology Letters 15(4), 611–613 (2003)

7 Performance Issues in Optical Burst/Packet Switching

D. Careglio (chapter editor), J. Aracil , S. Azodolmolky , J. García-Haro,
S. Gunreben, G. Hu, M. Izal, A. Kimsas, M. Klinkowski, M. Köhn,
E. Magaña, D. Morató, P. Pavón-Mariño, J. Perelló, J. Scharf, S. Spadaro,
I. Tomkos, A. Tzanakaki, and J. Veiga-Gontán

Abstract. This chapter summarises the activities on optical packet switching (OPS) and optical burst switching (OBS) carried out by the COST 291
partners in the last 4 years. It consists of an introduction, five sections with
contributions on five different specific topics, and a final section dedicated
to the conclusions. Each section contains an introductive state-of-the-art description of the specific topic and at least one contribution on that topic.
The conclusions give some points on the current situation of the OPS/OBS
paradigms.

7.1 Introduction

Optical Burst Switching (OBS) [84] and Optical Packet Switching (OPS) [16]
have arisen as an alternative to low-flexible wavelength switching network and are
still gaining considerations in the research community.

The principal design objective for an OBS/OPS network is that aggregated user data
is carried transparently as an optical signal, without O/E/O conversion. This optical
signal goes through the switches that have either none or very limited buffering capabilities. Besides, the control information is carried separately from the user data
either in time (OPS) or in space (OBS). In such a network the wavelengths are temporally utilised and shared between different connections. It increases network flexibility and its adaptability to the bursty characteristics of IP traffic.

An OBS/OPS network consists of a set of electronic edge nodes and optical
core nodes connected by WDM links (see Fig. 7.1). At the edge nodes, client
packets of the same forwarding equivalence class are assembled into containers
(called bursts in OBS and packets in OPS). This process is usually called burstification or packetisation. After transmission through the network towards their destination the containers are disassembled at the egress and the original client packets are forwarded to the client network. Each container is composed of a data
payload (usually also referred simply as burst or packet) and a header packet (HP).
The HP is generated when the burstification process is finished and carries all the
information necessary to discriminate the burst or packet inside the network, like
for instance, the traffic class or its length. Inside the network the control informa-

I. Tomkos et al. (Eds.): COST 291 – Towards Digital Optical Networks, LNCS 5412, pp. 189–235, 2009.

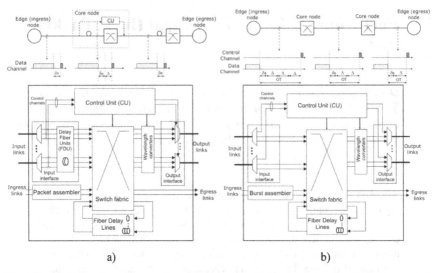

Fig. 7.1. a) OPS node and network architecture, b) OBS node and network architecture. δ_s is the switching time, Δ is the processing time, and OT is the offset time (only for OBS).

tion is processed electronically, whilst the data payload is transmitted all-optically, without optical to electrical conversion.

It has to be mentioned that in the case of OBS network, two different signalling protocols have been proposed adapting the ATM block transfer (ABT) standard designed for burst-switching ATM networks [47]:

- *Tell-and-Wait* (TAW) signalling based on delayed transmission [29]. The TAW protocol, which is recognised sometimes as a two-way signalling protocol, performs an end-to-end resources reservation with acknowledgment in advance of the burst transmission.
- *Tell-and-Go* (TAG) signalling based on immediate transmission [84]. The TAG protocol operates with a one-way signalling and it allocates transmission resources on-the-fly, a while before the burst payload arrives to a node.

The majority of research attentions are put on the one-way signalling model since two-way signalling protocols may present some concerns on the latency produced during the connection establishment process. For this reason this chapter only focus on an OBS network adopting the TAG signalling scheme, which is also the solution adopted in OPS networks.

According to this scheme, each core node must process on the fly the control information. In OPS network (Fig. 7.1(a)), the HP is usually time separated from the optical packet by a guard-time in the order of tens of nanoseconds which helps the extraction of the HP from the optical packet. In OBS network (Fig. 7.1(b)), the HP is delivered to the core node with some *offset time* prior to its burst data payload. While in the OBS network, the offset time is introduced in order to give time for both processing the control information and reconfiguring the switching matrix, in OPS this delay time is supplied by the fibre delay unit introduced at the input interface which delays the arrive of the optical packets.

Once received at the core nodes, the HP is processed in an electronic *controller*. The controller performs several functions, among others the burst *forwarding* and *resources reservation*. The forwarding function, which is related to the network *routing*, is responsible for determination of an output link (port) the data container is destined to. The resources reservation function makes a booking of a wavelength in the output link for the incoming data container. In case the wavelength is occupied by another burst a *contention resolution* mechanism, if exists, is applied. In case no resources are available for the incoming data, it is lost. After the data transmission is finished in a node the resources can be released for other connections.

Briefly, the main differences between OPS and OBS are:

- OPS uses short data containers (optical packets in the order of one to tens of microseconds), the HP (the control information) is attached at the head of the data packets and therefore both (control and data) use the same channel (i.e., in-band control), and finally the switching and control elements must be able to operate very fast (less than one microseconds).
- OBS uses large data containers (optical bursts in the order of tens to thousands of microseconds), the HP is transmitted out-of-band in a separate channel than the data bursts (but a close time relationship is required between control and data), and less time demanding are required for switching and control elements (tens to hundreds of microseconds).

It has to be noticed that the time demanding of the switching and control operations is a consequence of the length of the data containers; shorter data containers require faster operations in order to service the faster arrival rate and to optimize the utilisation of the channel capacity.

In summary the OBS/OPS paradigms support highly dynamic traffic in future networks. By switching on a burst/packet level in the optical data plane it provides on the one hand a much greater flexibility than a network based on circuit switching. With processing of information in the electrical domain, they avoid on the other hand severe technological challenges as for example optical signal processing.

The rest of the chapter summarises the research activities on OPS and OBS carried out by the COST 291 partners in the last 4 years. In the following, we include five sections with contributions on five different specific topics, namely OBS/OPS performance (Section 7.2), burstification mechanisms (Section 7.3), QoS provisioning (Section 7.4), routing algorithms (Section 7.5) and TCP over OBS networks (Section 7.6). Each section contains an introductive state-of-the-art description of the specific topic and at least one contribution on that topic.

Section 7.7 concludes the chapter with some discussions on the current situation of the OPS/OBS paradigms.

Some other aspects such as interoperability with control plane, physical layer constraints, burst switch architectures, test-beds implementation and verification, are not discussed in this chapter. A survey on OBS networks covering some of these issues is presented in [3].

7.2   OBS/OPS Performance

7.2.1   Introduction and State-of-the-Art

Two operations mainly determine the performance of the OBS/OPS networks: resource reservation and contention resolution.

The resources reservation process concerns the reservation of resources necessary for switching and transmission of data containers from input to output port. The resource reservation starts from the setup and finishes after the resource release. Both resources setup and release can be either explicit or estimated. Different resources reservation algorithms have been proposed adopting the above rules:

- *Just-In-Time* (JIT) [100] – performs an immediate resource reservation. It checks for the wavelength availability just at the moment of processing of header packet.
- *Horizon* [96] – performs estimated setup and resources release. It is based on the knowledge of the latest time at which the wavelengths are currently scheduled to be in use.
- *Just-Enough-Time* (JET) [105] – performs estimated setup and resources release. It reserves resources just only for the time of data transmission.

JET is one of the most efficient mechanisms, with improved data loss probability when comparing to other algorithms. A disadvantage is its high complexity compared to the O(1) runtime of Horizon and JIT [14].

The search of the resources can be based on several policies being the simplest ones based on random or round-robin. More advanced policies [101] are:

- *Latest Available Unscheduled Channel* (LAUC), which is a Horizon-type algorithm, keeps a track of the latest unscheduled resources and searches for a wavelength with the earliest available allocation;
- *Void-Filling* (VF), which is a JET-based algorithm, keeps a track of the latest unused resources and allows putting a data container into a time gaps before the arrival of a future scheduled one. VF algorithms achieve better performance than Horizon-based ones, however, at the cost of high processing complexity.

The resources available for the reservation depend on the capabilities of the nodes. Indeed, in case two or more containers pretend to use the same resource, a contention resolution must be applied. Two factors complicate the contention resolution: unpredictable and low-regular traffic statistics, and the lack of optical random access memories. The contention can be resolved with the assistance of following mechanisms:

- *Wavelength conversion* (WC) [20] – converts the frequency of a contending data container all-optically to other, available wavelength;
- *Deflection routing* (DR) [11] – forwards a data container spatially, in the switching matrix, to another output port;

- *Fibre delay line* (FDL) *buffering* [16] – operates in time domain and resolves the contention by delaying the departure of one of data containers by a specific period of time.

In case none of mechanisms can resolve the contention, the data container is dropped.

The wavelength conversion is natural way to resolve contention. A drawback of this mechanism, however, is high cost of WC devices, especially, in case of a full-wavelength conversion, which is performed in wide frequency range. Some solutions make use of limited or shared wavelength conversion capabilities (e.g., [26]).

Application of deflection routing is almost cost-less since no additional devices are necessary for this mechanism. On the other hand, it was shown that deflection routing can improves network performance under low and moderate traffic loads whilst it may intensify data losses under high loads [110]. Another drawback that has to be managed properly is the out-of-order arrival.

Even if one of the principal design objectives was to build a buffer-less network, the application of FDL buffering is considered as well. Both feed-forward and feed-back FDL buffer architectures can be used [45]. In [32] it was shown that combined application of FDL buffering with WC can significantly reduce data loss probability. Some of these results are illustrated in Section 7.6.

Several analytical studies have been proposed to model the behaviour of the resource reservation and contention resolution in OBS/OPS nodes (e.g., [2,9]). Section 7.2.2 studies the accuracy on the use of balking models to analytically estimate the blocking probabilities in OBS nodes that use Fibre Delay Lines (FDLs).

Section 7.2.3 compares the two different switch architectures for OPS nodes, namely Input-Buffered Wavelength Routed (IBWR) switch and Output Buffered (OB) switch.

To enhance the performance of the OBS networks, some hybrid approaches have been proposed employing more than one switching paradigm like Optical Burst Transport Network (OBTN) [34], Overspill Routing in Optical Networks (ORION) [97] or Optical Migration Capable Networks with Service Guarantees (OpMiGua) [6]. Section 7.2.4 presents a comparison between a generic OBS node and the OpMiGua node by means of a qualitative and quantitative analysis. In order to achieve a maximum of comparability both models are chosen as similar as possible and especially are fed with identical traffic.

7.2.2 On the Use of Balking for Estimation of the Blocking Probability for OBS Routers with FDL Lines

Burst blocking probability is the primary performance measure for OBS networks. Typical approach to reduce blocking probability is increasing the time during which an incoming request can be satisfied. This is usually made by storing the packet to be served in memory waiting for delivery at a later time. But since optical buffering is not available at the moment, nor it is a foreseeable technology that will appear in the close future, optical switch designers resort to alternate solutions such as the Fibre Delay Lines (FDLs). Due to the limited delay availability, a buffered burst may be dropped if the output port/wavelength occupation persists when the burst is to exit the FDL.

Typical approaches to this system assume N input and output ports c wavelengths per port and full wavelength conversion capability. Let us assume that the c wavelengths of an output port are occupied (namely the output port is blocked). An arrival to the system that finds the output port blocked will not enter an FDL *if the delay provided by the FDLs is not large enough to hold the burst during the system blocking time*; namely if the output port residual life is larger than the delay provided by the fibres. A queuing system in which arrivals decide on whether to enter the system based on the system state (number of users, current delay, etc) is called a *balking* system or a system with discouraged arrivals [37]. For instance, an $M/M/c/K$ system falls within this category, since arrivals will not enter the system if K customers are already inside it.

We describe the system as a continuous-time discrete Markov chain that represents the number of bursts in the output port (c servers and FDLs). The balking model incorporates the probability that a burst is dropped, i.e. the probability that a burst does not enter the system because the FDL is too short to hold the burst for the system residual life, into transition rates of states with index $i >= c$, as shown on Fig. 7.2.

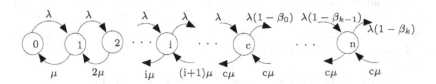

Fig. 7.2. *{Xt, t > 0}*, number of bursts in the output port

The probabilities β_k in a system with FDLs of length L are $\beta_{n-c} = P(T_n > L)$. They depend on T_n the residual life of state n which is the sum of the residual life of the blocked state and the departure time of every previous burst in the FDL. T_n can be calculated as a close expression for a Poisson-distributed arrival and burst length system. From this expression the steady state probabilities π_n for every state can be expressed as seen in (2). This is the model that has been proposed in [9,67].

On [72] we describe simulations performed to check the model on scenarios of 10 Gbps wavelengths in number c from 8 to 128. The burst average size was set to 15 kBytes, which is the average file size in the Internet as reported by [27], yielding a transmission time $E[X] = 12.288$ μs. Switching times will be assumed to be negligible, since SOA-based switches achieve switching times in the vicinity of nanoseconds [15,68,71]. Finally, each simulation run consists of 10^8 burst arrivals.

We compared simulation results to theoretical results from the model. We found discrepancies in blocking probability $P(blocking)$ versus the maximum FDL delay normalized by the time to transfer average burst $D_{max}/E[X]$ (see Fig. 7.3(a)). For low delay values it can be approximated accurately by the Erlang-B formula as expected. However, as D_{max} increases, theoretical blocking probability differs from simulation results.

The hypothesis of the balking model is checked to explain the discrepancy. The discrepancies can be traced to the calculation of β_k. It turns out that the probabilities β_k don't accurately model simulated values and this translates to theoretical state probabilities π_n which don't fit simulated values either. See Fig. 7.3(b) for example comparisons of theoretical β_k and π_n against simulation observed values, for a number of wavelengths equal to 64. Both values (β_k and π_n) take part in product form on the calculation of the loss probability. Fig. 7.3(b) also shows this product $\beta_k \pi_n$. The discrepancy in the discouraged arrival probability and state probabilities happen precisely for high occupancy states with small probabilities of occurrence. However, those are the states where losses take place. Therefore, the deviation from the analytical to the real values in that region of the state-space produces the misbehaviour of the loss probability shown in Fig. 7.3(a).

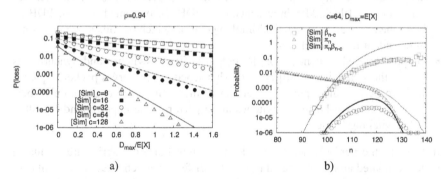

Fig. 7.3. Comparison of simulated and theoretical values, a) Burst dropping probability versus normalized FDL length, b) Comparison between the state probabilities (π_n) and the discouraged arrival probabilities (β_k)

Results in [72] show that the discrepancy between analytical and empirical results become more significant as the loss probability is decreased. Hence, the model becomes less accurate for realistic systems of WDM technology, with a higher output degree (number of wavelengths) and lower losses.

Thus we have shown that balking model accuracy depends on the ratio between fibre delay and service time. If the ratio is large then the balking model is not accurate to derive the blocking probability. On the other hand stronger discrepancies between analytical and simulation results are observed as the number of wavelengths per port increases. But precisely, the foreseeable technological evolution is towards hundreds of wavelengths.

7.2.3   A Performance Comparison of Synchronous Slotted OPS Switches

This contribution surveys the work in scheduling design and performance evaluation of OPS switching architectures, for synchronous slotted traffic. This means, switching nodes where traffic is composed of fixed size optical packets, which are aligned at switch inputs by means of synchronization stages. Results for fixed size and aligned traffic are a performance upper bound, when compared to the asynchronous and/or variable length traffic.

Two types of switching fabrics are studied: Input-Buffered Wavelength-Routed switches [116] (Fig. 7.4) and the OPS switching fabrics able to emulate output buffering (i.e. the KEOPS switch [38], the Output-Buffered Wavelength-Routed switch [116] or the space switch [15]).

IBWR switch is a more cost-effective and scalable architecture, when compared to output buffered fabrics, at a cost of a lower performance because of internal contention. The schedulers included in the comparison are:

- IBWR switch: The IBWR switch is evaluated with two parallel schedulers: (i) I-PDBM [86] scheduler which does not preserve packet sequence, and (ii) OI-PDBM scheduler, which preserves packet sequence at a cost of adding a further performance penalty [36]. Both of them are improvements to the Parallel Desynchronized Block Matching scheduler (PDBM), presented in [79]. PDBM-like schedulers allow a practical implementation which permits a response time independent from switch size.
- Output-buffered switches: For the output-buffered switches and synchronous traffic, the scheduler in [80] is used. This scheduler preserves packet sequence with no performance penalty, yielding to the optimum throughput/delay performance. Output-buffered switches are a performance upper bound for other OPS switching fabrics.

In [36, 79] the performance of IBWR and output buffered fabrics are evaluated under correlated and uncorrelated traffic, for different switch sizes. The results obtained show that the performance of the IBWR switch when packet order is not

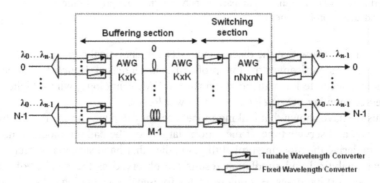

Fig. 7.4. Input-Buffered Wavelength-Routed switch (IBWR).

preserved (I-PDBM scheduler) is very close to the optimum performance given by output-buffered fabrics. A minor loss of performance appears when OI-PDBM scheduler is applied, which preserves packet order. Nevertheless, this performance loss is negligible even at medium and high loads, when the number of wavelengths per fibre is close to 32 or higher (that is, in Dense WDM networks). As an example, in most of the occasions, the same number of Fibre Delay Lines where required in IBWR switches and in output-buffered OPS architectures to achieve the target loss probability of 10^{-7}.

We conclude that the results endorse the application of the IBWR architecture in OPS networks, as a feasible competitor against less scalable output-buffered OPS architectures.

7.2.4   A Performance Comparison of OBS and OpMiGua Paradigms

While in the previous section aspects of OBS have been discussed, in this section OBS is compared with a hybrid optical network architecture named Optical Migration Capable Networks with Service Guarantees (OpMiGua) in order to determine which architecture is better suited for a given scenario. After introducing OpMiGua, we discuss qualitative differences and present results of a quantitative performance evaluation.

7.2.4.1   Optical Migration Capable Networks with Service Guarantees

OpMiGua inherently separates two different traffic classes [6]. High requirements concerning packet loss and jitter are granted by the so called Guaranteed Service class Traffic (GST). Traffic of this class is aggregated into bursts and transported in a connection oriented manner along preestablished end-to-end light paths and is given absolute priority. This ensures that there are no losses due to contention and delay jitter is minimized.

The other class with looser requirements is Statistically Multiplexed (SM) traffic. This is handled without reservations via packet switching. Losses due to contention and delay jitter due to buffering or deflection routing are allowed. Despite this inherent separation both traffic classes use sequentially the capacity of the same wavelength.

The architecture of a basic OpMiGua node is shown in Fig. 7.5. After entering the node on a wavelength SM and GST packets are separated in the optical domain according to a specific label, e.g., polarization. While GST packets are forwarded to a circuit switch, SM packets are directed to a packet switch. After traversing the respective switches GST and SM packets directed to the same output wavelength have to be multiplexed. Thus, by inserting SM packets in-between the gaps created by subsequent GST packets, the resource utilization is increased.

In order to maintain the absolute prioritization of GST packets, the switching decision for SM packets in the depicted scenario is aware of interfering GST packets on the output wavelengths within a sufficiently large time window [7].

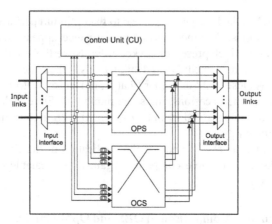

Fig. 7.5. OpMiGua node architecture

In the following we assume the packet switch as well as the circuit switch to be all-optical with full wavelength conversion but without any buffering. Also, we assume that the GST class is used for high priority (HP) and the SM class for low priority (LP) traffic. For OBS, we assume QoS differentiation for two traffic classes, i.e. high priority and low priority, by Offset Time Differentiation (see Section 7.4.2.1 for details on his behaviour).

7.2.4.2 Qualitative Comparison of OBS and OpMiGua

Comparing the two architectures, two main differences can be seen, that have an impact on the system performance. First, while in OBS all traffic is aggregated into bursts at the network ingress, in OpMiGua only the HP traffic is aggregated. Second, while in OBS all traffic shares all wavelengths, in OpMiGua each HP packet is transported on an end-to-end wavelength and only LP traffic can use all wavelengths – in the ingress as well as each core node.

In terms of delay, for reasonable load the delay of HP traffic is comparable in OBS and OpMiGua whereas the delay of LP traffic is higher in OBS. In OpMiGua, the delay of HP is due to three factors: delay in burst assembler, delay in each core node to have absolute priority of HP over LP, and delay due to the serialization of HP bursts into limited number of wavelengths; LP traffic is not aggregated in OpMiGua, thus it is only marginally delayed at the network ingress while the delay in core nodes depends only on the realization of the switching. In OBS, both HP and LP traffic classes are aggregated – thus delayed – and need to be delayed by the offset time; in contrast, the use of all wavelengths for HP bursts may reduce their waiting time.

In terms of delay jitter, it depends on the node architecture – e.g., whether processing delay is compensated by delay lines or by offset times – as well as on contention resolution strategies – whether FDLs and deflection routing is applied or not. Both aspects have impact on HP traffic as well as on LP traffic. Accordingly, in the network the delay jitter of HP traffic is usually higher in OBS than in OpMiGua whereas the delay of LP traffic is almost comparable.

In terms of network capacity, as in OpMiGua high priority traffic is only circuit switched, direct end-to-end wavelengths are necessary for each node pair exchanging HP traffic. Thus, a full mesh of wavelength channels is needed under the assumption that every node exchanges HP traffic with each other. In contrast, in an OBS network the lower bound is a single wavelength.

7.2.4.3 Quantitative Comparison of OBS and OpMiGua

Our approach for a quantitative comparison of the two architectures OBS and OpMiGua is to use simulation scenarios as similar as possible, which especially includes the traffic offered to both models. Traffic offered to the OBS and OpMiGua node is generated statistically identical traffic on packet level and fed afterwards to an architecture specific aggregation unit, which aggregates HP and LP packets if needed.

One commonly used metric for evaluation of architectures like OBS and OpMiGua is the packet or burst loss probability, which has the disadvantage of not considering differences in the length of lost units. We choose instead the bit loss probability (BLP) as metric, which specifies the lost traffic volume in comparison to total traffic. We consider for this metric both traffic classes in OBS and OpMiGua. However, in OpMiGua, HP traffic does not contribute to this metric as it is by definition lossless.

For the simulations we select a basic single node scenario with n incoming and outgoing fibres and w wavelengths per fiver. Traffic of both priority classes is equally distributed on all wavelengths with S giving the share of HP traffic with respect to the total traffic. Also, the traffic offered to the n output fibres is uniformly distributed. In case of OpMiGua each wavelength carries one HP connection. Packets are generated with exponentially distributed interarrival times and trimodal distributed length [17].

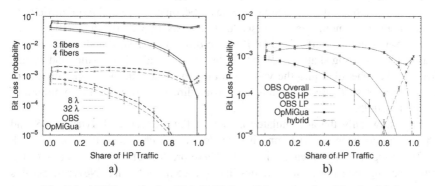

Fig. 7.6. a) BLP vs. S at load 0.6, b) BLP vs. S for n=4 and w=32 at load 0.6.

Traffic is aggregated per wavelength with a size threshold equivalent to a burst duration of 150 μs and a time threshold of 5 ms [51] (see section 7.3 for more details on burstification processes). The additional QoS offset of HP bursts in OBS we chose such that it is bigger than the maximum LP burst duration. This result in an absolute prioritization, but HP bursts may still be lost due to contention among themselves. Finally we use Just-Enough-Time (JET) and LAUC-VF as signalling and scheduling algorithm, respectively. For further details on the model, please refer to [89].

The dependency of BLP and S is shown in Fig. 7.6(a) for a fixed load of 0.6 in scenarios with 3 and 4 fibres and 8 and 32 wavelengths per fibre. At load 1 the mean generated traffic amount per time is equivalent to the maximum transmission capacity of the system. It can be seen that the BLP drops with increasing number of wavelengths. Furthermore the number of fibres has only a very small influence. Last but not least the BLP for OpMiGua is lower than that for OBS.

However, there are obvious differences in the behaviour of OBS and OpMiGua. The BLP of OpMiGua is monotonically decreasing with increasing S. This seems reasonable as the share of lossless HP traffic increases. Fragmentation of the available phases of output wavelengths due to HP traffic is not a real problem for the small LP packets.

All OBS curves show the same basic behaviour, but this is totally different to OpMiGua. Therefore it is exemplarily explained for the scenario $n=4$ and $w=32$, which is also depicted in Fig. 7.6(b). Furthermore, BLP is broken down into the parts caused by losses of LP and HP traffic ("OBS-LP" and "OBS-HP").

BLP for $S=0$ and $S=1$ should be nearly identical in case of OBS as the offset does not matter anymore if all bursts belong to the same traffic service class. The simulations clearly confirm this expectation.

For very small values of S the completion of HP bursts is mainly triggered by the timeout criterion, which results in small bursts. These small bursts fragment the phases during which a maximum size LP burst can be scheduled. This scheduling is not always possible and in comparison to $S=0$, where this fragmentation does not occur, the BLP is higher.

In the range $S=0.2$ to $S=0.8$ the BLP stays rather constant and originates only of LP losses. Although the LP share decreases it becomes more and more difficult to schedule the maximum size LP bursts due to increasing occupation by HP bursts.

For $S>0.8$ the LP part of the BLP traffic drops very fast. Besides the obvious reason of decreasing share of LP traffic, the LP bursts also get smaller and by this better to be scheduled into the voids. On the other hand an increasing amount of HP traffic is lost. These two trends in opposite directions result in the minimum of the BLP at 0.95.

Until now only the accumulated impact of the differences between OBS and OpMiGua has been observed and it is unclear to which extend the smoother HP traffic of OpMiGua influences the BLP. Therefore the OBS node is fed with HP traffic having the same characteristics like in case of OpMiGua. Nevertheless this hybrid scenario is rather theoretical, as it is impossible to guarantee this lossless HP traffic within an OBS network scenario.

The resulting BLP can also be seen in Fig. 7.6(b). While this BLP shows at small S more similarities to OBS, it finally behaves like OpMiGua and goes to zero. The sharp increase for $S>0$ is not as big as for OBS. The reason is that in this scenario less HP bursts are produced. However these bursts are longer as the HP traffic amount is still the same. Remaining differences to OpMiGua, which are in the order of one magnitude, are due to the aggregation of LP traffic.

7.2.4.4 Conclusions

With OBS and OpMiGua we compared two transport network architectures with QoS support for two traffic classes. Based on the current technological development status OBS has less stringent requirements, as switching is done on a bigger granularity.

With respect to delays, the predominant part (besides propagation) originates from aggregation in ingress nodes. Here OpMiGua might have a disadvantage in case of very bursty high priority traffic. On the other hand in OBS high priority traffic has an additional delay due to the offset between header control packet and burst.

Furthermore, for the investigated scenario OpMiGua is better suited. Although traffic generated for both models is statistically identical, traffic fed to the nodes itself shows differences due to absence of LP traffic aggregation and one single destination per wavelength for HP traffic in case of OpMiGua. Observed performance advantages of OpMiGua are caused by these two factors and the difference generally increases with higher HP traffic share.

7.3   Burstification Mechanisms

7.3.1.1 Introduction and State-of-the-Art

The architecture of a typical OBS edge router is depicted in Fig. 7.7. The switching unit forwards incoming packets to the burst assembly units. The packets addressed to the same egress node are processed in one burst assembly unit. There is one designated assembly queue for each traffic class.

Burstification (also known as burst assembly) algorithms can be classified as timer-based (e.g., [35,102]), size-based (e.g., [76,98]), and hybrid timer/size-based (e.g., [107]). In the timer-based scheme, a timer starts upon the arrival of the first packet to an empty queue, i.e. at the beginning of a new assembly cycle. After a fixed time (T_{Thr}), all the packets arrived in this period are assembled into a burst. In the threshold-based scheme, a burst is sent out when enough packets have been collected in the assembly queue such that the size of the resulting burst exceeds a threshold of S_{Thr} bytes. In the *hybrid* algorithm, a burst can be sent out when either the burst length exceeds the desirable threshold or the timer expires.

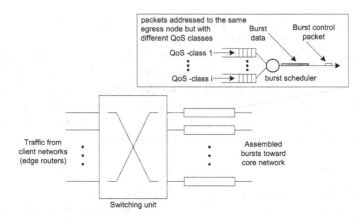

Fig. 7.7. Architecture of an OBS edge node.

Recently, it has been shown that the use of fixed thresholds in burstification algorithms may lead to some performance degradation since they are not flexible
enough to take into account the actual traffic situation. In fact, considering that
incoming traffic is in general strongly correlated traffic such as TCP or long-
range-dependent traffic [23,54,61], the burstification processes based on fixed
thresholds are not able to respond to the traffic changes accordingly. Several
adaptive burstification algorithms have been proposed to ameliorate this situation [12,85] which can better respond to traffic changes and can provide better
performance.

One example is illustrated in Section 7.3.2 where the case of a timer-based
burstfication algorithm is analyzed. Given a burstifier that incorporates a timer-
based scheme with minimum burst size, bursts are subject to padding in light-load
scenarios. Due to this padding effect, the burstifier normalized throughput may be
not equal to unity. The results, obtained using input traffic showing long-range
dependence, motivate the introduction of adaptive burstification algorithms, which
choose a timeout value that minimizes delay, yet they keep the throughput very
close to unity.

On the other hand, the burstification, which is executed at the edge nodes, can
substantially change the client traffic characteristics and lead to significant improvements to the network performance if the long-range dependence is alleviate.
A number of recent publications have studied the traffic characterization of the
burstification. The statistics for the size and interarrival time of bursts from the assembly are investigated in [22, 59]. The impact of burstification on the self-
similarity level of the data traffic is studied in [42, 48, 103, 107]. A complete
analysis is investigated in Section 7.3.3 where the impact of timer- and size-based
burstification algorithms on the self-similarity level of the output traffic is reported. Both static and adaptive algorithms are examined and the performance impact of the burstification algorithms in terms of burst assembly delay and its jitter
is assessed. The study has shown that the burst assembly mechanism at the OBS

edge router reduces the self-similarity level of the output traffic and that this reduction depends on the parameters of the algorithm. The results reveal that the
proposed adaptive burst assembly algorithm performs better comparing to its non-
adaptive counterpart.

7.3.2 Delay-Throughput Curves for Timer-Based OBS Burstifiers with Light Load

OBS proposals are in part motivated by the inability to switch optical paths fast
enough to be done on a per-packet basis. This problem is solved by gathering
bursts of packets to be switched to the same destination, but to keep a low
enough rate of switching a minimum burst size use to be proposed as well. This
leads to padding short bursts in order to keep this minimum size in timer-based
burst gatherers. Padding will not be likely to occur in medium to heavily loaded
OBS networks using a timer-based burstifier. However, a light load scenario
will potentially produce many bursts with a number of packets below the minimum burst size and padding will be necessary. But load fluctuations do happen
in highly loaded networks, during weekends or due to different busy hours at
different geographical locations and light-load epochs will be observed[1]. The
light-load will imply that when the timer expires, all packets awaiting transmission in the burst assembly queue are transmitted along with a padding space that
will add load to the network. Even if this load is not significant in the link that is
generating the burst it increases also load at other links and thus it should be
quantified.

  On [49] this effect is analyzed. The incoming traffic (bytes per time interval) is
modelled by a Fractional Gaussian Noise (FGN), which has been shown to model
accurately traffic from a LAN [74]. Note that in order to calculate the throughput
only the number of information bytes per burst matters and not the packet arrival
dynamics. Precisely, the FGN is a fluid-flow model that provides the number of
bytes per time interval only. While the small timescale traffic fluctuations are not
captured by the model, the long-range dependence from interval to interval is indeed accurately portrayed.

  According to our previous results in [48], for a timer-based burstifier, it turns
out that the traffic arriving per time interval T_0 is a Gaussian random variable X
with mean $\mu = \mu' T_0$ and standard deviation $\sigma = \sigma' T_0^H$ (being μ', σ' and H the
mean, standard deviation and Hurst parameter of the traffic arrival process at one
time unit time slots).

  The throughput of a given burstifier is defined as the ratio between the information bits and the total bits transmitted. If the minimum burst size is b_{min}, the
throughput will equal unity whenever $X > b_{min}$ and $E[X]/b_{min}$ if $X < b_{min}$. By using

---

[1] See for instance http://loadrunner.uits.iu.edu/weathermaps/abilene/ for daily variation of
    traffic in an Internet

a convenience variable $Y = min\{b_{min}, X\}$ the throughput can be expressed as $\rho = E[Y]/b_{min}$ and we derive in [49] an expression for ρ depending on input traffic parameters.

$$\rho = b_{min}^{-1}\left(\mu - \sigma \lambda(-\alpha)\right)\varphi(\alpha) + \left(1 - \varphi(\alpha)\right) \tag{7.1}$$

where $\varphi(x) = 1/\sqrt{2\pi}\, e^{-\frac{1}{2}x^2}$ and $\varphi(\alpha) = \int_{-\infty}^{x} \varphi(t)\,dt$ are the PDF and distribution function of a normalized Gaussian random variable.

To quantify the extra load that enters the OBS backbone because of the added padding we define a new convenience variable $Z = max\{b_{min}, X\}$ that denotes the bits generated by the burstifier. Z is a truncated Gaussian variable from which we derive (in [49]) an expression for the input rate to the OBS core introduced by the burstifier

$$R = \frac{E[Z]}{T_0} = T_0^{-1}\left[b_{min}\,\varphi(\alpha) + \left(\mu + \sigma\lambda(\alpha)\right)\left(1 - \varphi(\alpha)\right)\right] \tag{7.2}$$

Equations (1) and (2) are validated against high speed traffic from Abilene-I data set. The Abilene-I data set traces contain traffic from two OC-48 links, collected at US core router nodes and are provided by NLANR[2]. For the example we use 10 minutes worth of traffic from a 2.5Gbps link as a real-world traffic source for the burstifier. The trace selected shows an average traffic rate around 480Mbps which, assuming a 10Gbps wavelength in the OBS port, makes the utilization factor be approximately equal to 0.05. Fig. 7.8 shows equations compared to the burst process that would be generated by burstifiying the Abilene-I trace with several T_0 and b_{min} values. Similar results are obtained with synthetic FGN traffic generated with Random Midpoint Displacement algorithms that allows us to have results for broader H parameter range (Abilene-I traces have H values between 0.7 and 0.8.

Results show a negative gradient of the throughput with both the coefficient of variation (instantaneous variability) and Hurst parameter (long-range dependence). However, there is a timeout value that makes such gradient be equal to zero (as can be seen on Fig. 7.8). Such timeout value depends on the minimum burst size, the traffic load and, to a lesser extent, it also depends on the long-range dependence parameter H and the coefficient of variation c_v.

The above observation leads us to seek for an expression that provides the timeout value (T_0) for which the delay throughout curves flatten out to unity. This is beneficial to maximize the throughput at the minimum delay cost and also to decrease the network load. For the Abilene-I trace considered the increased traffic load due to padding is shown in Fig. 7.8. The effect of choosing a wrong timeout

---

[2] http://pma.nlanr.net/Traces/long/ipls1.html

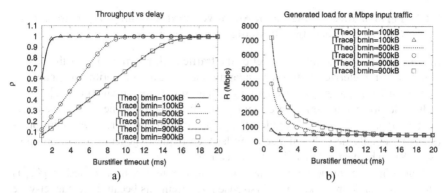

Fig. 7.8. Throughput-delay curve and input traffic to the OBS network for the Abilene-I trace.

value is very significant not only for the throughput, but also for the generated load to the OBS network.

Concerning the change rate of the traffic moments, other proposals based on link state estimation assume that the network load remains stable in timescales of minutes [92]. If that is the case, one could devise an adaptive burstifier that would offer minimum delay and maximum throughput for any given input traffic stream. The timeout value rate of change would be in the scale of minutes, which seems reasonable from a practical implementation standpoint.

In [49] we propose three different adaptive timeout algorithms and compare them for different values of the Hurst parameter H and coefficient of variation c_v. The proposed algorithms are trade-off of complexity versus accuracy. The simplest (L-estimate) requires to estimate the burstifier load $\hat{\mu}'$ and set timeout $T_0^L = b_{min}\big/\hat{\mu}'$. The chosen T_0^L is the number of sampling intervals needed to fill on average a size of b_{min} at the estimated rate. The basic assumption is that the influence of the second moment and H parameter is negligible.

Using estimators for first and second moments of the traffic arriving to the burstifier we can build more accurate algorithms (LV-estimate) or (LVH-estimate) using also estimations of H parameter of the arrival process. T_0^{LVH} or T_0^{LV} are chosen as the solutions of the nonlinear problem of minimizing T_0 subject to the condition that equation (1) gives throughput values above a desired threshold (i.e. $\rho\left(\hat{\mu}',\hat{\sigma}',\hat{H}'\right) > 0.95$).

Our trace-driven analysis of the Abilene backbone shows that, for most cases of real Internet traffic, first moment estimation is enough to provide a timeout value very close to the optimum. Thus, an adaptive timeout algorithm can be easily incorporated to timer-based burstifiers, with a significant benefit in burstification delay and throughput.

7.3.3   Performance Evaluation of Adaptive Burst Assembly Algorithms in OBS Networks with Self-Similar Traffic Sources

In this work the self-similarity level of the traffic both before and after the execution of a parameterized hybrid and adaptive burstification algorithm is analyzed. The burstification algorithm is an improvement of the one presented in [12].

In order to model the realistic input traffic from the client networks, the arriving and aggregated traffic is made of superposition of fractal renewal point process as it actually describes the self-similar web requests generated by a group of users [88]. The detailed model is described in [5].

Regarding the traffic volume measurement, the approach presented in [42] is adopted focusing on packet and burst-wise measurements because the packet-wise and burst-wise analysis is important on the performance of the electronic control units in core routers. The quantitative values for the Hurst parameter estimation are reported for the proposed adaptive burstification algorithm. The performance of the OBS edge node in terms of delay and delay jitter is also investigated.

7.3.3.1   Adaptive Burstification Algorithm

Within an OBS edge router, the incoming packets (e.g. IP packets) from the client networks will be forwarded to respective queues based on the destination address of egress OBS edge router and possibly the QoS parameters, where the burstification algorithms are used to generate the burst control packet and the data burst. Then the burst control packet and the optical burst will be scheduled to the transmitter and sent out to the core network.

The packet length distribution used in our study has been reported in [101] and has been modified to ignore the packets with size larger than 1500 bytes. The average packet length of the modified distribution is 375.5 bytes and reflects the realistic predominance of small packets in IP traffic.

The main disadvantage of such static burstification algorithm is that it does not take into account the dynamism of traffic and therefore they cannot respond to the traffic changes. This adversely impacts the network performance. Therefore adaptive burst assembly algorithms are proposed to ameliorate this situation. The main idea in these burstification algorithms is to adaptively change the value of the T_{Thr} and S_{Thr}. If we assume that the network uses a static routing algorithm, then according to the link capacity, for each burst assembly queue inside the edge router, we have the following inequality:

$$\sum_{i=1}^{N} \frac{avgBL_i}{T_{Thr,i}} \leq Bandwidth \tag{7.3}$$

where $avgBL_i$ represents the average burst length in the ith burst assembly queue, and the bandwidth of the link is given by $Bandwidth$. Since from (3) the value of T_{Thr} changes with the value of average burst length, we have to infer the value of

$avgBL_i$ from the traffic history. One possible approach is to take into account both the previous value of average burst length and the current sampled value ($SavgBL_i$) as expressed in the following expression [12]:

$$avgBL_i \leftarrow w_1\, avgBL_i + w_2\, SavgBL_i \qquad (7.4)$$

where w_1, w_2 are two positive weights ($w_1 + w_2 = 1$). Based on (3) and (4) the two threshold values for the adaptive burstification algorithm are computed as follows:

$$T_{Thr,i} = \alpha\, \frac{avgBL_i\, N}{Bandwidth} \qquad (7.5)$$

$$S_{Thr,i} = \begin{cases} avgBL_i & if\ avgBL_i > \beta\ E\left[L_p\right] \\ \beta\ E\left[L_p\right] & otherwise \end{cases} \qquad (7.6)$$

where α, β are burst assembly factors and $E[L_p]$ is expected packet length. In order to synchronize this adaptive burst assembly algorithm with the changes in TCP/IP traffic, we set the value of $w_2 > 0.5$. This will put more weight on the recent burst size. More specifically, when a long burst is sent out (high value of $AvgBL_i$) it is very probable that TCP will send out more packets in the sequel. Therefore it is better to increase the value of both time and size thresholds to deal more efficiently with the incoming traffic. Similarly as soon as the TCP traffic is terminated or initiating a slow start stage, by giving higher weight to w_2, we also dramatically decrease the time and size threshold values. The results that we will present in next section are obtained by setting $w_1 = 0.25$, $w_2 = 0.75$. More details of this adaptive burstification algorithm is presented in [5]. Note that $Tmin_{Thr}$ is given by the following equation:

$$Tmin_{Thr} = \frac{\beta\ E\left[L_p\right] N}{Bandwidth} \qquad (7.7)$$

We put a lower limit on T_{Thr} in order to keep the assembly period within a reasonable range and to prevent the burst length decreasing by too much.

7.3.3.2 Numerical Results

The simulation scenario consists of 12 client networks connected to an OBS edge router via a 10 Gbps link. The link between the OBS edge router and the core network is running at 40Gbps. The burst assembly algorithm is implemented within the OBS edge router. The incoming IP packets will be forwarded to the assembly queue associated with its egress edge router. We have defined three levels of traffic load (ρ) at the edge router: 0.3 (light load), 0.5 (medium load) and 0.7 (heavy load), which corresponds to 332889, 554816, and 776742 packets per second, respectively.

The simulation records each packet arrival at the OBS edge router regardless of the source client network and all the incoming packets comprise the aggregated input traffic. The twelve client networks are divided into four groups and the Hurst parameter of each group is set at $H=0.7, 0.75, 0.80$, and 0.85 respectively. Among the well-known Hurst parameter estimators, i.e. aggregated variance, R/S plot, periodogram, local Whittle and wavelet techniques [18], the wavelet analysis is used because it is robust to many smooth trends, non-stationarities, and high frequency oscillations [91]. In our simulation scenarios we have assumed that there is only one quality of service (QoS) class supported and the destination address of each IP packet is randomly selected from N egress edge routers within the core network. Thus there are N assembly queues in the OBS edge router. We choose $N=1, 10, 20$ in our simulation. All the traffic processes are measured at the time-scale of 100 μs. The simulation time is 6 seconds and owing to the sufficiently large queues in the OBS edge router, no packet loss is assumed.

The effect of adaptive burstification algorithm on the self-similarity level of the output traffic and burstification delay and delay jitter is studied in the following scenario. The T_{Thr} parameter of the burstification algorithm is estimated dynamically and the S_{Thr} parameter is also evaluated dynamically in favour of larger values for average burst length. Fig. 7.9 depicts the estimated Hurst parameter of the both aggregated input traffic and the optical output traffic, which is injected from edge router to the core network ($N = 1, N = 10$).

It can be seen that the burst-wise output traffic, which is the result of adaptive burstification algorithm, exhibits much lower level of self-similarity in terms of estimated Hurst parameter. In order to compare the hybrid and adaptive burstification algorithms in term of their effects on the self-similarity level of the output traffic, we set up another simulation scenario and the estimated Hurst parameter is depicted in Fig. 7.10. We have to mention that in order to make our comparison unbiased, we have focused on the distribution of bursts, which are generated according to time or size constraints and we have set the parameters for both hybrid [103] and adaptive algorithms in a way that both algorithms generate the same percentage of time-constrained and size-constrained bursts. In other words, both algorithms behave similarly as far as the distribution of bursts is concerned.

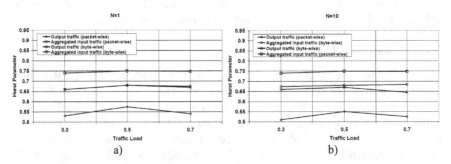

Fig. 7.9. Hurst parameter of input and output traffic for different values of load, a) $N = 1$, b) $N = 10$.

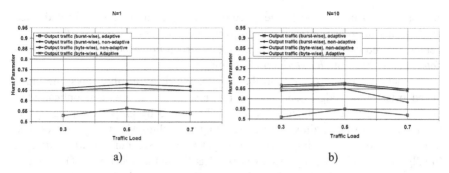

Fig. 7.10. Estimated H of output traffic for different values of load, a) N = 1, b) N = 10.

It can be observed that the byte-wise self-similarity level for both algorithms remains the same. However the burst-wise output traffic for the adaptive burstification algorithm, as expected, exhibits lower level of self-similarity in comparison to the non-adaptive (hybrid) burstification algorithm. This is due to the dynamic feature of algorithm, which adapts the value of both T_{Thr} and S_{Thr} to match dynamically with the incoming traffic.

It can be seen that the adaptive burstification algorithm performs noticeably better that its non-adaptive counterpart. In other words the burstification delay and its jitter in the adaptive algorithm are lower than the same metrics for the non-adaptive (hybrid) algorithm. This observation is valid mainly due to the mechanism that is employed in burstification algorithm. In the adaptive algorithm the T_{Thr} parameter is determined based on the weighted average of burst lengths. Thus T_{Thr} parameter tries to adapt itself according to the computed average burst length and also the recent value of burst length. Furthermore we also enforced a S_{Thr} in our burstification algorithm, which not only put a limit on burstification delay but also tries to synchronize with TCP/IP traffic as much as possible.

Summarizing, the obtained results show that the burstification algorithm at the OBS edges can be used as a traffic shaper to smooth out the burstiness of the input traffic as indicated by the noticeable reduction in the Hurst parameter. Comparing the traffic shaping capability, the adaptive outperform the non-adaptive (hybrid) algorithm in terms of reduction in Hurst parameter, burst assembly delay and burst assembly delay jitter.

7.4   QoS Provisioning

7.4.1   Introduction and State-of-the-Art

This section addresses the problem of quality of service (QoS) provisioning in OBS networks. The lack of optical memories results in quite complicated operation of OBS networks, especially, in case when one wants to guarantee a certain level of quality for high priority (HP) traffic. Indeed the quality demanding applications, like for instance real-time voice or video transmission, need for dedicated mechanisms in order to preserve them from low priority (LP) data traffic. In particular, the requirements concern to ensure a certain upper bounds on end-to-end *delay*, *delay jitter*, and *burst loss probability*.

The delays arise mostly due to the propagation delay in fibre links, the introduced offset time, edge node processing (i.e., burstification) and optical FDL buffering. The first two factors can be easily limited by properly setting up the maximum hop distance allowed for the routing algorithm. Also the delay produced in the edge node can be imposed by a proper timer-based burstification strategy. Finally the optical buffering, which in fact has limited application in OBS, introduces relatively small delays. Regarding the jitter, it depends on many factors and it is more complicated to analyze; nonetheless, since the delay can be easily bounded, its variations could be also limited accordingly. In this context the burst loss probability (BLP) metric is perhaps of the highest importance in OBS networks that operate with one-way signalling.

In a well-designed OBS network the burst losses should arise only due to resources (wavelength) unavailability in a fibre link. The probability of burst blocking in the link strongly depends on several factors, among others on the implemented contention resolution mechanisms, burst traffic characteristics, network routing, traffic offered to the network and relative class load. Since this relation is usually very complex the control of burst losses may be quite awkward in OBS networks.

Several components can contribute to QoS provisioning in OBS networks. In general, they are related to the control plane operation, through signalling (e.g., [21]) and routing (as e.g., in [10]) functions, and to the data plane operation both in edge nodes (e.g., [106]) and in core nodes (e.g., [52,53,112]). See Fig. 7.11 for a classification of the QoS mechanisms.

Although, a great number of QoS mechanisms have been proposed for OBS networks, still, only a few works study their comparative performance. In [112] some QoS scenarios with two different burst drooping principles applied, namely, a wavelength threshold-based and an intentional burst drooping are analyzed. Finally, the evaluation of different optical packet-dropping techniques is provided in [77]. In this direction Section 7.4.2 makes an extension to these

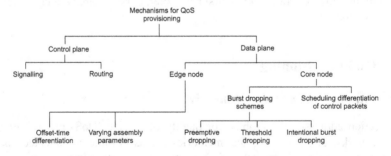

Fig. 7.11. Categories of QoS mechanisms in OBS networks.

studies. In particular, the performance of most frequently referenced QoS mechanisms, namely *offset time differentiation, full burst preemption* and *wavelength threshold-based dropping* are compared.

One of the more effective solutions, the *burst segmentation* mechanism [99], is analyzed in Section 7.4.3. The fact that a burst is composed by several packets makes it possible to drop *part* of a burst, so that the remaining packets may continue transmission in subsequent hops. Consequently, the use of burst segmentation provides significant throughput advantages.

7.4.2 Performance Overview of QoS Mechanisms in OBS Networks

7.4.2.1 Frequently Referenced QoS Mechanisms

In this study we focus on three mechanisms:

- *Offset time differentiation* (OTD), which is an edge node-based mechanism [106]. It assigns an extra offset-time to HP bursts in order to favour them during the resources reservation process (see Fig. 2.12a). The extra offset time, when properly setup, allows to achieve an absolute class isolation, i.e., the probability to block a HP class burst by a LP class burst is either inconsiderable or none.
- *Burst preemption* (BP), which is a core node-based burst dropping mechanism [52]. In case of the burst conflict, it overwrites the resources reserved for a LP burst by a HP one; the pre-empted LP burst is discarded (see Fig. 7.12b). In this work we consider a full preemption scheme, i.e., the preemption concerns the entire LP burst reservation.
- *Burst Dropping with Wavelength threshold* (BD-W), which is a core node-based burst dropping mechanism [112]. It provides more wavelength resources in a link to HP bursts than to LP bursts, according to a certain threshold parameter (see Fig. 7.12c). If the resource occupation is above the threshold, the LP bursts are discarded whilst the HP bursts can be still accepted.

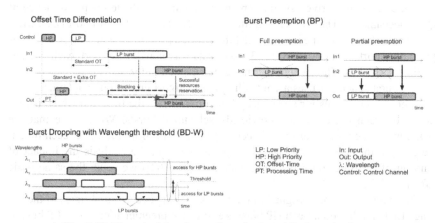

Fig. 7.12. The principle of operation of selected QoS mechanisms.

In order to gain some insight into the mechanism behaviour let us assume a Poisson burst arrival process and i.i.d. burst lengths. Under such an assumption, a burst loss probability in a link can be modelled with the Erlang loss formula (see e.g., [87]).

Both OTD and BP can be characterized by absolute class isolation. In the former, the extra offset time assures that the contention of HP bursts is only due to other HP burst reservations. In the latter, a HP burst can pre-empt whatever LP reservation and the loss of HP bursts is again only due to the wavelength occupation by other HP reservations. In both cases HP bursts compete among themselves in access to the resources and thus the HP class BLP can be estimated as $BLP_{HP} = Erlang(\alpha_{HP}\rho, c)$, where α_{HP} and ρ denote, respectively, the HP class relative load and the overall burst load and c the number of wavelengths.

The behaviour of BD-W depends greatly on its threshold (T_w) selection. Indeed, if $T_w = 0$ (i.e., no resources available for LP bursts), there is only HP class traffic accepted to the output link. Although, the mechanism achieves its topmost performance with regard to HP class and BLP_{HP} is the same as in OTD and BP, still, the LP class traffic is not served at all and $BLP_{LP} = 1$. Notice that in both OTD and BP the LP class traffic still has some possibilities to be served, in particular, if there can be found a free wavelength, not occupied by any earlier HP reservations (the OTD case), or the LP burst is not preempted (the BP case). Now, if we provide some wavelength resources for LP class traffic (i.e., $T_w > 0$), the performance of HP class will be worsening as long as HP bursts will have to compete with LP bursts. In the extreme case $T_w = c$, there is no differentiation between traffic classes and BD-W behaves as a classical scheduling mechanism. Accounting on this analysis, BD-W might require some regulation mechanisms in order to adjust the threshold value according to the required class performance and actual traffic load conditions.

7.4.2.2 Numerical Results

We set up an event-driven simulation environment to evaluate the performance of QoS mechanisms. The simulator imitates an OBS core node with no FDL buffering capability, full connectivity, and full wavelength conversion. It has 4 x 4 input/output ports and $c = \{4, 8, 16, 32, 64\}$ data wavelengths per port, each one operating at 10Gbps. The switching times are neglected in the analysis.

The burst scheduler uses a void filling-based algorithm. In our implementation, the algorithm searches for a wavelength that minimizes the time gap which is produced between currently and previously scheduled bursts. We assume that the searching procedure is performed according to a round-robin rule, i.e. each time it starts from the less-indexed wavelength. To avoid in the analysis the impact of varying offset times on scheduling operation (see [62]) we setup the same basic offset to all bursts.

The extra offset time assigned to HP bursts in OTD is equal to 4 times of the average LP burst duration. Each HP burst is allowed to preempt at most one LP burst if no free wavelength is available in BP. The preemption concerns a LP burst the dropping of which minimizes the gap produced between the preempting HP burst and the rest of burst reservations. We establish $T_w = 0.5c$ in BD-W so that LP class bursts can access at most the half of all the available wavelengths simultaneously.

The traffic is uniformly distributed between all input and output ports. In most simulations the offered traffic load per input wavelength is $\rho = 0.8$ (i.e., each wavelength is occupied in 80%) and the percentage of HP bursts over the overall burst traffic, also called HP class relative load α_{HP}, is equal to 30%.

The burst length is normally distributed (see e.g., [107]) with the mean burst duration $L = 32$ μs and the standard deviation $\sigma = 2 \ 10^{-6}$. In further discussion we express the burst lengths in bytes and we neglect the guard bands. Thus the mean burst duration L corresponds to 40 kbytes of data (at 10Gbps rate). The burst arrival times are normally distributed with the mean that depends on the offered traffic load and the standard deviation $\sigma = 5 \ 10^{-6}$.

We evaluate both a data loss probability, i.e., an effective lost of data due to the burst loss, and effective throughput, which represents the percentage of data burst served with respect to overall data burst offered.

All the simulation results have 99% level of confidence.

The results of BLP_{HP} presented in Fig. 7.13(a) confirm the correctness of theoretical argumentation provided in the previous section. In particular, we can see that the performance of both OTD and BP is similar without respect to the number of wavelength in the link.

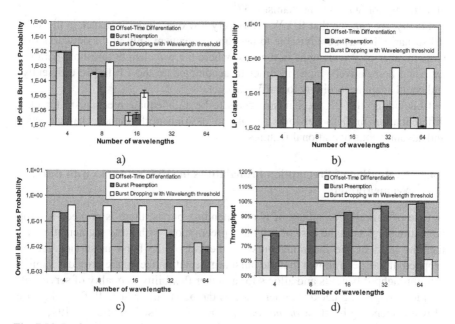

Fig. 7.13. Performance of QoS mechanism vs. link dimensioning ($\rho = 0.8$, $\alpha_{HP}=30\%$), a) HP class BLP, b) LP class BLP, c) overall BLP, d) effective data throughput.

Regarding BLP_{LP} and the overall burst loss and throughput performance (Fig. 7.13(b)-(d)), the results are slightly in the favour of BP when comparing to OTD. The explanation can be found in [62], where it is shown that the scheduling operation may be impaired by the variation of offset-times, the feature which is inherent to OTD mechanism.

Finally, we can see that the BD-W mechanism exhibits very poor performance. The reason is that BD-W has effectively fewer wavelengths available for the burst transmissions than the other mechanisms, whilst at the same time it attempts to serve the same volume of burst traffic.

7.4.2.3 Discussion and Conclusions

Both OTD and BP can be distinguished by their high performance.

Although, OTD is characterized by a relatively simple operation, as long as it does not require any differentiation mechanism in core nodes, still, this mechanisms may suffer from extended delays due to extra offset times. Also, the management of extra offset times with the purpose of providing absolute quality levels might be quite complex in the network.

On the other hand, there exist several proposals that extend the functionality of BP mechanism. Particular solutions focus on providing absolute quality guarantees to individual classes of service [112], improving resources utilization [99], and supporting a routing problem [64]. An inconvenient overhead in the data and control plane due to the preemption operation can be overcome with the assistance of a preemption window mechanism [57].

Finally, we can see that BD-W offers very low overall performance in the studied scenario. It may be advisable to use this mechanism only in the networks of a large number of wavelengths in the link, where the wavelength threshold parameter could be relatively high (in order to accommodate the LP traffic efficiently) and could adapt accordingly to traffic changes.

Concluding, the BP mechanism seems to be an adequate mechanism for QoS differentiation in OBS networks, thanks to its high performance characteristics and advantageous operational features.

7.4.3   Evaluation of Preemption Probabilities in OBS Networks with Burst Segmentation

In case of partial overlapping of two contending bursts, there is no need to drop the entire burst as in the case of full burst preemption; with burst segmentation, either the head of the incoming burst or the tail of the burst in service can be dropped. It has been shown that the burst segmentation technique provides significant throughput benefits and allows for a higher flexibility in quality of service allocation, by placing packets either towards the burst tail or head [99]. In the case of two contending bursts *with the same priority* a proposed solution (in [99]) is to drop the less amount of data. If the residual length of the burst in service is larger than the incoming burst length, then the burst in service wins the contention. The incoming burst is dropped (either entirely or partially -head-). If the residual length of the burst in service is smaller than the incoming burst length then the incoming burst wins the contention. The burst in service is segmented and the tail is dropped.

On [70] the preemption probability, or probability that the incoming burst wins the contention, is evaluated, within the same priority class. This probability is relevant for OBS network engineering for a twofold reason. First, since the incoming burst and the burst in service contend for the same resources, it is likely that they both follow the same route. Thus, due to tail dropping upon preemption, packet disordering may occur. Second, optical networks are limited by the so-called "electronic bottleneck". If preemption occurs, the optical switch must drop the tail of the burst in service and then switch the contending burst to the corresponding wavelength. This implies a processing cost not only in the optical domain but also in the electronic domain. Actually, additional signalling must be created to re-schedule bursts in the downstream nodes. Another control packet called "trailer" [99] is sent as soon as preemption happens in order to update scheduling information for the rest of OBS switches. Since this implies a processing cost, the likelihood of preemption becomes a relevant issue in OBS network performance.

Switching time is assumed to be negligible in comparison to the average burst length. Burst arrival can be assumed to be Poisson regardless of the possible long-range dependence of incoming traffic, as we have discussed in [48], but burst size will depend on the burst gathering algorithm and traffic input characteristics so several input size distributions will be considered.

Let's call (t_0, l_0) the arrival time and burst size of the first burst to arrive in a busy period, and (t_i, l_i) to subsequent bursts in the same busy period. We show in [70] that if there is a burst (t_*, l_*) in that busy period that wins the contention and preempts the first burst the time distribution of L_* is shifted to larger values in comparison to l_0. Intuitively, the preempting burst has a larger probability of high service times, in comparison to the burst in service.

$$P(l_* > x) > P(l_0 > x) \qquad \forall x > 0 \qquad (7.8)$$

From this theorem it turns out that preemption is less likely to occur for the burst that wins the contention than for the first burst in a busy period (burst 0). Hence, *the preemption probability reaches a maximum with the first burst in a busy period* and this probability is given by

$$P(L > A) = \int_0^\infty P(L > x)\, dF_A(x) \qquad (7.9)$$

where A is a random variable that provides the residual life of the server (wavelength). We derived it for several usual incoming burst length distributions in [70] and provided closed expressions for $P(L>A)$ for the case of exponential and Pareto-distributed burst lengths.

We verified by simulation upper bound preemption probabilities for several burst length distribution. For example, Fig. 7.14 shows the cases for Pareto and Gaussian distributions. Note as the utilization factor decreases, the busy periods tend to be shorter. Thus, the system behaviour is closer to the best case that was assumed for the upper bound derivation, i.e. pre-emption of the first burst in a busy period. Hence, the upper bound becomes closer to the simulation results.

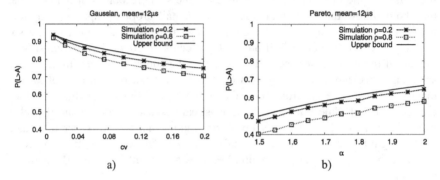

Fig. 7.14. Preemption probability, a) Gaussian and b) Pareto.

We have shown preemption probabilities are highly dependent on the burst length distribution. Hence, for the same traffic load, the burst assembly algorithm has a strong impact on the burst segmentation dynamics in the optical network core.

7.5   Routing Algorithms

7.5.1   Introduction and State-of-the-Art

In this section we concern on the problem of routing in optical burst switching networks (OBS). OBS architectures without buffering capabilities are sensitive to burst congestion. An overall burst loss probability (BLP) which adequately represents the congestion state of entire network is the primary metric of interest in an OBS network.

In general, routing algorithms can be grouped into two major classes: *non-adaptive* (when both route calculation and selection are static) and *adaptive* (when some dynamic decisions are taken) [93]. In static routing the choice of routes does not change during the time. On the other hand, adaptive algorithms attempt to change their routing decisions to reflect changes in topology and the current traffic. Adaptive algorithms can be further divided into three families, which differ in the information they use, namely *centralized* (or global), *isolated* (or local), and *distributive* routing. *Single-path* or *multi-path* routing corresponds, respectively, to the routing scenarios with only one or more paths between each pair of nodes available. If the decision of path selection in multi-path routing is taken at the source node, thus such routing is called *source routing*. A special case of multi-path routing is *deflection* (or alternative) routing. Deflection routing allows selecting an alternative path at whatever capable node in case a default primary path is unavailable.

Static shortest path routing based on Dijkstra's algorithm is the primary routing method frequently explored in OBS networks (e.g., [107]). In such routing, some links may be overloaded, while others may be spare, leading to excessive burst losses. Therefore several both non-adaptive and adaptive routing strategies, based on deflection, multi-path or single-path routing, have been proposed with the objective of the reduction of burst congestion.

Although deflection routing can improve the network performance under low traffic load conditions, still it may intensify the burst losses under moderate and high loads [110]. Indeed the general problem of deflection routing in buffer-less OBS networks is over-utilization of link resources, what happens if a deflected path has more hops than a primary path. Hence, since first proposals were based on the static route calculation and selection (e.g., [41]), in the next step the authors proposed an optimisation calculation of the set of alternative routes (e.g., [60,65]) as well as an adaptive selection of paths (e.g., [19]). The assignment of lower priorities to deflected bursts is another important technique which preserves from excessive burst losses on primary routes [11].

Multi-path routing represents another group of routing strategies, which aim at the traffic load balancing in OBS networks. Most of the proposals are based on a static calculation of the set of equally-important routes (e.g., [81]). Then the path selection is performed adaptively and according to some heuristic [75,95] or optimised cost function [66,94]. Both traffic splitting [4,63] and path ranking [46,104] techniques are used in the path selection process.

The network congestion in single-path routing can be avoided thanks to a proactive route calculation. Although most of the strategies proposed for OBS networks consider centralised calculation of single routes [111], still some authors focus on distributed routing algorithms [31,44]. Both optimisation [63] and heuristic [28] methods are used.

In literature they are present other routing strategies that give support to network resilience by the computation of backup paths [13,44] and to multicast transmission by duplicating [43,50].

In terms of network optimisation, since an overall BLP has a non-linear character [87], either linear programming formulation with piecewise linear approximations of this function [94] or non-linear optimisation gradient methods [40] can be used. Section 7.5.2 focuses on a multi-path source routing approach and applies a non-linear optimization of BLP with a straightforward calculation of partial derivatives to improve OBS network performance.

7.5.2 Optimization of Multi-Path Routing in Optical Burst Switching Networks

In a non-linear optimization problem we assume that there is a pre-established virtual path topology consisting of a limited number of paths between each pair of source-destination nodes. Using a gradient optimization method we can calculate a traffic splitting vector that determines the distribution of traffic over these paths. In order to support the gradient method we propose straightforward formulas for calculation of partial derivatives.

7.5.2.1 Routing Scenario

We assume that the network applies source-based routing, so that the source node determines the path of a burst that enters the network. Moreover, the network uses multi-path routing where a burst can follow one of the paths given between the pair of source-destination (S-D) nodes. We assume each node is capable of full wavelength conversion and thus there is no wavelength-continuity constraint imposed on the problem.

Selection of path p is performed according to a traffic splitting factor x_p. Constraints on the traffic splitting factor are the following: 1) x_p should be non-negative and less or equal to 1, and 2) the sum of traffic splitting factors for all paths connecting given pair of S-D nodes should be equal to 1.

The reservation (holding) times on each link are i.i.d. random variables with the mean equal to the mean burst duration. Bursts destined to given node arrive according to a Poisson process of (long-term) rate specified by the demand traffic matrix. Thus traffic offered to path p can be calculated as a fraction x_p of the total traffic offered between given pair of S-D nodes.

Here vector $x = (x_1, \dots , x_P)$, where P means the number of all paths, determines the distribution of traffic over the network; this vector should be optimized to reduce congestion and to improve overall performance.

7.5.2.2 Formulation and Resolution Method

A loss model of OBS network based on the *Erlang fixed-point approximation* was proposed in [87]. In particular, the traffic offered to link e is obtained as a sum of the traffic offered to all the paths that cross this link reduced by the traffic lost in the preceding links along these paths.

The formulation of [87] may bring some difficulty in the context of computation of partial derivatives for optimization purposes. Therefore in [56] we propose a simplified non-reduced link load model where the traffic offered to link e is calculated as a sum of the traffic offered to all the paths that cross this link. The rationale behind this assumption is that under low link losses, observed in a properly dimensioned network, the model in [87] can be approximated by our model [56]; in [56] we can see that the accuracy of the simplified model is very strict for losses below 10^{-2}.

Having calculated the traffic offered to each link, the main steps of the network loss modelling include the calculation of burst loss probabilities on links, given by the Erlang loss formula, loss probabilities of bursts offered to paths, and the overall burst loss probability B (see [56] for detailed formulae).

From this network loss model we define cost function $B(x)$ to be the subject of optimization. The optimization problem is formulated as to minimize $B(x)$ subject to the constraints imposed on the traffic splitting factor (discussed in SubSection 7.5.2.1). Since the overall BLP is a non-linear function of vector x the cost function is non-linear as well. A particularly convenient optimisation method is the Frank-Wolfe reduced gradient method (algorithm 5.10 in [83]); this algorithm was used for a similar problem in circuit-switched networks [40].

Gradient methods need to employ the calculation of partial derivatives of the cost function. The partial derivative of $B(x)$ with respect to x_p, where p means a path, could be derived directly from the network loss formulae by a standard method involving resolution of a system of linear equations. Such a computation, however, would be time-consuming.

Therefore instead in [56] we provide a straightforward derivation of the partial derivative that is based on the approach previously proposed for circuit switched networks [55]. We have managed to simplify the model described in [55] and make the calculation of partial derivatives straightforward, not involving any iteration. The calculation of gradient in our method, therefore, is not longer an issue.

It can be shown numerically that objective function $B(x)$ is not necessarily convex. Nevertheless, under moderate traffic loads we have observed that several repetitions of the optimization program always give us the same (with a finite numerical precision) near-optimal value of B.

7.5.2.3 Numerical Results

We evaluated the performance of our routing scheme in an event-driven simulator. In order to find a splitting vector x specifying a near-optimal routing we used solver *fmincon* for constrained nonlinear multivariable functions available in the *Matlab* environment. Then we applied this vector in the simulator.

The evaluation is performed for NSFnet (15 nodes, 23 links) and EON (28 nodes, 39 links) network topologies; different numbers of wavelengths (λs) per link are considered, each transmitting at 10Gbps. The optimized routing (OR) is compared with two other routing strategies: a simple shortest path routing (SP) and a pure deflection routing (DR). We consider 2 shortest paths per each source-destination pair of nodes; they are not necessarily disjoint. In SP routing only 1 path is available. Uniform traffic matrix and exponential burst inter-arrivals and durations are considered. All the simulation results have 99% level of confidence.

In Fig. 7.15 we show B as a function of offered traffic load for different routing scenarios. We see that the optimized routing can achieve very low losses, particularly, when compared with the shortest path routing. Analytical results ('OR-an' in the figure) correspond very well to simulation results. The optimization takes about

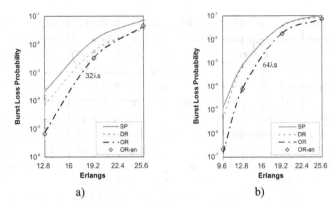

Fig. 7.15. Comparison of routing schemes a) NSFnet, b) EON.

23s and 1800s for NSFnet network (of 420 paths) and EON network (of 1512 paths), respectively, when using a non-commercial *Matlab* solver on a Pentium D, 3GHz computer.

7.5.2.4 Conclusions

In this Section we have proposed a non-linear optimization method for multi-path source routing problem in OBS networks. In this method we calculate a traffic splitting vector that determines a near-optimal distribution of traffic over routing paths. Since a conventional network loss model of an OBS network is complex we have introduced some simplifications. The references formulae for partial derivatives are straightforward and very fast to compute. It makes the proposed non-linear optimization method a viable alternative for linear programming formulations based on piecewise linear approximations of the cost function.

The simulation results demonstrate that our method effectively distributes the traffic over the network and the network-wide burst loss probability can be significantly reduced compared with the shortest path routing.

7.6   TCP over OBS Networks

7.6.1   Introduction and State-of-the-Art

TCP is today the dominant transport protocol in Internet, and it is expected to continue to be used. As TCP is not specifically designed for a particular technology, modifying the standard TCP can lead to a performance optimization in specific environments. In this direction, a great amount of novel TCP versions have been proposed for mobile networks, wireless mesh networks, and high speed networks such as optical switched networks. The development of TCP of the last years has covered three major topics: 1) making TCP more suitable for high speed environments and grid computing applications (e.g. Fast TCP and High Speed TCP), 2) making TCP more robust for non congestion events (e.g. TCP-NCR, TCP-PR and TCP-Aix), and 3) TCP for special environments and applications e.g. for wireless networks.

In the context of OBS network, the problem of having TCP has been widely studied in the literature. The design of novel specific TCP implementation are considered in [89,114,115]. Nonetheless, there is no consensus in whether a completely new TCP version is needed, and which new TCP version should be standardized within all the proposals. In order to have some benchmarking reference, TCP Reno 0 and TCP Sack [30] are generally considered as they are the most popular versions in current networks.

From the performance point of view, the main focus is put on the effect of the burstification delay on TCP behaviour [24,108,109]. In fact, the burstfication process can increase the value of the Round Trip Time and thus decrease the TCP throughput accordingly. At the same time, the high bandwidth delay product of the OBS networks contributes to enlarge faster the congestion window than in current networks and thus increase the TCP throughput [24,25]. As a consequence, the (timer and/or size) thresholds in the burstification process become important trade-offs to achieve high TCP performance [108].

On the other hand, TCP over OBS suffers of the so called False Timeout effect [109]. Due to the bufferless nature of OBS core network and the one-way signalling scheme, the OBS network is subject of random burst losses, even at low traffic loads. The random burst loss may be falsely interpreted as network congestion by the TCP layer, which is therefore forced to timeout and to decrease the sending window. Some mechanisms based on burst retransmission are proposed to alleviate the false timeout effect (e.g., [113]).

The effect of the packet reordering is addressed in Section 7.6.2 where a layered framework to measure the reordering introduced by contention resolution strategies in OBS networks is presented. In particular, characterization is based on the reordering metrics proposed by the IETF IPPM Working Group. The obtained results are twofold. First, they quantify the impact of burst reordering on TCP throughput performance, and secondly, they give insight into solving burst reordering by well dimensioned buffers.

7.6.2   Burst Reordering Impact on TCP over OBS Networks

7.6.2.1   Introduction

In this work, we follow a layered approach to study the viability of OBS as a carrier technology for TCP. Firstly, we quantify the introduced reordering at the OBS layer. With such purposes, we apply the reordering metrics presented by the IETF in [73], which provide us extensive information. First, they quantify the buffer size that should be placed at edge nodes to solve reordering at the OBS layer. This would permit the sending of already ordered packets to the IP layer, so that burst reordering would remain transparent to TCP. Second, in the case that reordering is left to the TCP layer, they provide information about the violation of the DUP-ACK threshold due to reordering, which allows TCP performance estimation.

It is widely known the effect of packet loss on TCP. In TCP Reno [1], the sender of a TCP session is notified of a packet loss by means of duplicate acknowledgments. In this context, the TCP fast retransmit algorithm is invoked once the duplicate acknowledgment threshold (DUP-ACK) is reached. As a result, the missing packet is retransmitted and the sender's congestion window is halved, which decreases TCP throughput significantly. A similar situation occurs whether a packet becomes reordered. Note, that in the event of reaching the DUP-ACK threshold, TCP may consider a reordered packet as lost, even though it is only delayed and it would later be received.

For the sake of generality, we quantify reordering in OBS networks under several contention resolution strategies. As in [33], we deal with Conv, Defl and FDL basic strategies and combinations of them. Because the order of application of each strategy is essential, combined strategies are named by a concatenation of the former's acronyms. In particular, performance of ConvFDL, ConvDefl, ConvFDLDefl and ConvDeflFDL is also here evaluated.

7.6.2.2 Scenario under Study

With evaluation purposes, we implement the 16-node COST 266 reference network [69]. For simplicity, all links have the same length of 200 km, which introduces a link propagation delay of 1 ms. Network resources are dimensioned according to a static traffic demand matrix, obtained from a 2006 European population model [69]. Particularly, a total demand of 9.9 Tbps is offered to the network which corresponds to 990 Erlangs for a 10 Gbps line rate. Then, wavelength capacity is distributed in the network, so that shortest path routing leads to equal blocking probabilities on all links (i.e., dimensioning according to the Erlang model [58]). In this context, different network load situations can be achieved by overdimensioning wavelength capacity by a given factor (denoted as *overdimensioning factor* in the figures).

Regarding traffic characteristics, the burst departure process follows a Poisson process and burst length is exponentially distributed with mean 100 kbit [32]. In turn, in OBS nodes, the number of add/drop ports is unlimited and the switching matrix is non-blocking. Besides, the delay for burst control packet processing is compensated by a short extra FDL of appropriate length at the input of the node.

With contention resolution purposes, we assume one FDL per node with a certain number of wavelengths. The length of this FDL equals the mean burst transmission time, defined as the time needed to transmit an average sized burst (i.e., 10 μs for 100kbit bursts over 10Gbps data links). Note that the wavelengths of this FDL are shared and the number of wavelength converters per node is unlimited. Nonetheless, if all wavelengths are occupied upon burst arrival in the FDL, the burst is discarded.

7.6.2.3 Simulation Results

In this section, we evaluate the performance of the strategies *Conv*, *ConvFDL*, *ConvDefl*, *ConvFDLDefl* and *ConvDeflFDL* in an OBS scenario. We focus not only on burst loss probability, but also on introduced reordering, which harms TCP performance as well. Note that a complete characterization of reordering becomes important, especially when assessing a protocol's viability over a given network. With such an objective, the IETF IPPM working group has recently standardized a set of metrics [73] to characterize reordering effects in generic packet networks (e.g., OBS networks). In this section, three of them are selected and further quantified.

Specifically, we evaluate reordering ratio, reordering extent and 3-reordering ratio, which provide a broad view of reordering in the scenario under study. To this end, we measure burst reordering between each demand source-destination pair and we provide global network statistics. Note, that if no wavelength conversion would have been feasible in the network, our conclusions on reordering would still be valid, as Conv is applied first in all schemes. In the evaluation, we assume 8 wavelengths in the FDLs mainly due to cost and hardware integration issues. Moreover, to avoid unnecessary load and high propagation delays in the network, we limit the number of deflections to 1. Previous works demonstrate that the improvements due to further deflections are marginal [32], as long as a reasonable amount of flexibility is allowed in the network. The results have been obtained using the event-driven simulation library IKRSimlib [8].

As can be seen in [82], for high and medium loads ConvDeflFDL introduces the highest reordering, followed by ConvFDLDefl, ConvDefl and ConvFDL. However, towards low loads, all strategies behave similarly in terms of reordering ratio. Particularly, the introduced reordering ratio by Conv alone was there not evaluated. In fact, when applying this strategy all bursts travel along the same path and no buffering is used. Therefore, no reordering is introduced. Concerning burst loss probability, it was distinguished that for high loads the performance of all the strategies that use deflection routing (i.e., ConvDefl, ConvFDLDefl and ConvDeflFDL) is poor, as they overload an already highly loaded network. Nonetheless, towards lower loads, deflection (alone or combined) decrease burst loss probability rapidly, as enough network resources become available. The majority of studies coincide that in a realistic OBS scenario, burst blocking probabilities should range from 10^{-3} to 10^{-6}. Particularly, in this operating range, all strategies introduced the same reordering to the network. However, ConvDeflFDL provided the best performance regarding burst loss probability. At a first sight, this leads to the conclusion that this strategy may provide the best compromise between burst losses and introduced reordering.

In addition, we analyze the possibility to restore burst order directly at the OBS layer. Then, already ordered packets could be sent to the IP layer, so that burst reordering would remain transparent to TCP. With these purposes, a possible solution is the placement of buffers on a per flow basis at OBS edge nodes. Such buffers would store incoming out-of-order bursts, waiting for the expected one to be received. In this context, the reordering extent metric provides information about the mean extent to which bursts are reordered. Therefore, this gives an idea of these buffers' size.

As depicted in [82], deflection routing technique introduces large extents, in the order of one thousand. In fact, deflected bursts transverse at least one more hop than those that go through the direct path. This accounts for an additional propagation delay of 1 ms, which is two orders of magnitude greater than the mean burst transfer time (10 μs in our scenario). Conversely, the use of buffering such as in ConvFDL introduces relatively low extents. Hence, these strategies would enable the restoration of the burst order directly at the OBS layer by means of small buffering capacities. It is noteworthy that towards low loads, the introduced extent by combined strategies tends to the former (e.g., towards low loads, ConvDeflFDL tends to ConvDefl). This is due to the fact that, in a low loaded network, contentions can be solved in the first attempt in most situations.

Until now, we have quantified the reordering ratio and introduced reordering extent for each contention resolution strategy under consideration. While the former provides a general view of what happens in the network, the latter evaluates the possibility to restore order directly in the OBS layer. Note that this information provides understanding about the origins of reordering and evaluates specific solutions to restore it. However, it does not illustrate the direct implication of reordering on TCP. It is our goal now to quantify the n-reordering burst ratio. To this end, we assume n = 3, which matches TCP Reno operation [1].

Referring again to [82], it was shown that 3-reordering ratio increases along with the overdimensioning factor. This could be due to several reasons. For low loads, deflected bursts have more possibilities to succeed, which would increase 3-reordering ratio. Moreover, for higher loads, since more reordering exists, this could decrease 3-reordering. For instance, let us assume a reordered burst. It may happen, that the following ones become also reordered, which could cause this one not to be 3-reordered. Further looking at the obtained results there depicted, it can be seen that buffering technique introduces less 3-reordering ratio than deflection, outperforming ConvFDL all the remainder strategies. For better illustration, absolute 3-reordering ratio was also evaluated in [82]. In fact, it quantifies the ratio of received packets, which become 3-reordered or more. The obtained results presented a behaviour inline with the reordering packet ratio. For high loads, differences between the strategies can be appreciated, outperforming ConvFDL the remainder ones. However, towards lower loads, in a more realistic OBS scenario, all strategies behave equally.

7.6.2.4 Impact of Burst Reordering on TCP Performance

In this section, we quantify the impact of burst reordering on final TCP throughput. Taking into account the already measured 3-reordering ratio at the burst layer in [82], we derive a worst case situation for 3-reordering packet ratio. Then, considering both burst reordering and burst loss pernicious effects, we provide a new figure of merit, called P_{FR}, which quantifies the probability to invoke *fast re-transmit* algorithm in TCP Reno. Finally, as the key point of this work, we estimate the theoretical TCP throughput over the scenario under study, which allows us to conclude on its viability.

For the n-reordering packet ratio, and according to the definition presented in [73], only the first packet contained in an n-reordered burst is considered as n-reordered. Intuitively, this leads to think that an upper bound for the n-reordering packet ratio is given when exactly 1 packet of the TCP flow under study is contained in each burst. One should remind, that the n-reordering packet ratio is measured on a per TCP flow basis. Therefore, only those packets belonging to the TCP flow under study are considered (bursts can be composed of more packets arriving from different TCP sessions). To validate this intuitive assumption, analytical derivations were provided in [82]. Particularly, it was concluded that the following equation must hold to ensure the worst case assumption

$$P\left(N_r \geq n_r\right) \geq \frac{1}{n_p} P\left(N_r \geq \left\lceil \frac{n_r}{n_p} \right\rceil\right), \qquad n_p, n_r \in N. \tag{7.10}$$

where $P(N_r \geq n_r)$ denote the Complementary Cumulative Distribution Function (CCDF) of a burst to become at least n_r-reordered and n_p stays for the number of packets of the same TCP flow per burst. In [82], it was obtained the CCDF of the burst n-reordering ratio for each strategy under study. Indeed, for $n_r = 3$ and $n_r \in N$, the gathered results accomplished in equation (10). This demonstrates that the assumption of having one packet of the same TCP flow under study to be contained in each burst truly contemplates the worst case scenario for the 3-reordering ratio.

This analysis allows us to estimate a worst case for the final TCP throughput, supposing that TCP runs over the network under study. According to the conclusion above, we assume that 1 packet per TCP flow is contained in each burst. In such a case, 3-reordering packet ratio coincides with the already measured 3-reordering burst ratio. Furthermore, we consider that upon contention a burst is entirely dropped. Thus, packet loss probability P_L equals to burst loss probability P_B. Note that if the receiver does not use selective acknowledgments and the sender uses the basic congestion control presented in [1], reordering has the same effect as packet loss. In fact, reordered packets which exceed the DUP-ACK threshold also trigger the fast retransmit algorithm (i.e., as if they would have been lost). Hence, whether $P(N_r* \geq n_r)$ identifies the CCDF function of a packet to become at least n_r-reordered, the probability to invoke fast retransmit algorithm can be stated as $P_{FR} = P(N_r* \geq n_r) + P_L$.

In Fig. 7.16(a), we depict the upper bound for P_{FR} for a DUP-ACK threshold set to 3. In particular, it is obtained as $P_{FR} = P(N_r \geq 3) + P_B$, using the results presented in [82]. As seen, for high loads, Conv and ConvFDL lead to better results, due to the lower reordering they introduce. However, for lower loads, all combined strategies provide similar performance. This is due to the fact that along this

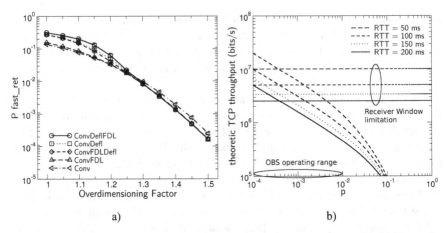

Fig. 7.16. a) Probability to trigger TCP fast retransmit (worst case scenario), b) Theoretic TCP throughput (bits/s) according to the model proposed in [78].

range, 3 reordering ratio dominates in front of P_L. The fact that Conv alone provides substantially worse performance demonstrates the need for additional contention resolution in OBS networks.

Up to now, several analyses have been proposed in the literature to model the steady state throughput of a TCP connection. In [78] a model, which considered both congestion avoidance phase and retransmissions caused by time out, is developed and an approximated formula for the throughput B_{TCP} of a TCP session is given.

Fig. 7.16(b) illustrates, for different RTT values, the theoretical TCP throughput according to this model. Mainly, it depicts B_{TCP} and the limitation due to the receiver limitation window, both function of p (the total packet loss probability along the path, or P_{FR} since, in this scenario, 3-reordering has the same effect as packet loss). In this way, given a certain p, the theoretical TCP throughput will be the minimum of both curves.

As mentioned earlier, OBS networks are usually dimensioned to achieve burst loss probabilities ranging from 10^{-3} to 10^{-6}. Looking at Fig. 7.16(a), a network dimensioned to achieve these values (from the results presented in [82], overdimensioning the network by 1.25 - 1.35) would experience P_{FR} values from 10^{-2} to 10^{-3}, depending on the strategy used. Observing now Fig. 7.16(b), we find that, for these p values, the performance of TCP is highly affected by the reordering introduced at the OBS layer. In fact, to assure the proper performance of TCP, p should be lower than 10^{-3}, so that the limiting factor would be the receiver advertised window, rather than the reordering introduced in the network. This demonstrates that reordering should be also considered when dimensioning an OBS network for TCP traffic. As seen, its impact on TCP is much more significant than P_L in the range of operation of typical OBS networks. Moreover, as far as TCP performance is concerned, almost all combined contention resolution strategies under study behave similarly. Although we mentioned earlier that *ConvDeflFDL* may outperform the remainder, such improvements are hidden by the fact that 3-reordering dominates in front of P_L.

7.6.2.5 Conclusions

In this section, we propose a layered framework to quantify the impact of burst reordering on TCP performance. First of all, we measure the reordering introduced by several contention resolution strategies. With such purposes in mind, we use the packet reorder metrics proposed by the IETF. Two different approaches to tackle reordering in an OBS scenario have been highlighted and subsequently evaluated. On the one hand, reordering can be solved directly at the OBS layer, by means of well dimensioned buffers. On the other hand, reordering can be left to higher layers, expecting this one to be solved by them.

For the former strategy, we quantify the size of the buffers which should be placed at OBS edge nodes on a per flow basis. Following this line, we find that deflection routing prohibits this solution, since the introduced extents are extremely high. Conversely, we demonstrate that buffering introduces significantly lower extents, which would, a priori, enable this strategy.

For the latter strategy, we focus on its impact on final TCP Reno performance. We propose a new figure of merit, named P_{FR}, which considers not only the pernicious effects from packet loss, but also the ones from caused by reordering. This allows us to conclude, based on the model proposed by [78] that the usual OBS operating range fits no more. On the contrary, network should be dimensioned taking into account not only burst loss probability, but also burst reordering introduced by contention resolution.

7.7 Conclusions

Ten years ago, the growth of the Internet and its bursty statistical characteristic were the main drivers to develop innovative data-centric optical transport networks. In this context the optical burst switching (OBS) and optical packet switching (OPS) technologies were proposed as promising network solutions overcoming the typical inefficiency of the circuit switching network. In fact they were designed with the aim of optimising the utilisation of the WDM channels by means of fast and highly dynamic resource allocation based on a statistical multiplexing scheme.

These ten years of research activities in OPS and OBS covered different, extensively and heterogeneous topics: novel switch architecture with no, partial or full wavelength conversion, multi-switching architecture, efficient scheduling algorithms, routing with traffic engineering capability, mechanisms to support QoS, novel TCP mechanism to enhance the random loss behaviour of the OPS/OBS networks, protection and restoration mechanisms, etc. Some of these topics have been reviewed in this chapter. An important issue which is a hot topic of current research activity is the deployment of control plane in OBS/OPS networks. As a solution, some studies have initiated to consider a common control plane based for example on the generalised MPLS protocol (GMPLS). Having a common control plane might be desired, in particular, in the context of coexistence of different switching technologies and of the network migration towards all-optical networks. Therefore, the loop can be closed allowing the continuous deployment of optical circuit switching, OBS and OPS.

Nonetheless, nowadays OPS/OBS are still not feasible since the majority of the required optical devices are not commercially available or even not proved in laboratory. This situation creates some slowdown interest in these fields. To move up and gain insight into OPS/OBS, a more strict cooperation between interdisciplinary areas is desired: researchers in photonic material, optical communication and optical networking should dedicate efforts in defining clear requirements, recommendations and guidelines and proposing viable solutions.

References

1. Allman, M., Paxson, V., Stevens, W.: TCP Congestion Control. RFC 2581 (1999)
2. Almeida jr., R.C., Pelegrini, J.U., Waldman, H.: A Generic-traffic Optical Buffer Modeling for Asynchronous Optical Switching Networks. IEEE Communications Letters 9(2), 175–177 (2005)
3. Aracil, J., et al.: Research in Optical Burst Switching within the e-Photon/ONe Network of Excellence. Optical Switching and Networking 4(1), 1–19 (2007)
4. Argos, C.G., de Dios, O.G., Aracil, J.: Adaptive Multi-path Routing for OBS Networks. In: 9th IEEE International Conference on Transparent Optical Networks (ICTON), Rome, Italy, pp. 299–302 (2007)
5. Azodolmolky, S., Tzanakaki, A., Tomkos, I.: Study of the Impact of Burst Assembly Algorithms in Optical Burst Switched Networks with Self-Similar Input Traffic. In: 8th IEEE International Conference on Transparent Optical Networks (ICTON), Nottingham, UK, pp. 35–40 (2006)
6. Bjørnstad, S., Hjelme, D.R., Stol, N.: A Highly Efficient Optical Packet Switching Node Design Supporting Guaranteed Service. In: European Conference on Optical Communication (ECOC), Rimini, Italy, pp. 110–111 (2003)
7. Bjørnstad, S., Hjelme, D.R., Stol, N.: A Packet-Switched Hybrid Optical Network with Service Guarantees. IEEE Journal on Selected Areas in Communications (Supplement on Optical Communications and Networking) 24(8), 97–107 (2006)
8. Bodamer, S., et al.: IND Simulation Library 2.3 User Guide Part I: Introduction (2004), http://www.ikr.uni-stuttgart.de/INDSimLib
9. Callegati, F.: Optical Buffers for Variable Length Packets. IEEE Communications Letters 4(9), 292–294 (2000)
10. Callegati, F., Cerroni, W., Muretto, G., Raffaelli, C., Zaffoni, P.: QoS Routing in DWDM Optical Packet Networks. In: WQoSR2004 co-located with QoFIS 2004, Barcelona, Spain (2004)

11. Cameron, C., Zalesky, A., Zukerman, M.: Prioritized Deflection Routing in Optical Burst Switching Networks. IEICE Transaction on Communications E88-B(5), 1861–1867 (2005)
12. Cao, X., Li, J., Chen, Y., Qiao, C.: Assembling TCP/IP Packets in Optical Burst Switched Networks. In: IEEE Global Communications Conference (Globecom), Taipei, Taiwan, pp. 2808–2812 (2002)
13. Chen, Q., Mohan, G., Chua, K.C.: Route Optimization for Efficient Failure Recovery in Optical Burst Switched Networks. In: IEEE HPSR 2006, Poznan, Poland (2006)
14. Chen, Y., Turner, J.S., Mo, P.-F.: Optimal Burst Scheduling in Optical Burst Switched Networks. IEEE/OSA Journal of Lightwave Technology 25(8), 1883–1894 (2007)
15. Chiaroni, D., et al.: First Demonstration of an Asynchronous Optical Packet Switching Matrix Prototype for Multi-Terabit-Class Routers/Switches. In: 27th European Conference on Optical Communication (ECOC), Amsterdam, Netherlands, pp. 60–61 (2001)
16. Chlamtac, I., et al.: CORD: Contention Resolution by Delay Lines. IEEE Journal on Selected Areas in Communications 14(5), 1014–1029 (1996)
17. Claffy, K., Miller, G., Thompson, K.: The Nature of the Beast: Recent Traffic Measurements from an Internet Backbone. In: International Networking Conference (INET), Geneva, Switzerland (1998)
18. Clegg, R.G.: A Practical Guide to Measuring the Hurst Parameter. In: Thomas N. (ed.) 21st UK Performance Engineering Workshop, School of Computing Science Technical Report Series, CS-TR-916, University of Newcastle (2005)
19. Coutelen, T., Elbiaze, H., Jaumard, B.: An Efficient Adaptive Offset Mechanism to Reduce Burst Losses in OBS Networks. In: IEEE Global Communications Conference (Globecom), St. Louis, MO, USA (2005)
20. Danielsen, S., Hansen, P., Stubkjear, K.: Wavelength Conversion in Optical Packet Switching. IEEE/OSA Journal of Lightwave Technology 16(12), 2095–2108 (1998)
21. de Miguel, I., González, J.C., Koonen, T., Durán, R., Fernández, P., Tafur Monroy, I.: Polymorphic Architectures for Optical Networks and their Seamless Evolution towards Next Generation Networks. Photonic Network Communications 8(2), 177–189 (2004)
22. de Vega Rodrigo, M., Götz, J.: An Analytical Study of Optical Burst Switching Aggregation Strategies. In: 3rd International workshop on Optical Burst Switching (WOBS), San Jose, CA, USA (2004)
23. de Vega Rodrigo, M., Spadaro, S., Remiche, M.-A., Careglio, D., Barrantes, J., Götz, J.: On the Statistical Nature of highly-aggregated Internet Traffic. In: 4th International Workshop on Internet Performance, Simulation, Monitoring and Measurement, Salzburg, Austria (2006)
24. Detti, A., Listanti, M.: Impact of Segments Aggregation on TCP Reno Flows in Optical Burst Switching Networks. In: IEEE Infocom, New York, NY, USA, pp. 1803–1812 (2002)
25. Detti, A., Listanti, M.: Amplification Effects of the Send Rate of TCP Connection through an Optical Burst Switching Network. Optical Switching and Networking 2(1), 49–69 (2005)
26. Dogan, K., Akar, N.: A Performance Study of Limited Range Partial Wavelength Conversion for Asynchronous Optical Packet/Burst Switching. In: IEEE International Conference on Communications (ICC), Ankara, Turkey, pp. 2544–2549 (2006)

27. Downey, A.: Evidence for long-tailed Distributions in the Internet. In: ACM Sigcomm Internet Measurement Workshop, San Francisco, CA, USA (2001)
28. Du, Y., Pu, T., Zhang, H., Quo, Y.: Adaptive Load Balancing Routing Algorithm for Optical Burst-Switching Networks. In: OSA Optical Fiber Communication Conference and Exhibit (OFC), Anaheim, CA, USA (2006)
29. Duser, M., Kozlovski, E., Kelly, R.I., Bayel, P.: Design Trade-offs in Optical Burst Switched Networks with Dynamic Wavelength Allocation. In: 26th European Conference on Optical Communications (ECOC), Munich, Germany (2000)
30. Floyd, S., Mahdavi, J., Mathis, M., Podolsky, M.: An Extension to the Selective Acknowledgement (SACK) Option for TCP. RFC 2883 (2000)
31. Gao, D., Zhang, H.: Information Sharing based Optimal Routing for Optical Burst Switching (OBS) Network. In: OSA Optical Fiber Communication Conference and Exhibit (OFC), Anaheim, CL, USA (2006)
32. Gauger, C.: Performance of Converter Pools for Contention Resolution in Optical Burst Switching. In: Optical Networking and Communications (OptiComm), Boston, MA, USA, pp. 109–117 (2002)
33. Gauger, C., Köhn, M., Scharf, J.: Comparison of Contention Resolution Strategies in OBS Network Scenarios. In: 6th International Conference on Transparent Optical Networks (ICTON), Wroclaw, Poland, pp. 18–21 (2004)
34. Gauger, C., Mukherjee, B.: Optical Burst Transport Network (OBTN) – A Novel Architecture for Efficient Transport of Optical Burst Data over Lambda Grids. In: IEEE High Performance Switching and Routing (HPSR), Hong Kong, P.R. China, pp. 58–62 (2005)
35. Ge, A., Callegati, F., Tamil, L.: On Optical Burst Switching and Self-Similar Traffic. IEEE Communication Letters 4(3), 98–100 (2000)
36. González-Castaño, F.J., Rodelgo-Lacruz, M., Pavón-Mariño, P., García-Haro, J., López-Bravo, C., Veiga-Gontán, J., Raffaelli, C.: Guaranteeing Packet Order in IBWR Optical Packet Switches with Parallel Iterative Schedulers. Accepted for publication to European Transactions on Telecommunications journal
37. Gross, D., Harris, C.M.: Fundamentals of Queueing Theory, 2nd edn. John Wiley and Sons, Chichester (1985)
38. Guillemot, C., et al.: Transparent Optical Packet Switching: the European ACTS KEOPS Project Approach. IEEE Journal of Lightwave Technology 16(12), 2117–2134 (1998)
39. Gunreben, S., Hu, G.: A Multi-layer Analysis on Reordering in Optical Burst Switched Networks. IEEE Communications Letters 11(12), 1013–1015 (2007)
40. Harris, R.J.: The Modified Reduced Gradient Method for Optimally Dimensioning Telephone Networks. Australian Telecommunications Research 10(1), 30–35 (1976)
41. Hsu, C., Liu, T., Huang, N.: Performance Analysis of Deflection Routing in Optical Burst-Switched Networks. In: 21st IEEE Infocom, New York, NY, USA (2002)
42. Hu, G., Dolzer, K., Gauger, C.: Does Burst Assembly really Reduce the Self-Similarity? In: OSA Optical Fiber Communication Conference and Exhibit (OFC), Atlanta, March 2003, pp. 124–126 (2003)
43. Huang, X., She, Q., Zhang, T., Lu, K., Jue, J.P.: Small Group Multicast with Deflection Routing in Optical Burst Switched Networks. In: 5th International workshop on Optical Burst Switching (WOBS), San Jose, CA, USA (2006)
44. Huang, Y., Heritage, J., Mukherjee, B.: Dynamic Routing with Preplanned Congestion Avoidance for Survivable Optical Burst-Switched (OBS) Networks. In: OSA Optical Fiber Communication Conference and Exhibit (OFC), Anaheim, CL, USA (2005)

45. Hunter, D.K., Chia, M.C., Andonovic, I.: Buffering in Optical Packet Switches. IEEE/OSA Journal of Lightwave Technology 16(12), 2081–2094 (1998)

46. Ishii, D., Yamanaka, N., Sasase, I.: Self-learning Route Selection Scheme using Multipath Searching Packets in an OBS Network. OSA Journal of Optical Networking 4(7), 432–445 (2005)

47. ITU-t Recommendation I.371, Traffic control and congestion control in B-ISDN (2000)

48. Izal, M., Aracil, J.: On the Infuence of Self Similarity on Optical Burst Switching Traffic. In: IEEE Global Communications Conference (Globecom), Taipei, Taiwan, pp. 2308–2312 (2002)

49. Izal, M., Aracil, J., Morató, D., Magaña, E.: Delay-Throughput Curves for Timer-based OBS Burstifiers with Light Load. IEEE/OSA Journal of Lightwave Technology 24(1), 277–285 (2006)

50. Jeong, M., Qiao, C., Vandenhoute, M.: Distributed Shared Multicast Tree Construction Protocols for Tree-shared Multicasting in OBS Networks. In: 11th International Conference on Computer Communications and Networks (ICCCN), Miami, Florida, USA, pp. 322–327 (2002)

51. Junghans, S.: Pre-Estimate Burst Scheduling (PEBS): An Efficient Architecture with low Realization Complexity for Burst Scheduling Disciplines. In: International Conference on Broadband Networks (Broadnets), Boston, Massachusetts, USA, pp. 1124–1128 (2005)

52. Kaheel, A., Alnuweiri, H.: A Strict Priority Scheme for Quality-of Service Provisioning in Optical Burst Switching Networks. In: 8th IEEE Symposium on Computers and Communications (ISCC), Turkey (June 2003)

53. Kaheel, A., Alnuweiri, H.: Quantitative QoS Guarantees in Labeled Optical Burst Switching Networks. In: IEEE Global Communications Conference (Globecom), Dallas, TX, USA (2004)

54. Karagiannis, T., Molle, M., Faloutsos, M., Broido, A.: A Nonstationary Poisson View of Internet Traffic. In: IEEE Infocom, Hong Kong, P.R. China (2004)

55. Kelly, F.P.: Routing in Circuit-Switched Networks: Optimization, Shadow Prices and Decentralization. Advanced Applied Probability 20, 112–144 (1988)

56. Klinkowski, M., Pioro, M., Careglio, D., Marciniak, M., Sole-Pareta, J.: Non-linear Optimization for Multipath Source-Routing in OBS Networks. IEEE Communications Letters 11(12), 1016–1018 (2007)

57. Klinkowski, M., Careglio, D., Morató, D., Solé-Pareta, J.: Preemption Window for Burst Differentiation in OBS. In: OSA Optical Fiber Communication Conference and Exhibit (OFC), San Diego, CA, USA (2008)

58. Köhn, M., Gauger, C.: Dimensioning of SDH/WDM Multilayer Networks. In: 4th ITG Symposium on Photonic Networks, pp. 29–33 (2003)

59. Laevens, K.: Traffic Characteristics inside Optical Burst Switched Networks. In: Opticom, Boston, MA, USA, pp. 137–148 (2002)

60. Lee, S., Kim, H., Song, J., Griffith, D.: A Study on Deflection Routing in Optical Burst-Switched Networks. Photonic Network Communications 6(1), 51–59 (2003)

61. Leland, W., Taqqu, M., Willinger, W., Wilson, D.: On the Self-Similar Nature of Ethernet Traffic (extended version). IEEE/ACM Transactions on Networking 2(1), 1–15 (1994)

62. Li, J., Qiao, C., Xu, J., Xu, D.: Maximizing Throughput for Optical Burst Switching Networks. In: IEEE Infocom, Hong Kong, P.R. China, pp. 1853–1863 (2004)

63. Li, J., Mohan, G., Chua, K.C.: Dynamic Load Balancing in IP-over-WDM Optical Burst Switching Networks. Computer Networks 47(3), 393–408 (2005)
64. Li, J., Yeung, K.L.: Burst Cloning with Load Balancing. In: OSA Optical Fiber Communication Conference and Exhibit (OFC), Anaheim, CL, USA (2006)
65. Long, K.-P., Yang, X., Huang, S., Chen, Q.-B., Wang, R.: An Adaptive Parameter Deflection Routing to Resolve Contentions in OBS Networks. In: Boavida, F., Plagemann, T., Stiller, B., Westphal, C., Monteiro, E. (eds.) NETWORKING 2006. LNCS, vol. 3976, pp. 1074–1079. Springer, Heidelberg (2006)
66. Lu, J., Liu, Y., Gurusamy, M., Chua, K.: Gradient Projection based Multi-path Traffic Routing in Optical Burst Switching Networks. In: IEEE High Performance Switching and Routing workshop (HPSR), Poznan, Poland (2006)
67. Lu, X., Mark, B.L.: Performance Modeling of Optical-Burst Switching with Fiber Delay Lines. IEEE Transactions on Communications 52(12), 2175–2182 (2004)
68. Ma, X., Kuo, G.-S.: Optical Switching Technology Comparison: Optical MEMS vs. other Technologies. IEEE Communications Magazine 41(11), 16–23 (2003)
69. Maesschalck, S., et al.: Pan-European Optical Transport Networks: An Availability-based Comparison. Photonic Network Communications 5, 203–225 (2003)
70. Magaña, E., Morató, D., Izal, M., Aracil, J.: Evaluation of Preemption Probabilities in OBS Networks with Burst Segmentation. In: IEEE International Conference on Communications (ICC), Seoul, Korea (2005)
71. Masetti, F., et al.: Design and Implementation of a Multi-terabit Optical Burst/Packet Router Prototype. In: OSA Optical Fiber Communication Conference and Exhibit (OFC), Anaheim, CA, USA (2002)
72. Morató, D., Aracil, J.: On the Use of Balking for Estimation of the Blocking Probability for OBS Routers with FDL Lines. In: Chong, I., Kawahara, K. (eds.) ICOIN 2006. LNCS, vol. 3961, pp. 399–408. Springer, Heidelberg (2006)
73. Morton, A., Ciavattone, L., Ramachandran, G., Shalunov, S., Perser, J.: Packet Reordering Metrics. RFC 4737 (2006)
74. Norros, I.: On the Use of Fractional Brownian Motion in the Theory of Connectionless Networks. IEEE Journal on Selected Areas in Communications 13(6), 953–962 (1995)
75. Ogino, N., Arahata, N.: A Decentralized Optical Bursts Routing based on Adaptive Load Splitting into Pre-calculated Multiple Paths. IEICE Transactions on Communications E88-B(12), 4507–4516 (2005)
76. Oh, S., Kang, M.: A Burst Assembly Algorithm in Optical Burst Switching Networks. In: OSA Optical Fiber Communication Conference and Exhibit (OFC), Anaheim, CA, USA, pp. 771–773 (2002)
77. Overby, H., Stol, N.: QoS Differentiation in Asynchronous Bufferless Optical Packet Switched Networks. Wireless Networks 12(3), 383–394 (2006)
78. Padhye, J., Firoiu, V., Towsley, D.F., Kurose, J.F.: Modeling TCP Reno Performance: A Simple Model and its Empirical Validation. IEEE/ACM Transactions on Networking 8(2), 133–145 (2000)
79. Pavon-Marino, P., Garcia-Haro, J., Jajszczyk, A.: Parallel Desynchronized Block Matching: A Feasible Scheduling Algorithm for the Input-Buffered Wavelength-Routed Switch. Computer Networks 51(15), 4270–4283 (2007)
80. Pavon-Mariño, P., Gonzalez-Castaño, F.J., Garcia-Haro, J.: Round-Robin Wavelength Assignment: A new Packet Sequence Criterion in Optical Packet Switching SCWP Networks. European Transactions on Telecommunications 17(4), 451–459 (2006)

81. Pedro, J.M., Monteiro, P., Pires, J.J.O.: Efficient Multi-path Routing for Optical Burst-Switched Networks. In: 6th Conference on Telecommunications (ConfTele), Peniche, Portugal (2007)

82. Perelló, J., Gunreben, S., Spadaro, S.: A Quantitative Evaluation of Reordering in OBS Networks and its Impact on TCP Performance. In: 12th conference on Optical Network Design and Modeling (ONDM 2008), Vilanova i la Geltru, Spain (2008)

83. Pioro, M., Medhi, D.: Routing, Flow, and Capacity Design in Communication and Computer Networks. Morgan Kaufmann, San Francisco (2004)

84. Qiao, C., Yoo, M.: Optical burst switching (OBS) – a new Paradigm for an Optical Internet. Journal of High Speed Networks (Special Issues on Optical Networks) 8(1), 69–84 (1999)

85. Rajaduray, R., Ovadia, S., Blumenthal, D.J.: Analysis of an Edge Router for span-constrained Optical Burst Switched (OBS) Networks. IEEE/OSA Journal of Light-wave Technology 22(11), 2693–2705 (2004)

86. Rodelgo-Lacruz, M., Pavón-Mariño, P., González-Castaño, F.J., García-Haro, J., López-Bravo, C., Veiga-Gontán, J.: Enhanced Parallel Iterative Schedulers for IBWR Optical Packet Switches. In: Tomkos, I., Neri, F., Solé Pareta, J., Masip Bruin, X., Sánchez López, S. (eds.) ONDM 2007. LNCS, vol. 4534, pp. 289–298. Springer, Heidelberg (2007)

87. Rosberg, Z., Vu, H.L., Zukerman, M., White, J.: Performance Analyses of Optical Burst Switching Networks. IEEE Journal on Selected Areas in Communications 21(7), 1187–1197 (2003)

88. Ryu, B., Lowen, S.: Fractal Traffic Models for Internet Simulation. In: 5th IEEE International Symposium on Computer Communications (ISCC), Antibes, Juan les Pins, France, pp. 200–206 (2000)

89. Scharf, J., Kimsas, A., Köhn, M., Hu, G.: OBS vs. OpMiGua – A Comparative Performance Evaluation. In: Proceedings of the 9th International Conference on Transparent Optical Networks (ICTON 2007), Rome, pp. 294–298 (2007)

90. Shihada, B., Ho, P.-H., Zhang, Q.: A Novel False Congestion Detection Scheme for TCP over OBS Networks. In: IEEE Global Telecommunications Conference (Globecom), Washington, DC, USA, pp. 2428–2433 (2007)

91. Stoev, S., Taqqu, M.S., Park, C., Marron, J.S.: On the Wavelet Spectrum Diagnostic for Hurst Parameter Estimation in the Analysis of Internet Traffic. Computer Networks 48, 423–445 (2005)

92. Tan, S.K., Mohan, G., Chua, K.C.: Link Scheduling State Information based on Set Management for Fairness Improvement in WDM Optical Burst Switching Networks. Computer Networks 45(6), 819–834 (2004)

93. Tanenbaum, A.S.: Computer Networks, 2nd edn. Prentice-Hall, Englewood Cliffs (1988)

94. Teng, J., Rouskas, G.: Traffic Engineering Approach to Path Selection in Optical Burst Switching Networks. OSA Journal on Optical Networking 4(11), 759–777 (2005)

95. Thodime, G., Vokkarane, V., Jue, J.: Dynamic Congestion-based Load Balanced Routing in Optical Burst-Switched Networks. In: IEEE Global Communications Conference (Globeom), San Francisco, CA, USA (2003)

96. Turner, J.S.: Terabit Burst Switching. Journal of High Speed Networks 8(1), 3–16 (1999)

97. Van Breusegem, E., Cheyns, J., Colle, D., Pickavet, M., Demeester, P.: Overspill Routing in Optical Networks: a new Architecture for Future-proof IP over WDM Networks. In: Optical Networking and Communications (OptiComm), Dallas, TX, USA, pp. 226–236 (2003)

98. Vokkarane, V.M., Haridoss, K., Jue, J.P.: Threshold-based Burst Assembly Policies for QoS Support in Optical Burst-Switched Networks. In: Optical Networking and Communications (OptiComm), Boston, MA, USA, pp. 125–136 (2002)

99. Vokkarane, V.M., Jue, J.P.: Prioritized Burst Segmentation and Composite Burst Assembly Techniques for QoS Support in Optical Burst-Switched Networks. IEEE Journal on Selected Areas in Communications 21(7), 1198–1209 (2003)

100. Wei, J.Y., McFarland, R.I.: Just-in-time Signalling for WDM Optical Burst Switching Networks. IEEE Journal of Lightwave Technology 18(12), 2019–2037 (2000)

101. Xiong, Y., Vandenhoute, M., Cankaya, H.: Control Architecture in Optical Burst Switched WDM Networks. IEEE Journal on Selected Areas in Communications 8(10), 1838–1851 (2000)

102. Xu, J., Qiao, C., Li, J., Xu, G.: Efficient Channel Scheduling Algorithms in Optical Burst Switched Networks. In: IEEE Infocom, San Francisco, CA, USA, pp. 2268–2278 (2003)

103. Xuw, F., Yoo, B.S.J.: Self-Similar Traffic Shaping at the Edge Router in Optical Packet Switched Network. In: IEEE International Conference on Communications (ICC), New York, NY, USA, pp. 2449–2453 (2002)

104. Yang, L., Rouskas, G.N.: Adaptive Path Selection in Optical Burst Switched Networks. IEEE/OSA Journal of Lightwave Technology 24(8), 3002–3011 (2006)

105. Yoo, M., Qiao, C.: Just-enough-time (JET): A High Speed Protocol for Bursty Traffic in Optical Networks. In: IEEE/LEOS Summer Topical Meetings, Montreal, Canada, pp. 26–27 (1997)

106. Yoo, M., Qiao, C., Dixit, S.: Optical Burst Switching for Service Differentiation in the Next-Generation Optical Internet. IEEE Communications Magazine 39(2), 98–104 (2001)

107. Yu, X., Li, J., Cao, X., Chen, Y., Qiao, C.: Traffic Statistics and Performance Evaluation in Optical Burst Switched Networks. OSA/IEEE Journal of Lightwave Technology 22(12), 2722–2738 (2004)

108. Yu, X., Qiao, C., Liu, Y., Towsley, D.: Performance Evaluation of TCP Implementations in OBS Networks. Tech. Rep. 2003-13, CSE Dept., SUNY, Buffalo (2003)

109. Yu, X., Qiao, C., Liu, Y.: TCP Implementations and False Timeout Detection in OBS Networks. In: IEEE Infocom 2004, Hong Kong, P.R. China, pp. 774–784 (2004)

110. Zalesky, A., Vu, H., Rosberg, Z., Wong, E., Zukerman, M.: Modelling and Performance Evaluation of Optical Burst Switched Networks with Deflection Routing and Wavelength Reservation. In: IEEE Infocom, Hong Kong, P.R. China, pp. 1864–1871 (2004)

111. Zhang, J., et al.: Explicit Routing for Traffic Engineering in Labeled Optical Burst-Switched WDM Networks. In: IEEE International Conference on Computational Science (ICCS), Krakow, Poland (2004)

112. Zhang, Q., Vokkarane, V.M., Jue, J.P.: Biao Chen: Absolute QoS Differentiation in Optical Burst-Switched Networks. IEEE Journal on Selected Areas in Communications 22(9), 1781–1795 (2004)

113. Zhang, Q., Vokkarane, V.M., Wang, Y., Jue, J.P.: Analysis of TCP over Optical Burst-Switched Networks with Burst Retransmission. In: IEEE Global Telecommunications Conference (Globecom), St. Louis, MI (2005)

114. Zhang, Y., Li, L., Wang, S.: TCP over OBS: Impact of Consecutive Multiple Packet Losses and Improvements. Photonic Network Communications (in print)

115. Zhang, Y., Wang, S., Li, L.: B-Reno: A New TCP Implementation Designed for TCP over OBS Networks. In: Future Generation Communication and Networking (FGCN), Jeju Island, Korea, pp. 185–190 (2007)
116. Zhong, W.D., Tucker, R.S.: Wavelength Routing-based Photonic Packet Buffers and their Applications in Photonic Packet Switching Systems. Journal of Lightwave Technology 16(10), 1737–1745 (1998)

8   Multi-layer Traffic Engineering (MTE) in Grooming Enabled ASON/GMPLS Networks

M. Köhn (chapter editor), W. Colitti, P. Gurzì, A. Nowé, and K. Steenhaut

Abstract. The Automatically Switched Optical Networks (ASONs) and the Generalized Multi Protocol Label Switching (GMPLS) control plane are envisaged to play an important role in the next generation Internet. They provide optical networks with intelligence and automation and they enable the Multi-layer Traffic Engineering (MTE) paradigm. In this chapter we review different aspects of Multi-layer Traffic Engineering that have been investigated in COST action 291. We introduce integrated routing schemes and show fundamental performance metrics. Furthermore, we discuss mechanisms for multi-layer traffic engineering beyond routing that either improve the overall performance of the network or increase the fairness among different users. Finally, we summarize results of an performance evaluation that shows the impact of fundamental traffic and network characteristics on the performance.

8.1   Introduction

The increasing usage of the Internet around the world has been leading to a massive growth in traffic volume and dynamics to be transported by network backbone. To cope with this, highly flexible and dynamic IP-over-WDM solutions were envisioned a few years ago that provide virtually unlimited bandwidth and support dynamics not only in the electrical layer but also in the optical layer. During the dot-com-bubble, the need for circuit-switched connections of smaller bandwidth than full wavelengths was mostly neglected and greenfield solutions with large optical cross connects seemed close to reality. However, today carriers look much more closely at new business opportunities as well as operating and capital expenditure (OPEX/CAPEX) reductions and thus an important requirement is the bandwidth provisioning at sub-wavelength granularity.

Automatically Switched Optical Networks (ASONs) are optical transport networks with dynamic connection capability [1]. Controlled by the Generalized Multi Protocol Label Switching (GMPLS) [2], an ASON assist the IP/MPLS layer in traffic engineering the network by automatically allocating capacity. The cooperation between IP/MPLS layer and ASON/GMPLS layers is called Multi-layer Traffic Engineering (MTE). The MPLS and GMPLS control plane similarity and the MTE paradigm have brought the IP and optical domains in a tighter relationship. A platform containing both IP/MPLS and ASON/GMPLS domains are envisaged to be the preferred infrastructure for the future Internet [3]. It allows an

I. Tomkos et al. (Eds.): COST 291 – Towards Digital Optical Networks, LNCS 5412, pp. 237–252, 2009.

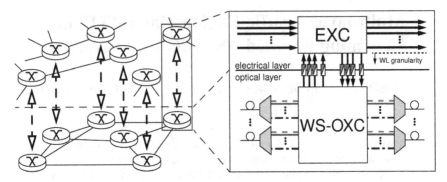

Fig. 8.1. Network and Node Architecture

operator to execute routing algorithms in a multi-layer and integrated fashion. Integrated routing takes into consideration both layers' states when accommodating traffic and consequently improves the resource optimization while meeting the user's needs.

Such multi-layer networks consist of multi-layer nodes with switches on the electrical layer as well as cross connects on the optical WDM layer. The principal network architecture we use throughout this work is shown in Fig. 8.1. It consists of nodes that are interconnected by fibres or fibre trunks. In the nodes, the electrical layer consists of a non-blocking electrical switch (EXC). This is either a packet switch or a TDM switch with switching capabilities for all granularities. In the optical layer, after demultiplexing the wavelengths are either connected directly to the output multiplexer or via a transponder to the EXC. For this, a wavelength selective non-blocking optical cross connect (OXC) operating on wavelength granularity is used which is capable of setting up and tearing down on demand transparent lightpaths from any source node to any destination node through the network. With this, the virtual topology on the electrical layer, i.e., the capacity connecting two nodes can be adapted during operation by adding and removing links. The EXC and OXC are interconnected by a limited number of tunable transponders at a given line rate.

In the following we address the different aspects of traffic engineering in such networks. In Section 8.2 we present a new integrated routing scheme and evaluate its performance. In Section 8.3, we introduce improvements to traffic engineering schemes which have been investigated during COST Action 291. Finally, we discuss application scenarios for such networks with respect to key traffic and network characteristics in Section 8.4.

8.2   Routing and Grooming in Multi-layer Networks

In literature, several algorithms for routing and grooming have been proposed in the last years, e.g. Necker et al. [4], Zhu et al. [5], Kodialam et al. [6], Zhu et al. [7].

They all try to minimize the blocking probability but do not explicitly consider the current load situation of the network.

In the following, we first review the two basic algorithms commonly used in literature for reference. We then introduce our new algorithm and show its performance.

8.2.1 Basic Schemes

Integrated MTE routing can accommodate new services either on the existing virtual topology or on newly established lightpaths [8]. The two basic policies are:

- **Optical Layer First (OLF) policy.** The system first attempts to establish a new direct lightpath between source and destination nodes. If a new lightpath cannot be established due to physical resource shortage (i.e. unavailability of wavelengths and ports) the system tries to aggregate the traffic over the existing virtual topology. The OLF policy tries to exploit the physical resources as much as possible. The advantage of this policy is a lower blocking probability and a higher QoS in terms of packet loss and end-to-end delay. In fact, the communication often occurs on direct end-to-end lightpaths without traversing intermediate electronic IP routers. Such components, in fact, are considered to be the bottleneck of the communication as a consequence of the optical-electronic-optical signal conversion and queuing delay. Nevertheless, the OLF policy likely installs a high quantity of lightpaths which are not optimally used and therefore it results in a non optimal resource utilization level.
- **IP Layer First (ILF) policy.** The system first attempts to groom an LSP request over the existing virtual topology. If there is not sufficient bandwidth to find a path, a new lightpath establishment is triggered. Unlike the OLF policy, the ILF policy tries to exploit the available capacity as much as possible by aggregating traffic on the existing virtual topology. The advantage for an operator is the more optimized use of the capacity at the expenses of a higher blocking probability. Since the communication occurs over paths consisting of more than one lightpath, this policy results in a higher packet loss and end-to-end delay which can degrade the QoS.

8.2.2 Adaptive Integrated Multi-layer Routing

The comparison between the two multi-layer routing policies illustrated in Section 8.2.1 highlights that ILF is a policy optimal from the operator's perspective while OLF is the one that better meets the user's needs. This implies that a strategy that uses both routing policies would achieve a fair compromise between resource optimization and QoS. In this section we describe a policy for adaptive integrated routing in ASON/GMPLS networks. The policy decides the multi-layer routing policy according to the load condition. The scheme is called Network State Dependent (NSD) and has been proposed in [8].

8.2.2.1 Network State Dependent (NSD) Strategy

The NSD strategy is a combination of the OLF and ILF policies. More precisely, it is based on the idea that the OLF policy should be used under high load conditions on the IP/MPLS layer (i.e. virtual topology) while the ILF policy should be preferred when the load condition is high on the optical layer (i.e. physical topology). This means that the routing decision is based on the resource utilization level experienced on both the virtual and physical topologies. In fact, a high resource utilization level on the virtual topology is an indication of a possible congestion due to a high number of LSPs aggregated on the lightpaths. In this case the system should facilitate the installation of new lightpaths in order to add additional capacity to the IP/MPLS layer. While, a high resource utilization level on the physical topology indicates that the system has established a high number of lightpaths and therefore is running out of physical resources. In this case, LSP grooming should be favoured in order to better optimize the capacity on the virtual topology.

To measure the resource utilization on the physical and virtual topologies, we introduce two indexes described in subparagraphs 8.2.2.1.1 and 8.2.2.1.2, respectively. The strategy's operations during the routing decision are then summarized in subparagraphs 8.2.2.1.3.

Resource Usage Index on the Physical Topology Ipt(tk)

This index is a measure of the optical layer's resources being used at a certain instant t_k in which a new LSP request arrives at a node. The quantities contributing to the index $I_{pt}(t_k)$ are the total number of busy ports $P_{busy}(t_k)$ and the total number of busy wavelengths $W_{busy}(t_k)$ at the instant t_k. These are the parameters giving an indication on the number of lightpaths being installed in the system. In fact, a node having no available ports is unable to be source or destination node of a lightpath, while a fibre without free wavelengths is unable to establish new lightpaths. The formula calculating the physical resource usage index is reported in (6.1).

$$I_{PT}(t_k) = \frac{fP_{busy}(t_k) + f^{-1}W_{busy}(t_k)}{fNP + f^{-1}M_{fibres}W} \tag{6.1}$$

In (1), N is the total number of nodes, P the total number of ports per node, M_{fibres} the total number of fibres, W the number of wavelengths per fibre and f is the ratio between W and P. The presence of the factor f is due to the influence that the minimum number between P and W has on the number of lightpaths that can be installed. More precisely, when $P>W$ the maximum number of lightpaths is limited by the number of wavelengths and therefore less importance should be given to the number of ports. When $P<W$ the maximum number of lightpaths is limited by the number of ports and therefore we give less importance to the number of wavelengths.

Resource Usage Index on the Virtual Topology IVT(tk)

$I_{VT}(t_k)$ indicates the logical layer's resources being used at a certain instant t_k in which a new LSP request arrives at a node. The quantity contributing to this index is the total capacity being used among the already established lightpaths, $C_{used}(t_k)$. In fact, by calculating the resource usage index on the virtual topology the system has an indication about the level of exploitation of the virtual topology. The formula representing the $I_{VT}(t_k)$ index is reported in (6.2).

$$I_{VT}(t_k) = \frac{C_{used}(t_k)}{M_{lightpaths}(t_k)C_{tot}} \qquad (6.2)$$

In (1), $M_{lightpaths}(t_k)$ indicates the total number of lightpaths being established at the instant t_k and C_{tot} is the amount of lightpath capacity.

Routing Decision Phase

The routing decision phase is summarized in Fig. 8.2.

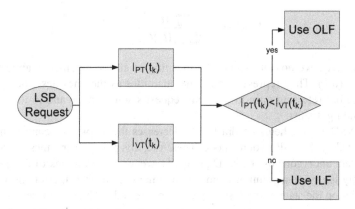

Fig. 8.2. Routing decision phase in the NSD strategy.

As illustrated in Fig. 8.2, when a node receives a new LSP request at an instant t_k, it evaluates and compares the two resource usage indexes. The result of the comparison decides the suitable policy to execute. If the resource usage index related to the physical topology is the higher one, the system accommodates the LSP request by using the ILF policy. If the resource usage index related to the virtual topology is the highest one, the system chooses the OLF policy to route the request. The aim of the NSD strategy is to balance the resource utilization on the optical and virtual topologies in order to avoid congestion. However, the proposed strategy requires every node to have full knowledge about the network state when accommodating a connection. This means that every node has to flood LSAs to inform the other nodes about the number of its available ports, the number of available wavelengths on its outgoing fibres and the used capacity on its outgoing lightpaths. Even though the OSPF-TE protocol guarantees that such information is advertised over the entire network, there can be scalability problems in wide network topologies. To overcome this problem, we propose a heuristic that is based on the NSD policy but requires a lower amount of control information exchanged. The proposed heuristic is described in the following section.

8.2.2.2 NSD_source Heuristic

In the NSD_source heuristic the routing decision is based only on the resource usage experienced by the LSP request's source node. This means that the node accommodating the request calculates the two resource usage indexes taking into consideration only its number of available ports, the number of available wavelengths on its outgoing fibres and the used capacity on its outgoing lightpaths. Therefore, (6.1) and (6.2) are replaced by the following equations, respectively:

$$I_{PT}(t_k) = \frac{fP_{busy}^{source}(t_k) + f^{-1}W_{busy}^{source}(t_k)}{fP^{source} + f^{-1}M_{fibers}^{source}W} \qquad (6.3)$$

$$I_{VT}(t_k) = \frac{C_{used}^{source}(t_k)}{M_{lightpaths}^{source}(t_k)C_{tot}} \qquad (6.4)$$

For simplicity, we do not explain the meaning of all the parameters appearing in (6.3) and (6.4). They represent the same quantities as the ones described in (6.1) and (6.2) but referring only to the LSP request's source node and to its outgoing fibres and lightpaths.

The NSD_source heuristic drastically decreases the amount of control information needed by the LSP's source node. Therefore, this policy overcomes the scalability problem undergone by the NSD policy. However, the drawback of this policy is that likely the routing solution is not the optimal one. In fact, the routing decision only relies on the state of the LSP request's source node, which is not representative of the complete network state. Since the amount of control information needed strictly depends on the network topology, the operator should find a fair tradeoff and balance the quantity of control flooding needed with the quality of the solution.

8.2.3   Simulation Study

This section discusses our simulation study. Information on simulation settings and network topology can be found in [8]. The performance of the proposed strategy has been evaluated and compared with the performance obtained for the policies OLF and ILF and also with a *random* strategy. When accommodating a request with the random strategy, the source node of the LSP request randomly selects either the OLF or the ILF policy as routing policy.

The performance has been evaluated in terms of blocking probability and resource usage, presented in subparagraphs 8.2.3.1 and 8.2.3.2, respectively.

8.2.3.1 Blocking Probability

The blocking probability is defined as the ratio between the number of rejected LSP requests and the total number of generated LSP requests.

The blocking probability is illustrated in Fig. 8.3.

As expected, the ILF policy results in a blocking probability significantly higher than the one obtained for the OLF policy. The gap between the two related curves reaches values of 5%. This is due to the fact that the OLF policy establishes a high amount of lightpaths and therefore provides the virtual topology with a huge amount of capacity. Moreover, the physical resources do not saturate as a consequence of the high number of wavelengths and ports used. This has been necessary to use high traffic loads and to test the network under overload conditions.

The blocking probability experienced by the system when using the proposed NSD strategy is almost similar to the one measured for the OLF policy. For traffic loads between 180 and 220 Erlangs it is slightly higher, while for loads higher than 220 Erlangs it becomes slightly lower. This is a significant result if we also consider the improvement that the NSD policy obtains in terms of resource usage (see paragraph 8.2.3.2): the NSD policy combines the advantages of both ILF and OLF policy. In fact, it achieves the blocking probability obtained by the OLF policy while balancing the resource usage experienced in the physical and virtual topology.

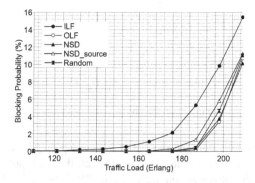

Fig. 8.3. Blocking probability against traffic load.

As anticipated in paragraph 8.2.2.2, the heuristic NSD_source reduces the amount of control information needed at the expenses of a higher blocking probability with respect to the NSD policy. In Fig. 8.3, the gap between the two curves reaches the value of more than 2% for traffic loads of about 200 Erlangs. This is the proof that the system is not always able to find the best routing solution and therefore the resources are not used in an optimal way. Consequently, a higher number of LSP requests needs to be blocked. This is confirmed by the blocking probability experienced by the random policy. It remains in the middle between the blocking probabilities measured for the NSD and the NSD_source policies. This means that the random choice between the ILF and OLF policies achieves a use of the resources which is better optimized with respect to the policy that uses a limited amount of information (NSD_source). It is also interesting to notice that until a traffic load of about 190 Erlangs, the random policy and the NSD policy have the same values of blocking probability. This suggests that the random policy could be a convenient solution for an operator, but not under relatively high load conditions. In more loaded networks, in fact, it is more convenient to have more control on the use of the resources. This is only achieved by the NSD strategy.

8.2.3.2 Resource Usage

The resource usage has been evaluated by measuring the average resource usage indexes on the physical and virtual topologies, introduced in paragraphs 8.2.2.1.1 and 8.2.2.1.2, respectively. To further demonstrate the benefits of the proposed strategy, we also report the average number of established lightpaths and the average number of used wavelengths.

The resource usage indexes on the physical and virtual topologies are reported in Fig. 8.4 and 8.5, respectively.

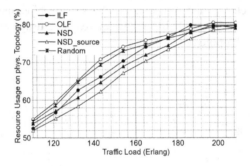

Fig. 8.4. Resource usage on the physical topology against traffic load.

Fig. 8.4 and 8.5 demonstrate that the use of only one policy between OLF and ILF is not a convenient choice for an operator. In fact, the OLF policy has a high resource usage on the physical topology (about 75% for traffic loads of 160 Erlangs), as a consequence of the high amount of established lightpaths; while the ILF policy results in a high amount of used resources on the virtual topology (about 85% for traffic loads of 160 Erlangs), as a consequence of the higher quantity of LSPs multiplexed on the same lightpaths. A high resource usage can bring the system in an overloading state and therefore to an increase of the blocking probability when the traffic increases.

On the virtual topology (Fig. 8.5), the proposed NSD strategy experiences a resource usage which is higher than the one reported for the OLF policy but significantly lower than the one experienced the ILF policy. On the physical topology (Fig. 8.4), the NSD strategy achieves a resource usage which is lower than both ILF and OLF policies, under any traffic load condition. This reinforces our thesis

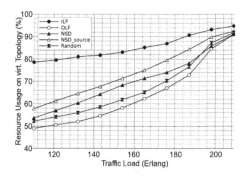

Fig. 8.5. Resource usage on the virtual topology against traffic load.

that the proposed hybrid policies improve the network performance with respect to a pure ILF or OLF based network.

The NSD_source obtains a slightly lower resource usage on the physical topology with respect to the one measured for the NSD strategy (Fig. 8.4); while the resource usage is slightly higher on the virtual topology (Fig. 8.5). In both cases the difference between the two curves is limited between about 1% and 2%. It is straightforward that if we only look at the resource usage, the NSD_source heuristic represents a better choice compared to the NSD strategy. In fact, the two policies achieve similar values of the resource usage but the NSD_source heuristic needs a significantly lower quantity of control information. However, the blocking probability of the NSD_source policy shows values that are about 2% higher than the ones measured for the NSD strategy. This can represent a rather significant difference and therefore the operator needs to find a more convenient compromise between control information needed and blocking probability when he chooses an information limited based strategy like NSD_source.

Fig. 8.4 also shows the drawbacks of the random policy. It undergoes a resource usage on the physical topology which has similar values as the one observed for the OLF policy. Unlike the NSD strategy, the random policy follows the low blocking probability experienced by the OLF policy but does not provide any improvements in terms of resource usage. As a consequence, the random policy does not represent a convenient choice for an operator.

To reinforce the demonstration of the benefits of the proposed NSD strategy, we also report the average number of installed lightpaths and of used wavelengths. They are illustrated in Fig. 8.6 and 8.7, respectively.

In line with the resource usage reported in Fig. 8.4 and 8.5, we can observe in Fig. 8.6 and 8.7 that the NSD and NSD_source policies result in a significantly lower number of established lightpaths and of used wavelengths. It is straightforward that the ILF policy should not be considered in this analysis due to its unacceptably high blocking probability. The difference among the curves is more visible in the region where the blocking probability is very low (for traffic loads between 120 and 160 Erlangs). This is indeed the region of higher interest for an operator because he can accommodate the highest amount of traffic.

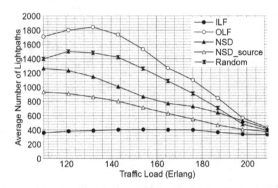

Fig. 8.6. Average number of established lightpaths against traffic load.

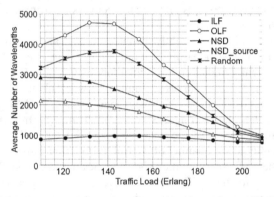

Fig. 8.7. Average number of used wavelengths against traffic load.

8.3 Improvements for Multi-layer Routing and Grooming Schemes

Routing and grooming is only one part of traffic engineering in a multi-layer network. Beyond solely controlling the traffic in a multi-layer network by routing and grooming schemes, extensions are feasible, that either try to further reduce the blocking probability or to improve other important characteristics of the network. Examples of such extensions are rerouting of established connections if resources are needed to setup new connections (e.g. [9]).

In COST action 291, we worked on two types of extensions and investigated them in depth. First, we developed a new approach to rearrange connections in order to optimize the usage of network resources and to reduce the blocking probability. Second, we generalized an existing connection admission control (CAC) scheme that improves fairness among connections. In the following, we introduce both work shortly and explain the most important results.

8.3.1 Online Optimization at Connection Teardown

Analyzing the connection setup processes in a dynamic network, one can see that in general the selection of a path for a new connection can only be optimal for the instant of connection setup. This is due to the fact that at this instant of time neither the holding time of the connection itself nor other connections which arrive during the holding time are known. Accordingly, all algorithms try to find a path for a new connection based on heuristics.

In a low loaded network these algorithms usually select a path which is optimal for the entire connection holding time. The reason is the small number of e.g. detours. Thus the amount of resources occupied by a certain connection is usually always minimal. Furthermore, due to the low load the probability that a new connection is blocked due to a detour of another connection is low.

In contrast, in a medium or highly loaded network often detours are needed leading to a waste of resources. Especially in high load situations, high blocking probabilities can be observed due to positive feedback [10].

This problem tries to solve our extension which is applicable to any routing and grooming scheme. We reduce the occupied capacity by rearranging detoured connections onto shorter paths. We rearrange connections always if capacity is freed, i.e., if a connection is torn down. The algorithm works as follows:

As soon as a connection is torn down, a set of established connection candidates is created. This set contains candidates, which could use the resources freed by the terminated connection. Among these connections, those are selected for rearrangements which most improve the network state. For this, several questions must be solved, e.g.:

- A metric must be defined to measure the network state. This metric must increase if it can be expected that connections arriving in future see a reduced blocking probability.
- A scheme must be defined to select the set of candidates. This scheme should select with low computational complexity candidates based on simple rules.

Basic results of an investigation are published in [11]. In this publication, we use candidate set consisting of all connections originating and terminating at the terminating connection's source and destination node, respectively. Furthermore, the terminating connection and all candidate connections must occupy the same bandwidth. For measuring the network state, we compare two metrics. First, we use the total grooming capacity occupied for grooming in electrical nodes. Second, we use the total link capacity occupied in the electrical layer.

In general, the algorithm reduced the blocking probability significant while only up to 10% of the connections have been touched. Nevertheless, it has to be mentioned that the routing and grooming scheme and the selection scheme and the metric of the rearrangement algorithm have to be well aligned in order to maximize the performance gain. Detailed results will be published.

8.3.2 Admission Control for Improving Fairness

While the commonly known routing and grooming schemes are usually designed for minimizing the overall blocking probability or maximizing resource efficiency, fairness is not considered. The term fairness reflects that with respect to a certain service attribute, e.g. the required bandwidth, all independent connection requests having the same requirement will experience the same service quality, e.g., the same blocking probability. Furthermore, different service requirements lead to a well defined differentiation in the service quality. Since from both perspectives— the user's as well as the operator's perspective—fairness is an important aspect, an additional mechanism has to be provided in order to ensure fair handling of connections. This is usually at the cost of penalizing the overall network performance. So such a mechanism has to be carefully designed and optimized.

In general, fairness applies to different aspects, e.g. distance, or required bandwidth. For bandwidth fairness, in [12] a CAC algorithm has been presented for TDM/WDM networks. This algorithm provides fairness among connections of different bandwidth granularities. The authors define a network to be fair if the blocking probability of a number of connections is independent from the granularities requested. This means the blocking probability of N_1 connections of a bandwidth of B_1 is equal to N_2 connections of bandwidth B_2 as long as $N_1 \cdot B_1 = N_2 \cdot B_2$. It works as follows:

If for a connection request a path with sufficient free capacity is available, the CAC has to decide whether the connection request is rejected for fairness reasons or can be accepted. For this, it classifies all arriving connection requests according to their bandwidth j. For each j it monitors the actual blocking probability p_j. Furthermore, it derives the so called blocking probability per unit line speed $q_j = 1 - \sqrt[j]{1 - p_j}$. This inherently assumes that the small capacity connections are independent of each other. It further derives from the actual overall blocking probability the so called target blocking probability P_j.

The CAC accepts a connection request of bandwidth j if the blocking probability of this bandwidth, i.e. p_j, is greater than its target blocking probability P_j. If and only if p_j is smaller than P, it randomly rejects the connection request with a rejection probability $Q = 1 - p_j / P$.

Based on this description, it can be seen that this algorithm can be generalized to any class-based systems. For this transfer, we identified two functions that have to be adapted accordingly: First, a scheme for classifying connection requests to a certain class and, second, a scheme for determining the target blocking probabilities per class. In the following, we explain these functions for the case of distance fairness.

In case of distance fairness, we classify the connections according to the distance between the connections endpoints. In a multi-layer network this can be translated to different mathematical definitions ranging from the airline distance to the hop distance in the actual virtual topology. As simulation studies have shown that the distance dependant behaviour is usually correlated to the hop distance in the physical topology, we use the length of the shortest path between source and destination node in the optical layer.

The calculation of the target blocking probability we derive from the definition of fairness. We consider two fairness definitions.

- First, assuming a nation-wide network, the blocking probability shall be equal for all connections independent of the distance. In this case, the target blocking probability P_j is the mean blocking probability P. The blocking probabilities per class are not normalized, i.e. $q_j = p_j$.
- Second, in international networks it may be required that similar to the bandwidth fairness definition N_1 connection requests with a distance of D_1 hops have the same blocking probability as N_2 connection with a distance of D_2 hops if $N_1 \cdot D_1 = N_2 \cdot D_2$. Analog to the bandwidth fairness, we normalize the blocking probabilities per unit hop length.

Beyond this generalization, we developed a new formula to calculate the rejection threshold. The threshold as proposed is in scenarios with low overall blocking probability too aggressive.

The blocked connections can be separated into two groups: First, connections that are rejected by the network as no free path is available and second, connections that are blocked by the CAC. We use the probabilities $P_{j,NW}$ and $P_{j,CAC}$ for a connection request of bandwidth j being blocked by the network or rejected by the CAC, respectively.

The target of the CAC system is to control the normalized blocking probability of a connection class, i.e., the sum of the network blocking probability and blocking probability due to rejection by the CAC, such that a given target blocking probability is reached: $P_{j,target} = P_{j,NW} + P_{j,CAC}$. With the probability $1 - P_{j,NW}$, a connection request is not blocked by the network and has to be handled by the CAC. There, it is either accepted if $p_j > P_{j,target}$ holds or with the probability Q randomly blocked. So, the blocking probability due to (random) rejects by the CAC can be calculated by $P_{j,CAC} = Q \cdot (1 - P_{j,NW})$. Using this, the required rejection probability Q can be calculated.

We investigated these new schemes in several scenarios [13]. The results of this investigation show that the algorithm improves the fairness in all scenarios at cost of overall blocking probability. This degradation depends on the method which is used to determine the rejection probability, but can be reduced to far below one order of magnitude.

8.4   Evaluation of Traffic and Network Patterns

In this section we summarize results of a detailed performance evaluation of two basic multi-layer architectures with respect to the traffic composition and network capacities.

The first network and node architectures have been introduced above. The second network and node architecture are derived from this architecture by replacing the dynamic switching OXC by a static optical layer, i.e., after demultiplexing the wavelengths are either connected directly to the output multiplexer or via a transponder to the EXC. The fixed transponders terminate the wavelengths at a given line rate. This means that in contrast to a single-layer network with only a electrical layer, wavelengths can transparently pass through the node and only those wavelengths that have to be terminated in the node are converted to the electrical domain while all the other wavelengths bypass the electrical domain without being modified and without occupying any transponders or switching resources there. However, as those optical bypasses either have to be installed in advance or reconfigured manually, the network is only dynamic in the electrical layer. Although dynamic switching of connections is also possible due to the functionalities in the electrical layer, this architecture is referenced as static.

Comparing the two multi-layer architectures, several major differences can be observed. With respect to functionality, in the dynamic architecture a new degree of freedom has been added as additional lightpaths can be established and unused lightpaths can be removed. Also, the used components vary. In the static case the wiring in the optical layer is fixed. So the wavelengths of the transponders are predetermined and thus, fixed transponders can be used. In the dynamic architecture, the wavelengths of the transponders are selected on demand and thus tunable transponders have to be used. Still, tunable transponders are more expensive than fixed transponders but prices are declining and due to e.g. more efficient stock keeping, operators are willing to pay the higher price. Another difference lies in the OXC. While in the static case either no such component or only a simple wiring panel is needed, this component is more complex in the dynamic architecture. Depending on characteristics like switching speed or number of wavelengths, OXCs are a high cost factor.

In both network and node architectures, the interconnection between the nodes as well as between the cross connects have an impact on the overall performance. Specifically, in the static scenario the number of lightpaths connecting a node pair has to be dimensioned. With this, the demands for the optical layer are known in advance and can be used for dimensioning of fibres. For this, the so called routing and wavelength assignment problem (RWA) must be solved [14]. In the dynamic scenario also the optimal number of transponders and fibres for the nodes and links has to be determined. As the lightpaths are established on demand, heuristics have to be used to calculate the optimized fibre topology. To be as general as possible while keeping the number of degrees of freedom on a reasonable level, we assume the optical layer to be infinitely large. This is reasonable as the cost of a multi-layer transport network is mostly dominated by the transponder cost [15]. With this assumption, no influences of specific RWA algorithms and specific dimensioning schemes must be considered. Compared to a scenario with limited resources in the optical layer, the difference will be rather small in the static scenario as the path of a wavelength in the optical layer has no influence on the routing in the electrical layer. In the dynamic case, decreasing the number of fibres will lead to worse results due to blocking in the optical layer if the dimensioning is too tight.

We performed simulation studies in the fictitious 16-node reference network of COST 266 [16] with 23 bidirectional links. We considered two total network capacities and two traffic mixes: a low capacity network dimensioned for a total offered traffic of 4.9 Tbps, and a high capacity network dimensioned for a total offered traffic of 49 Tbps. These networks have a mean number of 2.0 and 20.0 lightpaths per node pair, respectively.

The traffic mixes both consist of requests of 1/64 and 1/16 of a wavelength but differ with respect to the maximum bandwidth granularity. While one mix contains 20% of full wavelength connections, the highest granularity in the second mix is a quarter of a wavelength.

The results show, that the traffic mix has a significant impact on the performance of static and dynamic multi-layer networks [17]. A network with a dynamic switching optical layer can cover changes in the traffic mix very well, however, at the cost of increased complexity of required components as well as of routing. Assuming the traffic mixes used in this paper represent today's and future requirements, it can be a solution to first introduce nodes with a transparent static optical plane and when high bandwidth requests come into the network extend the nodes by a dynamic switching optical layer. However, the performance benefits can not be the only driver to bring dynamic optical cross connects into the field. Still, other value adding features e.g. restoration of entire fibres or lightpaths will be an argument.

References

1. ITU-T Rec. G.8080/Y1304, Architecture for the Automatic Switched Optical Networks (ASON) (2001)
2. Mannie, E. (ed.): Generalized multi-protocol label switching architecture. draft-ietf-ccamp-gmpls-architecture-07.txt (May 2003)
3. Oki, E., Shiomoto, K., Shimazaki, D., Imajuku, W., Takigawa, Y.: Dynamic Multi-layer Routing Schemes in GMPLS-Based IP + Optical Networks. IEEE Communication Networks 43 (2005)
4. Necker, M., Gauger, C., Bodamer, S.: A new efficient integrated routing scheme for SDH/SONET-WDM multilayer networks. In: Proceedings of the Optical Fibre Communication Conference (OFC), vol. 2, pp. 487–488 (2003)
5. Zhu, K., Mukherjee, B.: On-Line Approaches for Provisioning Connections of Different Bandwidth Granularities in WDM Mesh Networks. In: Proceedings of the Optical Fibre Communication Conference, OFC (2002)
6. Kodialam, M., Lakshman, T.V.: Integrated dynamic IP and wavelength routing in IP over WDM networks. In: Proceedings of Twentieth Annual Joint Conference of the IEEE Computer and Communications Societies (INFOCOM 2001), pp. 358–366 (2001)
7. Zhu, H., Zang, H., Zhu, K., Mukherjee, B.: A novel generic graph model for traffic grooming in heterogeneous WDM mesh networks. IEEE/ACM Transactions on Networking 11(2), 285–299 (2003)

8. Colitti, W., Gurzì, P., Steenhaut, K., Nowé, A.: Adaptive Multilayer Routing in the Next Generation GMPLS Internet. In: Second Workshop on Intelligent Networks: Adaptation, Communication and Reconfiguration, Bangalore, India, 10 January (2008)

9. Doumith, E.A., Gagnaire, M.: Traffic Routing in a Multi-layer Optical Network Considering Rerouting and Grooming Strategies. In: Proceedings of IEEE Global Telecommunications Conference (IEEE GLOBECOM '06), pp. 1–5 (2006)

10. Nakagome, Y., Mori, H.: Flexible routing in the global communication network. In: Proceedings of the 7th International Teletraffic Congress, ITC 7 (1973)

11. Köhn, M.: A new efficient online-optimization approach for SDH/SONET-WDM multi layer networks. In: Proceedings of the Optical Fibre Communication Conference, OFC (2006)

12. Thiagarajan, S., Somani, A.K.: Capacity Fairness of WDM Networks with Grooming Capabilities. Optical Networks Magazine 2, 24–31 (2001)

13. Köhn, M.: Improving fairness in multi service multi layer networks. In: Proceedings of the 7th International Conference on Transparent Optical Networks (ICTON), pp. 53–56 (2005)

14. Gargano, L., Vaccaro, U.: Routing in all-optical networks: Algorithmic and graphtheoretic problems. In: Althöfer, I., et al. (eds.) Numbers, Information and Complexity, pp. 555–578. Kluwer Academic Publishers, Boston (2000)

15. Bodamer, S., Späth, J., Glingener, C.: An efficient method to estimate transponder count in multi-layer transport networks. In: Proceedings of IEEE Global Telecommunications Conference (GLOBECOM '04), pp. 1780–1785 (2004)

16. de Maesschalck, S., Nederlof, L., Vaughn, M., Wagner, R.E.: Traffic Studies for Fast Optical Switching in an Intelligent Optical Network. Photonic Network Communications 8(3), 285–307 (2004)

17. Köhn, M.: Comparison of SDH/SONET-WDM Multi-layer Networks with Static and Dynamic Optical Plane. Optical Switching and Networking 2(4), 249–259 (2005)

9 Network Resilience in Future Optical Networks

L. Wosinska (chapter editor), D. Colle, P. Demeester, K. Katrinis,
M. Lackovic, O. Lapcevic, I. Lievens, G. Markidis, B. Mikac, M. Pickavet,
B. Puype, N. Skorin-Kapov, D. Staessens, and A. Tzanakaki

Abstract. Network resilience is an issue of deep concern to network operators being eager to deploy high-capacity fibre networks, since a single failure in the network could result in significant losses of revenue. The importance of network reliability will keep pace with the steadily increasing
network capacity. For very-high-capacity future optical networks, carrying
multitudes of 10 Gbit/s channels per fibre strand, a failure of optical connection will interrupt a vast amount of services running on-line, making the
connection availability a factor of great significance. Therefore the ultra-
high capacity future optical networks will face a challenge of providing
very efficient and fast survivability mechanisms. In this chapter we review
the terminology and basic resilience techniques along with the results of research work on optical network survivability performed in the frame of
COST291 cooperation. Our research work was focused on reliability performance improvement and on recovery in multilayer optical networks.

9.1 Introduction

Backbone networks carrying Internet (IP) traffic, possibly enhanced with Multi-
Protocol Label Switching (MPLS) functionality, are supported by Optical Transport Networks (OTNs) that provide transmission links between IP routers. By applying Wavelength Division Multiplexing (WDM), OTNs are capable of carrying
many independent channels, carried on different wavelengths, over one single optical fibre. This allows the network to transport huge amounts of data and provide
communication services that play a very important role in many of our daily social
and economical activities. For instance, strategic corporate functions show an increasing dependence on communication services.

Communication networks can be subject to both unintentional failures, caused
by natural disasters, wear out and overload, software bugs, human errors, etc and
intentional interruptions due to maintenance. As core communication networks
also play a vital military role, key telecommunication nodes were favoured targets
during the Gulf War, and could become a likely target for terrorist activity. For
business customers, disruption of communication can suspend critical operations,
which may cause a significant loss of revenue, to be reclaimed from the telecommu-

I. Tomkos et al. (Eds.): COST 291 – Towards Digital Optical Networks, LNCS 5412, pp. 253–284, 2009.

nications provider. In fact, availability agreements now form an important component of Service Level Agreements (SLAs) between providers and customers.

In the cutthroat world of modern telecommunications, network operators need a reliable and maintainable network in order to hold a leading edge over the competition. Fast and scalable network recovery techniques are of paramount importance in order to provide the increasingly stringent levels of reliability these network operators are demanding for their future networks.

A multilayer transport network typically consists of a stack of single-layer networks. There is usually a client-server relationship between the adjacent layers of this stack. Each of these network layers may have its own (single-layer) recovery schemes. As will be shown in the following sections, it is important to be able to combine recovery schemes in several layers in order to cope with the variety of possible failures in an efficient way and to benefit from the advantages of the schemes in each layer. It is worth mentioning that implementing a multilayer recovery strategy does not mean that all the recovery mechanisms will be used at every layer.

As Internet traffic is continuously shifting and changing in volume over time, for instance due to diurnal traffic fluctuation and overall traffic growth, there is ongoing research towards creating optical networks with the flexibility to reconfigure transmission according to traffic demands. This requires the possibility to set up and tear down OTN layer connections that implement logical links in the higher network layer in real-time, which has led to the concept of intelligent optical networks (IONs). In addition to allowing the network to adapt to changing traffic demands, this flexibility in setting up lightpaths on demand turns restoration into a viable recovery option.

In the following sections we introduce some basic terminology, describe the common resilience techniques, and address the issues related to improvement of the network reliability performance. We also discuss three generic approaches for providing recovery in multilayer networks (more specifically in IP-over-OTN networks) namely single-layer recovery schemes in multilayer networks, static multilayer recovery schemes and the dynamic multilayer recovery strategies. Some quantitative studies and comparisons between the different methods will reveal the advantages and disadvantages of each approach.

9.2   Terminology

In this section we provide the set of definitions used in this chapter [1].

Fault or **failure** represents a catastrophic event that causes the disruption of communications. For instance a failure might be a fibre disruption, an interface failure, or a device failure.

Resilience, survivability and **fault-tolerance** are synonyms and represent network's ability to continue to provide service in the presence of failures. The general aim of **resilience** is to make network failures transparent to users.

We define **protection** as the use of pre-assigned capacity to replace a failed transport entity. It can be very fast (typically, traffic should not be interrupted for more than 60 ms [2] in the event of a failure for SONET/SDH networks). Three fundamental types of protection mechanisms are: 1+1, 1:1 and 1:N protection (so called shared protection).

Link protection is the mechanism that automatically switches the traffic on the failed fibre link to a pre-assigned protection path between the nodes adjacent to the failed link.

Path protection is the mechanism that automatically switches the traffic from the failed path through a pre-assigned protection path. The backup path should be link or node disjoint with the primary path.

Restoration is based on the use of capacity available between nodes to replace a failed transport entity. Restoration is also called **network protection** and is based on alternative routing.

Network integrity is defined as the ability of a network to provide the desired QoS to the services, not only in normal (i.e., failure-free) network conditions, but also when network congestion or network failure occurs.

Physical topology is the real network, composed of optical links and photonic nodes.

Virtual topology (or **logical topology**) is the view of the network available to the higher layer switches.

Network survivability is a network's ability to continue to provide service in the presence of failures.

Instantaneous availability $A(t)$ is the probability that an item is in up state at a given instant of time, t [3].

Instantaneous unavailability $U(t)$ is the probability that an item is in down state at a given instant of time, t. $U(t)=1-A(t)$ [3].

(Asymptotic) availability or steady-state availability A is the limit, if this exists, of instantaneous availability when the time tends to infinity. Under certain conditions, for instance constant failure rate and constant repair rate, the asymptotic availability can be expressed as [3]: $A= MUT/(MUT+MDT)$ or $A=MTTF/(MTTF+MTTR)$, where MDT = the mean down time, MUT = the mean up time, $MTTF$= expectation of time to failure and $MTTR$=expectation of time to restoration (repair).

(Asymptotic) unavailability or **steady-state unavailability** is equal to: $U=1-A$.

Note: if not specified in this paper the term (un)availability means asymptotic or steady-state (un)availability.

Reliability is the probability that an item can perform a required function under stated conditions for a given time interval.

Connection availability (asymptotic or steady-state availability of the connection).is the probability that there is at least one optical path available between the considered network ports.

Network port (or **ingress/egress point**) is a point on the node, which gives access to the photonic network. Connected to the port may be a higher level switch, e.g. an SDH cross-connect, an ATM node, or an IP router.

Optical link is a bidirectional physical connection between two nodes. Optical links consist of fibres and fibre amplifiers.

Optical path (or lightpath) is a bidirectional wavelength channel between two network ports which may consist of one or more optical links and optical network nodes.

9.3   Basic Resilience Techniques and Failure Management

Network survivability schemes can be classified in two forms, i.e., protection and restoration.

Protection refers to pre-provisioned failure recovery. Protection schemes are typically fast and recovery time below 50 ms can be offered. Varying protection levels can be provisioned, ranging from 1+1, 1:1, to 1:N depending on user demands and budgetary constraints. If the network is protected on the 1+1 or 1:1 basis the single failure network survivability is 100%. Protection, however, can be quite expensive due to need of duplicating network equipment.

Restoration refers to rerouting the traffic around the failure if there are resources available. The alternative route is computed after occurrence of the failure searching for the available resources and therefore restoration uses to take longer time than protection where the spare resources are pre-computed and reserved. If no network resources are available upon failure the restoration is not possible. The average number of blocked calls over time shows the connection survivability.

Failure management deals with the countermeasures taken to compensate for vulnerabilities in the network, which include prevention, detection and reaction mechanisms [4]. Prevention schemes can be realized through hardware (e.g. strengthening and/or alarming the fibre), transmission schemes (e.g. coding schemes), or network architecture and protocols.

Detection mechanisms are responsible for identifying and diagnosing failures, locating the source and generating the appropriate alarms or notification messages to ensure successful reaction. Due to the constraints inherent in optical performance monitoring, these tasks are more difficult than in electrical networks. Methods to locate and recover from various component faults are proposed in [2]. In [5] and [6], efficient centralized failure location algorithms are proposed which process alarms received from various monitoring equipment to find a small set of potential failure locations. In [7], the authors propose a model for monitoring and localizing crosstalk attacks, and show that it is not necessary to place monitoring equipment at all the nodes in the network. In [8], a distributed algorithm based on message-passing is proposed to help localize propagating attacks, as well as component faults.

The third aspect of failure management is reaction to failures. Reaction mechanisms restore the proper functioning of the network by isolating the failure source, reconfiguring the connections, rerouting and updating the security status of the network [4]. In the presence of attacks, it is crucial that reaction mechanisms quickly isolate the source to preclude further attacks. Restoration techniques can use preplanned backup paths or reactive rerouting schemes, which can be slower but more efficiently utilize network resources. Reacting quickly and efficiently is crucial, not only due to the high data rates, but also to prevent from triggering a plethora of higher level reaction mechanisms. Automatic reaction mechanisms to handle component faults are proposed in [2]. In [8], distributed attack localization algorithms are applied to network restoration to achieve automatic protection switching and loopback in ring networks.

9.4   Resilient Network Performance Improvement, Evaluation Methods and Parameters

The increase in the complexity of optical network architectures causes the need for suitable network static design strategies. Given a specific recovery technique, the resilient network static design consists in allocating network resources, so that all (or some, in case of resilience differentiation) connections can be recovered from the considered failure event.

The various proposed solutions can be classified into two main groups: heuristic methods and exact methods. The former returns suboptimal solutions that in many cases are acceptable and have the advantage of requiring a limited computational effort. The latter, mainly Integer Linear Programming (ILP) based approaches, are much more computationally intensive and do not scale well with the network size.

While static design deals with the case of a known set of permanent connection requests, recently, studies on dynamic traffic routing are gaining more attention. In the case of dynamic traffic, optical networks optimized for a specific set of static connections may also be used to support on-demand lightpath provisioning.

9.4.1 Availability Calculation in Optical Network

As the complexity of a network increases, analytical availability calculation becomes more and more time and space consuming. Often it is very hard or even impossible to include all parameters from a real network in the analytical availability calculation. Especially, failure dependencies between network components make analytical approach inapplicable. Additional motivation for analysis of the influence of dependent failures comes from the fact that failure dependencies are often neglected in availability calculations assuming that dependency has no significant influence on availability performance. Failure dependencies in real redundant network structures decrease availability. In cases where it is obvious that dependencies exist, at least it should be proved if the availability is significantly affected. For example, influence of failure dependence could be checked in the ring network where ingress and egress links in some node share a common duct. Each failure caused by digging could affect more than one cable and probably all cables in the duct. The data from the field show that failures caused by digging are the most frequent fault events.

In order to evaluate the availability of complex network structures with possible dependent failures Monte Carlo simulation is used.

9.4.1.1 Availability Model

Markov availability model is applied in both analytical and simulation approach. In analytical approach availability expression for a structure is derived from transition probabilities between states of the network. In the model to be applied in simulation for each component a separate two-state Markov model is used. A working state of a component is changed to non-working state by occurrence of failure and the opposite transition occurs by repair action. Entire network state is evaluated from component states according to logical expressions that describe the relationship between component events (failure/repair) and total network state (working or non-working state). Basic parameters for each Markov availability model are component failure (λ) and repair rates (μ). It is assumed that both rates are constant. This approximation reflects the real behaviour of electronic and photonic components' failures during operation period of time. Constant rates lead to exponentially distribute times to failure (*TTF*) and times to repair (*TTR*). As the rates are invariant in time, simulation procedure can cover unlimited period of real time. Dependencies of failures are modeled by marking any pair or group of components that show any partial or total grade of dependency.

9.4.1.2 Monte Carlo Simulation

When used for network availability calculation, Monte Carlo simulation is used to generate times to failures and repairs of components in the network. Each *TTF* and *TTR* is derived from random number generator with defined probability density function (PDF) related to chosen component. Generated random numbers are uniformly distributed in [0, 1] interval. For example, in order to obtain an exponential from a uniform distribution following transformation is used:

$$t = -\frac{1}{\lambda}\ln(1-x) \tag{9.1}$$

where x is random number.

Statistical data related to occurrence of specific component failure are collected during component life test or by measuring *TTF*s for deployed systems. By monitoring real optical links one can distinguish between failures of cable and optical/electronic devices. By monitoring maintenance data from the field PDF for *TTR* can be estimated. Mean time to failure and mean time to repair can be calculated as the mean value of corresponding PDFs.

Each component changes randomly up and down state. Impact of each component state change is analyzed and decision should be done if the network state will be affected by component state change. Network total uptime T_{up} and downtime T_{down} are cumulatively calculated. When the simulation is completed the network availability A is calculated as:

$$A = \frac{T_{up}}{T_{up} + T_{down}} \tag{9.2}$$

Unavailability $U=1-A$, which is the complement of A, is used more frequently as a more suitable measure.

Event-driven simulation is assumed. An *event* can be produced just in two cases: at time point when a component failure occurs or at time point when some repair action is completed. Single simulation *iteration* is triggered by an event. It is assumed that only one component changes its state at time. A component can change its state from working to non-working or vice versa. At the beginning of the simulation all components are assumed to be in working state. Events, *TTF* or *TTR* for each component are generated according to the PDFs. A current component event occurs at the specified time that was set up during previous component event. All component produced events (*failures* and *repairs*) are put on the common simulation heap. The heap is used in the way that the next event to be selected for processing is taken as the earliest event in the heap. At the very beginning of the simulation all events on the heap represent failures of components. The first event is taken from the heap and corresponding component's state is changed from working to non-working state. At any other time point during the simulation when an event is selected from the heap, the change of the component state depends on type of the event: repair event changes component state from non-working to working state and failure event produces an opposite transition.

In the case where dependency of failures appears with the probability d within a set of components, the following procedure is carried out: when the iteration is triggered by one of component events from the set, first it should be decided whether in this iteration the dependent failure occurs or not. The presence of dependency is generated randomly with probability d. If the dependency of failures is selected then all components from dependent set are changing their state.

In every iteration, the impact of component state change on the network state is analyzed. In redundant structure a single component change may or may not cause the overall change of the network state. If the network state is changing, network uptime T_{up} or network downtime T_{down} are cumulatively increased. When the simulation is completed, the network availability A is calculated using (9.2). Note that network uptime and downtime obtained from simulation are treated in the same way as experimental data from monitoring successful/unsuccessful operation of a real system.

Availability calculation based on Monte Carlo simulation also introduces a *simulation error* but this error can be controlled by the number of simulation iterations. Unfortunately, better simulation accuracy requires the increase of simulation time. In addition, time complexity of simulation depends on the number of network elements and the level of network redundancy. In high redundant network total outcome of some network entity is very rare event and many of single or multiple element failures, including dependent failures, should be simulated before total outcome occurs.

9.4.1.3 Case Study: Ring Network with Dependent Failures

In this section, a simple case study of a ring network with dependent failures is analyzed. Ring topology is shown in Fig. 9.1. Topology consists of three nodes and links between them. A length is assigned to each link (in kilometers). Bolded sections of the links adjacent to the nodes A, B and C share a common duct and their failures are mutually dependent. In this case study the length of the section with dependent failures is 100 m. Mutually dependent parts of the links are denoted as (A_1, A_2), (B_1, B_2), (C_1, C_2). This model reflects the real cable deployment when two cables that belong to separate ring sections use the same duct; in most cases at the exit from the building where ring node is located. According to data from the field [9]

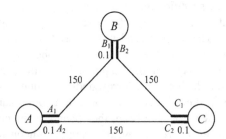

Fig. 9.1. Ring network

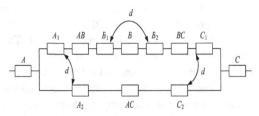

Fig. 9.2. Availability model for the connection A-C

70% of the total cable failures are caused by cable cuts. In other words, when a failure occurs there is a 70% probability that a cable cut failure occurred. We assume that a cable cut affects all cables in the duct. Thus, we assume that there is 70% dependence between failures of cables that share the same duct. If there were data from the field with more details on failure dependencies between fibres in a cable and between cables in a duct, these data could be used in simulation.

Availability model for the connection between nodes A and C in the analyzed ring topology is shown in Fig. 9.2. Nodes A and C are assumed perfect in order to avoid their impact on connection availability. In real network the nodes have high availability comparing to those of optical cables. 1+1 protection is assumed so the availability model consists of two paths: primary path over links A_2, AC and C_2, and spare path that passes node B. Dependence between failures of availability model elements is denoted with d. The grade of dependence between failures can vary in the range of maximum 100%, when failures are totally dependent, to minimum 0%, when failures are independent.

Availability of the connection A-C can be calculated analytically for the case when failures of the components are independent using following formula:

$$A_{A-C} = A_{A1} A_{AB} A_{B1} A_B A_{B2} A_{BC} A_{C1} + A_{A2} A_{AC} A_{C2} - A_{A1} A_{AB} A_{B1} A_B A_{B2} A_{BC} A_{C1} A_{A2} A_{AC} A_{C2} \quad (9.3)$$

Components availability data, taken from [10] and shown in Table 9.1, are used both for analytical availability calculation and as input parameters for the Monte Carlo simulation. Components failure rates are expressed in FIT (Failures In Time). The FIT rate of a component is the number of failures that can be expected in 10^9 hours of operation.

Table 9.1. Components Availability Data

Component Type	λ [FIT]	MTTF [h]	MTTR [h]
Nodes	1000	1 000 000	6
Cable [1/km]	100	10 000 000	21

Analytical and simulation results are presented in Table 9.2.

Table 9.2. Path A-C Unavailability Results

Dependence [%]	Calculation method	Unavailability (U) x 10^{-5}
-	Analytical	200.71
-	MC simulation	197.39
70	MC simulation	476.90
100	MC simulation	620.63

In the first column of the Table 9.2 there are dependence grades between failures in the section of cables that share a common duct. The comparison of simulation and analytical unavailability figures for the connection *A-C* for the case where failures are assumed independent, show a small difference caused by simulation error. When dependences between failures of 70% and 100% are considered, unavailability of connection *A-C* increases for 140% and 210% respectively compared to the case when failures are independent.

Table 9.3 shows how the simulation error depends on the number of simulation iterations. It typically took about a minute for a million of iterations to be computed.

Table 9.3. Simulation Error

Number of iterations ($\times 10^6$)	Simulation error [%]
30	8.31
150	1.90
250	1.65

9.4.2 Recovery Time

Due to the extremely high data rate in WDM networks, one should minimize the time between the occurrence of failure and the time at which the rerouting is completed. The fault notification time is dependent on the transmission speed in the communication media as well as the speed of the processing of the notification message by the intermediate nodes. For a given media such as fibre optic based network, this time can be reduced by ensuring that intermediate nodes will give priority treatment to notification messages necessary to trigger a recovery action. The design and implementation of fast signalling mechanisms that can reduce the total time in the recovery process is an important issue [11].

Self-organization is a phenomenon where order spontaneously emerges from disorder through the local interactions of distributed individual entities. Such systems have certain functionality and often form characteristic structures, typically small-world and/or scale-free forms [12]. Small worlds are topological structures that have short average path lengths, while exhibiting high clustering. It has been shown that small worlds can be achieved from highly-clustered lattices simply by replacing a few links at random with shortcuts between distant nodes [13], dramatically reducing the average path length. This concept could be applied to optical networks to improve recovery after failures, as well as other dynamic processes. Namely, high-speed short cuts between distant parts of a network could potentially enable faster system-wide communication, thus aiding dynamic processes such as synchronization, control and management.

Scale-free topologies have degree distributions which follow a power law [14]. This means that a few of the nodes are of very high degree, often referred to as hub nodes, while the large majority nodes are of very little degree. Such structures are robust to random failures but can be very vulnerable to attacks on hub nodes. However, if the hub nodes are sufficiently protected, these topologies could help maintain a highly robust system.

In transparent optical networks, lightpaths are established and torn down due to new connection requests and/or failures. A supervisory plane is maintained on a separate wavelength from the data channels to exchange monitoring and other failure management information which is opto-electronically processed at each node. In case of failure, the end nodes of affected lightpaths are notified via the supervisory plane in order launch their restoration mechanisms. These mechanisms most often involve rerouting failed lightpaths over backup paths. In the mean time, failure management tries to locate and isolate the failure on the basis of alarms collected from the monitoring equipment. Recovery time depends on the speed of restoration, as well as failure isolation in case of propagating attacks.

The supervisory plane topology is equal to the physical interconnection of fibres which, due to geographical considerations, forms a lattice-like structure. Such structures are usually highly clustered, but exhibit high average path lengths making communication between distant parts of the network fairly slow. According to the work of Watts and Strogatz, adding a few shortcuts could drastically reduce average path length and, consequently, speed-up distant communication. Of course, adding new fibre across distant parts of the optical network is not feasible due to the huge cost involved in laying down new fibre. However, if we establish a few high-speed *transparent* shortcuts (i.e. supervisory lightpaths) with no OEO conversion at intermediate nodes, and superposition them onto the existing supervisory plane, we can form a small world. An example of such a supervisory plane for the 14-node NSF network is shown in Fig. 9.3(a).

We have developed algorithms to create such structures in [15], which not only form small worlds but also investigate the potential to create scale-free topologies. We show that by strategically adding a small number of transparent shortcuts, we

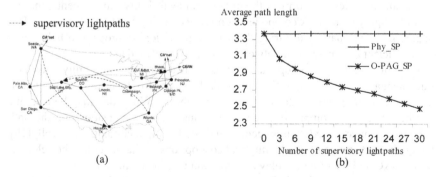

(a)   (b)

Fig. 9.3. (a) An example of a small world supervisory plane for the 14-node NSF network; (b) The average path length of the topologies obtained by the O-PAG algorithm compared to the physical topology of the 29-node basic European network from [16].

can significantly reduce the average path length and, thus, speed up failure management processes. An example of the reduction of average path length achieved by one of the proposed algorithms, *O-PAG_SP*, for the 29-node European basic network with 48 bidirectional edges from the COST Action 266 project [16] in comparison with the standard supervisory plane topology, *Phy_SP*, is shown in Fig. 9.3(b). While establishing shortcuts with *O-PAG_SP*, special attention was paid to reducing the time needed for end nodes of lightpaths to be notified in case of failure in order to reduce recovery time. By reducing the average path length in general, we also reduce the time needed by failure management to collect alarms from monitoring equipment in the network. Once such a supervisory plane is established, it is desirable that it self-maintains the desired structure in the presence of changes in the network. We propose a self-organizing algorithm to maintain such topologies in [17], subject to changes in data lightpaths, monitoring equipment and/or unexpected failures. The algorithm is based on periodic queries sent between nodes to evaluate the potential of connecting to other distant node via supervisory lightpaths.

We have been investigating the possibilities of developing a small-world scale-free supervisory plane aimed at speeding-up communication between distant parts of the network for faster recovery. We have developed algorithms to create and maintain such structures which show very promising results.

9.4.3 Network Performance Improvement through Differentiated Survivability

The trade-offs, such as the balance between overall cost and degree of resilience in shared vs. dedicated protection [18] play an important role in network design and operation. Quality of service awareness has gained vital importance in service provisioning with the rollout of applications that impose quality requirements on data transfer. In order to fulfill these requirements the underlying networking technology must be capable of offering end-to-end transport services at satisfactory availability levels. To meet the end-to-end availability requirements, the impact of employed network components on service quality has to be evaluated. This impact must be quantified during the service provisioning process to help determine efficient assignments of network resources to traffic demands.

Most of the literature on optical circuit-switched network availability makes the fundamental assumption that the failure probability of optical node equipment is negligible when compared to link failure probability. While this may be true in a number of cases, a comprehensive analysis must take into consideration the optical cross-connect (OXC) architecture and switching technology as well [19]. Given the wide range of availability and cost options available today, the selection of the OXC architecture may play an important role in certain networks.

In [19] the impact of optical node failures on wavelength-division-multiplexed networks is investigated, where reliable end-to-end optical circuits are provisioned dynamically. At the node level, the optical cross-connect (OXC) equipment availability measure is estimated using component level availability models. At the network level, end-to-end optical circuits are provisioned only when the level of connection availability required by the application can be guaranteed. With the objective of yielding efficient utilization of the network resources, i.e., fibres and OXCs, circuit redundancy is achieved by means of shared path protection (SPP) switching, in combination with differentiated reliability (DiR). The resulting optimal routing and wavelength assignment problem is proven to be NP-complete. To produce suboptimal solutions in polynomial time, a heuristic technique is presented, which makes use of a time-efficient method to estimate the end-to-end circuit availability in the presence of multiple (link and node) failures. Using the proposed heuristic, a selection of representative OXC architectures and optical switching technologies is examined to assess the influence of the node equipment choice on the overall network performance.

9.4.3.1 Algorithm Specification

In this section an algorithm for path-based protection is proposed based on implementing the backup multiplexing technique under dynamic traffic demands where existing lightpaths cannot be rerouted and future lightpath requests are not known. The use of the backup multiplexing technique is selected in order to facilitate efficient resource sharing. In this framework different routing and wavelength assignment schemes that considerably enhance the spare capacity utilization are investigated and proposed. Through the proposed novel wavelength assignment scheme (that dedicates a consecutive number of wavelengths to protection lightpaths) a significant performance improvement compared to commonly used techniques is observed. In addition, traffic demands are assigned three classes of service with regards to network recovery and adopt the concept of resilience priority classes to maximize network resource utilization. The three types of lightpaths considered are: 1) high priority protected lightpaths, 2) unprotected lightpaths and 3) low priority preempted lightpaths. A high priority protected lightpath has a working path and a diversely routed backup path. Both the working and the backup lightpaths are identified before the provisioning of the working path according to the backup multiplexing scheme. An unprotected lightpath is not protected with a backup path and upon any failure along the lightpath a dynamic restoration mechanism is initiated to provide an alternative route without any guarantees. Finally low priority preempted lightpaths are unprotected lightpaths that can use the backup routes of the high priority lightpaths. In case of high priority lightpath failure, preemption of this low priority traffic takes place.

The online version of the Routing and Wavelength Assignment (RWA)/resilience problem is solved, i.e. traffic requests arrive and get served sequentially without knowledge of future incoming requests [20, 21]. This makes this contribution valid for usage both in the network design and – most importantly – the traffic engineering field. In addition it is assumed that only a single link could fail at any instance of time and re-routing of already established connections is not allowed.

Last, the model does not take into consideration any wavelength conversion capability of the network and thus wavelength continuity across any path is a tight constraint in the problem definition.

The proposed algorithm is suitable to support differentiated services with regards to survivability taking into consideration the following three classes of service:

- A premium class (class-1) offering one dedicated primary path plus one shared but diversely routed backup path
- A standard class (class-2) providing for one unprotected but dedicated primary path that can be restored dynamically in case of failure and
- A low-priority class (class-3) offering a single path that may share links with class-1 backup paths and can be pre-empted in the case of a class-1 primary path failure to allow for activation of the backup mechanism.

The routing and wavelength assignment problems are solved in two separate steps. Routing is implemented based on the Dijkstra's algorithm to compute a primary and a backup path for a given demand. The wavelength assignment algorithm assigns wavelengths to the primary and backup paths favouring resource sharing between the current demand and the already established requests.

After the initialization phase, in which the algorithm collects network topology information (i.e. number of nodes, number of links, wavelengths per fibre, network connections, backup path wavelength assignment scheme) and constructs the required matrices to monitor the network state (Al, Bl and Rl), connection requests arrive for random source and destination pairs. First, independent of the request's service class, a primary lightpath is established through the primary lightpath computation phase. This phase consults the Rl matrix and assigns costs to the network links based on the following approach: if a link has no free wavelengths, its cost is set to infinite and it is not considered by the Dijkstra algorithm for the path computation. If available wavelengths exist on the link, the cost is set to be inversely proportional to the number of spare wavelengths, thus offering a degree of load balancing. After weights are assigned to the network links, the widest shortest path routing algorithm is run on the weighted graph, calculating a number of shortest paths and selecting the one that traverses the minimum number of hops and for which at least one common free wavelength exists on all its links. If no path is found, the connection is blocked. If at least one path is found, a list of possible wavelengths that can be allocated to it is identified and the first wavelength is chosen (assuming that they are sorted in increasing order) to form the primary lightpath.

After the primary lightpath is ready to be established, the Al and Rl matrixes are updated to reserve the appropriate wavelength and the algorithm proceeds to the examination of the request's service class. If the request belongs to class-2 and preemption is enabled, Bl matrix is also updated to allow sharing of the allocated wavelength from future backup paths of class-1 traffic that has the authority to preempt class-2 lightpaths.

If the established demand requires a backup path (class-1), the flow control moves to the backup computation phase. Here the available bandwidth Sl(a) consisting of the residual bandwidth (Rl) and the portion of the backup bandwidth (γl) that can be shared as described earlier is first identified excluding the links utilized by the primary path. Then based on this available bandwidth (Sl(a)), for each wavelength an auxiliary graph is generated representing the current network state. For this new topology formulation link costs are assigned based on the following strategy: On the links for which the wavelength under consideration belongs to γl a zero weight is assigned and if it belongs to Rl a unit cost is assumed. On the other hand links on which the wavelength is already allocated (by primary lightpaths) are not considered in the auxiliary graph and cannot be used for the backup calculation. An attempt to find a lightpath for each wavelength follows. If no lightpath is found for any wavelength, the connection is blocked due to backup path blocking, requesting from the algorithm to roll back the updates of Al and Rl previously performed by the primary path computation phase. In case of multiple backup lightpaths computations the algorithm must allocate one, based on the selected wavelength assignment scheme. If the random pick (RP) wavelength assignment scheme is selected the lightpath is chosen randomly from the set of the available lightpaths. For the last fit (LF) scheme the lightpaths with minimum cost are identified and the last one (when sorted in increasing order) is selected, whereas for the first fit (FF) the first one from the minimum cost lightpaths is allocated. In the final step of the algorithm Bl and Rl are updated for the links which residual bandwidth is used.

9.4.3.2 Performance Study

The results presented in this section, which are extensively presented and discussed in [20], are generated based on the Pan-European test network defined by COST 239 [22] that comprises 11 nodes and 26 links (Fig. 9.4). Links are considered bidirectional and if a link failure occurs the traffic flow in both directions will be disrupted. Lightpaths comply with the wavelength continuity constraint and connection requests are equally likely to have any of the network nodes as source or destination. Also we assume that calls arrive one by one and their holding time is long enough to consider that accepted calls do not leave (incremental traffic). A connection is blocked if either a primary or a backup path can not be established. The results shown in the following figures are the average values over 20 independent repetitions of the described experiment configuration.

The performance of three wavelength assignment schemes i.e. Last Fit, Random Pick and First Fit when applied for the backup lightpath establishment is investigated. First fit is the wavelength assignment scheme used for the primary path establishment through all simulation results presented.

In Fig. 9.5 the average blocking probabilities for Last Fit , First Fit and Random Pick are compared for uniform fibre capacities of C=8 and C=16 wavelengths. LF wavelength assignment scheme provides improved network performance compared with FF of around 4% and 2% for high network loads for 16 and 8 channels per fibre respectively. In addition the LF significantly outperforms RP since it can offer a blocking improvement of 14% and 8% for the two different fibre capacity parameters. These observations can be explained by the difference in the restoration capacity occurring from the various wavelength assignment schemes. A relevant analysis [20] has shown that the Last Fit wavelength assignment algorithm maximizes the backup path link reuse although a small number of links are dedicated for backup paths that are used more than once compared to the case of the Random Pick algorithm. The increase in restoration capacity of the Last Fit over the Random Pick scheme constitutes the main reason of the lower blocking probability of the Last Fit scheme. Last Fit is a simple and fast wavelength assignment scheme able to increase considerably the backup link reuse by dedicating a small but consecutive portion of the wavelength band to backup paths, allowing a large amount of the precious residual bandwidth for the primary paths that are allocated based on a First Fit scheme.

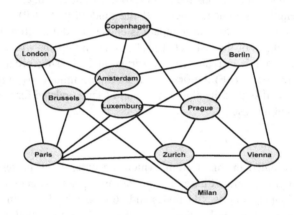

Fig. 9.4. Pan-European test network COST 239

Fig. 9.6 illustrates the results obtained by considering the coexistence of both class-1 and class-2 traffic, with the option of either enabling or disabling preemption. More specifically two scenarios are compared in this setting: one for which class 1 traffic comprises 50% and one for which class 1 traffic comprises 80% of the total requests respectively. These two cases are compared with the case, in which all the traffic is considered as class 1 traffic. The benefit offered by the preemption enabled scheme is up to 12%, when half of the incoming traffic is assigned as class 1 and up to 8% when 80% is set us class 1.

For the non preemptive scheme the benefit reduces to 5% and 3% respectively indicating the superiority of the preemptive approach in terms of network performance. This improvement offered by the preemptive scheme is at the expense of the reliable provisioning of low priority traffic, which can be tolerated for many

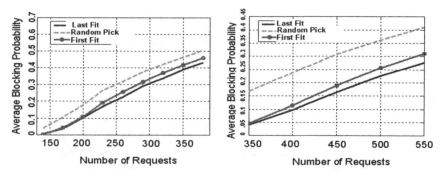

Fig. 9.5. Network performance for the three backup path wavelength assignment schemes and for different fibre capacity (a) C=8, (b) C=16

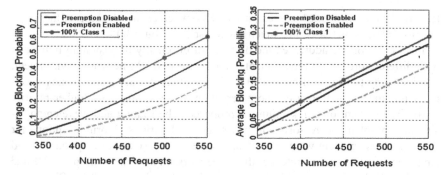

Fig. 9.6. Average blocking probability when (a) 50% and (b) 80% of the requested connections are assigned as class 1 traffic and LF scheme is used for C=16

non-real time applications. The preemptive scheme although utilizing a smaller number of links compared to the non preemptive case provides an increase in the link reuse percentage since it allows the low priority class-2 traffic to be shared among the backup paths of the higher priority traffic. When no preemption is allowed the number of possible shared paths is significantly reduced since only 50% of the total demands require backup paths resulting in inefficient backup resource utilization with considerable impact on the network performance.

Finally, in 4.7 we analyze the blocking probabilities of the different classes co-existing in the network when preemption is allowed. In Fig. 9.7(a) 80% of the total traffic is considered as class 1 and 20% as class 2. The blocking probability of the class 1 traffic is high compared to the low priority traffic (a difference of about 10% is observed) although the overall blocking is reduced when considering this differentiation scheme. In Fig. 9.7(b) the same percentage of class 1 and class 2 demands is assumed and almost the same blocking probability is observed for the two classes, causing a higher reduction in the overall blocking probability. Also in this case the blocking probability of high priority traffic is reduced considerably at

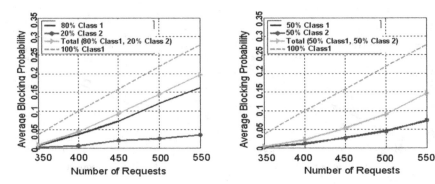

Fig. 9.7. Analyzing the blocking probabilities of the different classes in the network when (a) 80% and (b) 50% of class1 traffic is requested.

least for heavier network loadings (around 8%) whereas the blocking of the lower priority traffic is increased in a much smaller scale (about 4%).

In this section we addressed the problem of efficiently provisioning lightpaths with different protection requirements in a dynamic WDM network environment. The incoming traffic is differentiated to classes of service according to survivability requirements; additionally, the preemption of low priority by higher priority traffic in the event of a link failure is proposed. The routing and wavelength assignment problems were solved in two stages and various algorithm options for wavelength assignment were evaluated. In case of the use of pre-emption detailed simulation results demonstrate significant network improvement of up to 12% and considerable decrease in the blocking probability of the high priority traffic.

9.5   Security Issues in Transparent Optical Networks

Security issues in optical networks are of prime importance due to the high capacity these networks offer. Namely, a single failure can lead to tremendous data loss. Transparent optical networks (TONs) are dynamically reconfigurable networks which establish and tear-down all-optical connections, called lightpaths, between pairs of nodes. These connections can traverse multiple links in the physical topology and yet transmission is entirely in the optical domain. Although transparency has many attractive features, such as speed and insensitivity to data rate and protocol format, performance monitoring is much more difficult since it must be performed in the optical domain. Furthermore, malicious signals can propagate from the source to other parts of the network without loosing their attacking capabilities due to the lack of regeneration at intermediate nodes. To ensure secure network operation, the optical network employs a failure management system, designed to deal with failures and security threats. Failure management information, such as alarms from monitoring equipment, is exchanged via a set of supervisory channels. We propose a self-organizing approach to arranging these supervisory channels in such a way as to speed-up failure recovery-time.

In general, failures include both component faults and deliberate attacks on the optical network. Component faults include single or multiple component malfunctions which can be a consequence of natural fatigue, improperly installed or configured equipment, or external influence (e.g. power loss). Some common component faults which can degrade network performance, such as fibre cuts and transmitter, receiver, and optical amplifier faults, are given in [2]. Attacks, on the other hand are malicious attempts to interfere with the secure functioning of the optical network. Various attacks have been described in [4], [23], [24] and [25]. While faults only affect the connections passing directly through them, attacks can spread and propagate throughout the network. As such, rerouting mechanisms which can tolerate hardware failure do not necessarily protect against attacks since the re-routed attacked signal may carry attacking capabilities itself [7]. Furthermore, while faults usually occur due to the aging of the equipment, attacks can occur at any time during the life span of the network and can also appear sporadically [23]. Thus, attacks are much harder to locate and isolate.

Here we classify attacks according to the components whose vulnerabilities they exploit. Gain competition in optical amplifiers is a common target for attackers. Namely, an amplifier has a finite amount of gain available (a limited pool of upper-state photons), which is divided among the incoming signals. Thus, by injecting a high-power signal within the amplifier passband, an attacker can deprive other signals of power while increasing its own, allowing it to propagate through the network causing service degradation or even service denial. An example is shown in Fig. 9.8(a) where the attacker deprives User 3 of adequate gain. If this high-power signal is injected on a wavelength other than the legitimate data channels (out-of-band jamming), but still within the amplifier passband, it can cause cross-modulation which can be used to tap a data signal [4]. Tapping attacks enable unauthorized users to gain access to data either for eavesdropping or traffic analysis purposes. Light can also be injected onto a legitimate data wavelength (in-band jamming), not only causing gain competition in amplifiers, but degrading the signal on that particular wavelength by raising its signal-to-noise ratio (SNR).

Another major vulnerability in optical networks is caused by optical switching nodes which can exhibit significant crosstalk effects. This happens when part of an input signal leaks onto one or more unintended output signals and causes interference. An attacker can exploit this by injecting a very strong input signal which, in addition to causing interference, can make the attacked signal acquire attacking capabilities itself. Thus, such a crosstalk attack can propagate though the network, affecting links and nodes that are not even traversed by the original attacking signal [8]. An example of such an attack is shown in Fig. 9.8(a). We can see that the attacker is able to attack User 2 (via User 1) even though they do not traverse any common components. Not only can this cause wide-spread service disruption, but it makes identifying and localizing the source much more difficult. Another efficient tapping attack achieved by exploiting crosstalk [4] is shown in Fig. 9.8(b).

Here an attacker requests a legitimate data channel but does not send any data on it. Consequently, the channel carries only leakage it picks up via crosstalk. This weak leakage signal is then amplified into a strong tapped signal and delivered directly to the attacker.

Besides amplifiers and switches, unshielded optical fibres can also be exploited by an attacker with physical access to the fibre. A simple overt attack, which can be considered a fault by failure management, is achieved by cutting the fibre. A more covert attack, which is much harder to locate, can be achieved by slightly bending the fibre to tap part of a signal. It is also possible to inject a jamming signal onto the fibre causing service disruption. A particularly malicious attacker can combine the two by tapping a signal and then injecting noise at the tapping point to achieve both eavesdropping and degradation of the SNR on the attacked channel (correlated jamming). Additionally, long distances and high-power signals can introduce nonlinearities in the fibre causing crosstalk effects between wavelengths. Furthermore, dispersion and attenuation characteristics can be changed by warming the fibre and thus degrade the transmission quality.

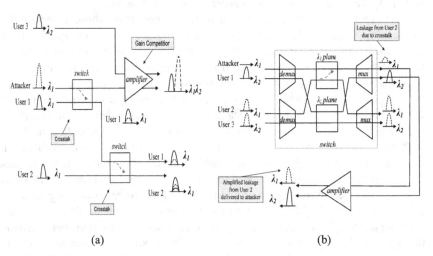

(a)   (b)

Fig. 9.8. An example of (a) gain competition and propagating crosstalk attacks, and a (b) combined tapping attack.

9.6 Multilayer Resilience

In this section three generic approaches for providing recovery in multilayer networks (more specifically in IP-over-OTN networks) will be discussed: single-layer recovery schemes in multilayer networks, static multilayer recovery schemes and dynamic multilayer recovery strategies. Some quantitative studies and comparisons between the different methods will reveal the advantages and disadvantages of each approach.

9.6.1 Single Layer Recovery in Multilayer Networks

This section discusses how recovery functionality can be introduced in multilayer networks by applying single-layer recovery schemes. The concepts and discussions are focused on a two-layer network, but are mostly generic and therefore applicable to any multilayer network.

9.6.1.1 Survivability at the Bottom Layer

In this approach, recovery of a failure is always done at the bottom layer of the multilayer network. In an IP-over-OTN network for example, this implies that the 1+1 optical protection scheme, or any other recovery scheme which is deployed at the OTN layer, attempts to restore the affected traffic in case of a failure.

This strategy has the benefit that only a simple root failure has to be treated, and that the number of required recovery actions is minimal (the recovery actions are performed on the coarsest granularity). In addition, failures do not need to propagate through multiple layers before triggering any recovery action.

However, this recovery strategy cannot handle problems that occur due to failures in a higher network layer. Moreover, if a node failure occurs in the OTN layer (being an OXC failure), the OTN layer recovery mechanism will only be able to restore the affected traffic that transits the failed bottom-layer node (being the OXC). The co-located higher-layer IP router will become isolated due to the failure of the OXC underneath, and thus all traffic treated by this IP router cannot be restored in the lower (optical) layer.

This is illustrated in Fig. 9.9. We label the top level nodes lower case and the bottom layer nodes upper case. The considered network carries two traffic flows between client layer nodes *a* and *c*. One traffic flow (*a-d-c*, indicated with a full line) transits the client-layer node *d* (using two logical links *a-d* and *d-c*), while the other traffic flow (*a-c*, indicated with a dashed line) uses a direct logical link from *a* to *c*, and only transits the server-layer node *D*. Now let's assume that a failure occurs in the bottom layer, for example the failure of node *D*. The server layer cannot recover the first traffic flow *a-d-c*. This is due to the fact that the client-layer node *d* becomes isolated due to the failure of *D*, which is terminating both logical links *a-d* and *d-c*. This failure can only be resolved at the higher layer. The second traffic flow *a-c, however,* is routed over a direct logical link between nodes *a* and *c*. This logical link transits only the failing node *D* in the bottom layer, which means that this traffic flow can be restored by the bottom-layer recovery scheme (dotted line on figure).

9.6.1.2 Survivability at the Top Layer

Another strategy is to provide the survivability at the top layer of the network. In our example of an IP-over-OTN network, this could for instance be the IP restoration technique or MPLS-based restoration [84]. The main advantage of this strategy is that it can cope with higher layer failures as well. A major drawback is,

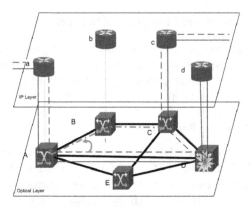

Fig. 9.9. Survivability at the bottom layer

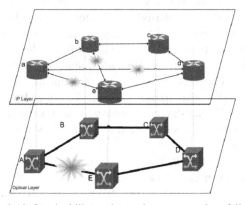

Fig. 9.10. Survivability at the top layer - secondary failures

however, that it typically requires a lot of recovery actions, due to the finer granularity of the flow entities at the top layer.

A single root failure in the lower layer can introduce a complex scenario of secondary failures in higher network layers. This is illustrated in Fig. 9.10, where the failure of an optical link in the bottom layer corresponds with the simultaneous failure of three logical IP links in the top layer. These three logical IP links are part of a Shared Risk Link Group (SRLG) [85]. The recovery scheme in the top layer will have to recover from three simultaneous link failures, a quite complex failure scenario, in clear contrast with a recovery scheme at the bottom layer which would only have to cope with a single link failure.

Another disadvantage of recovery at the top layer only is that traffic injected directly in the lower layer (e.g. wavelength channels directly leased by a customer) can not be recovered by the optical network operator, even if the failure happens in the optical layer itself.

9.6.2 Interworking between Layers

In the previous section some strategies are discussed that apply a single-layer re-covery mechanism in order to provide survivability in the multilayer network. The advantages of these approaches can be combined by running recovery mechanisms in different layers of the network as a reaction to the occurrence of one single network failure. More generally speaking, the choice in which layer(s) to recover the affected traffic due to a failure will depend on the circumstances, for example on the failure scenario that occurred.

This interworking between layers however requires some rules or coordination actions in order to ensure an efficient recovery process. These rules strictly define how layers and the recovery mechanisms within those layers react to different fail-ure scenarios, and form a so-called escalation strategy. Several escalation strategies are discussed in this section: uncoordinated, sequential, and integrated escalation.

9.6.2.1 Uncoordinated

The easiest way of providing an escalation strategy is to simply deploy recovery schemes in the multiple layers without any coordination at all. This will result in parallel recovery actions at distinct layers. Consider again the two-layered net-work (Fig. 9.11), with, for instance, the failure of the physical link A-D in the server layer. This failure of the physical link will also affect the corresponding logical link a-d in the client layer, and hence affects the considered traffic flow a-d-c. Since the recovery actions in both layers are not coordinated, both the recov-ery strategy in the client layer and the recovery strategy in the server layer will at-tempt recovery of the affected traffic. This implies that in the client layer the traf-fic flow from a to c is rerouted by the recovery mechanism of the client layer, resulting in a replacement of the failed path a-d-c by for instance a new path a-b-c. At the same time, the server layer recovers the logical link a-d of the client layer topology by rerouting all traffic on the failing link A-D through node E. It is clear that in this example recovery actions in a single layer would have been sufficient to restore the affected traffic.

The main advantage of the uncoordinated approach is that this solution is sim-ple and straightforward from an implementation and operational point of view. However, Fig. 9.11 shows the drawbacks of this strategy. Both recovery mecha-nisms occupy spare resources during the failure, although one recovery scheme occupying spare resources would have been sufficient. Usually during failure-free conditions spare resources are used to accommodate low priority traffic during failure-free conditions, but this so-called 'extra traffic' must be pre-empted when the spare resources are needed to recover from a failure. Hence, a repercussion of the uncoordinated approach is that more extra traffic than necessary is potentially disrupted. The situation can even be worse, consider for example that the server layer reroutes the logical link a-d over the path A-B-C-D instead of A-E-D, then both recovery mechanisms need spare capacity on the links A-B and B-C. If these higher layer spare resources are supported as extra traffic in the lower layer, then there is a risk that these client layer spare resources are pre-empted by the recovery action in the server layer, resulting in "destructive interference". Or in other words, none of the two recovery actions was able to restore the traffic, since the client layer reroutes the considered flow over the path *a-b-c*, which was disrupted by the server layer recovery. The research done in [86] illustrates that these risks may exist in real networks: the authors prove that a switchover in the optical domain may trigger traditional client layer protection. Moreover, such a multilayer recovery strategy can have significant repercussions on the overall network stability.

In [87], the authors show a real life example of network convergence problems that follow the impetuous use of the uncoordinated approach in an IP-over-OTN network, where the OTN layer features 1+1 path protection. They observe IP network convergence times after the occurrence of a link failure in the OTN layer. Although protection in the optical layer recovers a link within 20 ms, the recovery of the IP traffic that was transiting the link takes over 60 s in some cases.

In summary, although simple and straightforward, just letting the recovery mechanisms in each layer run without a coordinating escalation strategy has its consequences on efficiency, capacity requirements and even ability to restore the traffic.

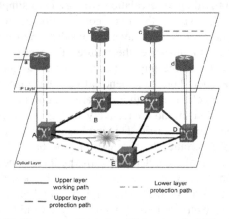

Fig. 9.11. The uncoordinated multilayer survivability strategy

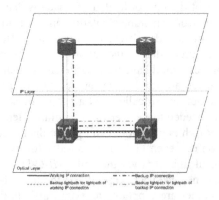

Fig. 9.12. Duplicated protection

9.6.2.2 Sequential Approach

A more efficient escalation strategy, in comparison with the uncoordinated approach, is the sequential approach. Here the responsibility for recovery is handed over to the next layer when it is clear that the current network layer is not able to perform the recovery task. For this escalation strategy two questions must be answered: in which layer to start the recovery process, and when to escalate to the next layer. Two approaches exist, the bottom-up escalation strategy and the top-down escalation approach, each having different variants.

Bottom-up Escalation

With this strategy, the recovery starts in the lowest-detecting layer and escalates upwards. The advantage of this approach is that recovery actions are taken at the appropriate granularity: first the coarse granularities are handled, recovering as much traffic as soon as possible, and recovery actions on a finer granularity (i.e., in a higher layer) only have to recover a small fraction of the affected traffic. This also implies that complex secondary failures are handled only when needed.

An issue that must be handled in the bottom-up escalation strategy is how a higher network layer knows whether it is the lowest layer that detects the failure (so it can start with the recovery) or has to wait for a lower layer instead. This issue is tackled in Section 9.5.2.3.

Top-down Escalation

With top-down escalation it is the other way around. Recovery actions are now initiated in the highest-possible layer, and the escalation goes downwards in the layered network. Only if the higher layer cannot restore all traffic, actions in the lower network layer are triggered. An advantage of this approach is that a higher layer can more easily differentiate traffic with respect to service types and so it can try to restore high priority traffic first. A drawback of this approach however is that a lower layer has no easy way to detect on its own whether a higher layer was able to restore the traffic (an explicit signal is needed for this purpose). So here the implementation is somewhat more complex and not currently implemented. There is also a problem of efficiency, since it is very well possible that for example 50 % of the traffic carried by a wavelength channel in an optical network is already restored by a higher network layer recovery mechanism, hence protecting this wavelength in the optical layer as well is only useful for the other 50% of the carried traffic.

9.6.2.3 Implementation of an Escalation Strategy

The actual implementation of these escalation strategies is another issue. Two possible solutions are described here (for the ease of explanation, the bottom-up escalation strategy is assumed in what follows).

A first implementation solution is based on a *hold-off timer* of T_{HO} seconds. Upon detection of a failure, the server layer starts the recovery immediately, while the recovery mechanism in the client layer has a built-in hold-off timer that must expire before initiating its recovery process. In this way, no client recovery action will be taken if the failure is resolved by the server-layer recovery mechanism before the hold-off timer expires. The main drawback of a hold-off timer is that recovery actions in a higher layer are always delayed, independent of the failure scenario. The challenge of determining the optimal value for T_{HO} is driven by a trade-off between recovery time versus network stability and recovery performance.

The second escalation implementation overcomes this delay by using a *recovery token signal* between layers. This means that the server layer sends the recovery token (by means of an explicit signal) to the client layer from the moment that it knows that it cannot recover (all or part of) the traffic. Upon reception of this token, the client layer recovery mechanism is initiated. This allows limiting the traffic disruption time in case the server layer is unable to do the recovery. A disadvantage, compared to the hold-off timer interworking, is that a recovery token signal needs to be included in the standardization of the interface between network layers.

9.6.3 Multilayer Survivability Strategies

9.6.3.1 Static Recovery Techniques

Multilayer survivability involves more than just coordinating the recovery actions in multiple layers. There is also the issue of the spare resources, and how they have to be provided and used in an efficient way in the different layers of the network. One way or another the logical (spare) capacity assigned to the recovery mechanisms that are deployed at higher network layers, must be transported by the lower layer. There are several ways to realize this.

The most straightforward option is called duplicated protection, and is depicted in Fig. 9.12 for a point-to-point example. (Note that we made abstraction from the physical disjointedness of working and backup path in this conceptual example. The extension towards larger networks and the introduction of physical disjointedness is straightforward.)

Each working IP link is transported via a lightpath in the OTN layer. To cope with OTN layer failures, the lightpath is protected by a backup lightpath. To cope with IP layer failures, the IP link is protected by a spare IP link (to be transported via the OTN layer as well). Moreover, if the spare capacity that is provisioned in the logical IP network is simply protected again in the underlying optical layer (backup lightpath for lightpath of spare IP link), we are coping with duplicated protection. Despite the reduced complexity, this is a rather expensive solution. Hence, investing in duplicated protection is very debatable and probably only meaningful in a few exceptional network scenarios.

A first possibility to save investment in physical capacity is carrying the spare capacity in the logical higher-layer network allocated to the higher-layer network recovery techniques, as unprotected traffic in the underlying network layer(s) (see Fig. 9.13 for the IP-over-OTN example).

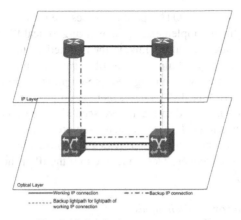

Fig. 9.13. Logical spare unprotected

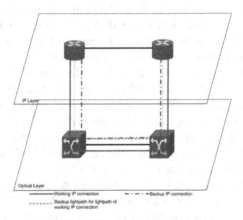

Fig. 9.14. Common pool strategy

This strategy, called *logical spare unprotected*, still allows protecting against any single failure: a cut of the bottom fibre (carrying the lightpath of the working IP link) would trigger the optical network recovery, while a failure of one of the outer router line cards would trigger the IP layer network recovery. A prerequisite for such a scenario is that the optical network supports both protected and unprotected lightpaths. It is crucial to guarantee that a single network failure is not able to affect simultaneously a working IP link and the unprotected lightpath carrying the IP spare capacity protecting that same link. Otherwise, the spare IP capacity would also become unavailable for recovery of the failure, and the recovery process would fail.

One step beyond simply carrying the spare capacity of the logical higher network layers as unprotected traffic in the underlying layer is to allow pre-empting this unprotected traffic by the network recovery technique of the lower network layer. This is the *common pool strategy* [88], and an example is given in Fig. 9.14

for an IP-over-OTN network. OTN spare resources are provisioned for the optical protection of the lightpath implementing the working logical IP link. The lightpath implementing the spare logical IP link is then routed along the same (optical) spare resources. In case of a failure of the fibre carrying the working logical IP link, the optical protection will be triggered, pre-empting the lightpath implementing the spare logical IP link. In that case, there is no problem in pre-empting this lightpath since it is not needed in the failure scenario. However, the pre-emption of lightpaths carrying logical spare capacity requires additional complexity. In summary, the common pool strategy provides a pool of physical spare capacity that can be used by the recovery technique in either the IP or the optical layer (but not simultaneously).

9.6.3.2 Dynamic Recovery Techniques

In the previous section, static multilayer recovery strategies have been discussed. They are called static, because at the time of a failure the logical network topology (in an IP-over-OTN network, this is the IP layer topology) remains unchanged (static). As such, the logical network must be provided with a recovery technique and the required spare resources for survivability reasons.

Dynamic multilayer survivability strategies differ from such static strategies in the sense that they actually use logical topology modification for recovery purposes. This requires the possibility to set up and tear down lower layer network connections that implement logical links in the higher network layer in real-time. Optical networks will therefore be enhanced with a control plane, which gives the client networks the possibility to initiate the set-up and tear-down of lightpaths in the optical layer. This is used to reconfigure the logical IP network in case of a network failure. This approach has the advantage that the logical network spare resources should not be established in advance in the logical IP network and thus the underlying optical network should not care about how to treat these client layer spare resources. In the optical layer, however, spare capacity still has to be

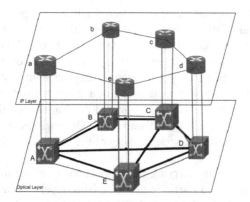

Fig. 9.15. Scenario before failure

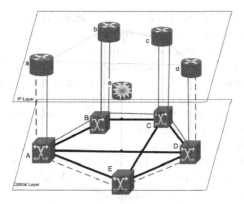

Fig. 9.16. Scenario after failure

provided to deal with lower layer failures such as cable cuts or OXC failures. Enough capacity is also needed in the optical layer to support the reconfiguration of the logical IP network topology and the traffic routed on that topology.

An illustration of a dynamic reconfiguration of the logical higher-layer topology in case of failures is given in Fig. 9.15 and Fig. 9.16 for an IP-over-OTN network. Initially, the traffic flow from router *a* to router *d* is forwarded via the intermediate router *e*. To this end the logical IP network contains the IP links *a-e* and *e-d*, implemented by the lightpaths *A-E* and *E-D* in the OTN network. When router *e* fails, routers *a* and *d* will detect this failure, and use the User-Network Interface (UNI) to request the optical layer for a tear-down of the links *a-e* and *e-d*. The resulting free capacity in the optical layer can be used to set up a direct logical IP link from router *a* to router *d*. This is requested to the underlying optical network by requesting the set-up of the lightpath between OXCs *A* and *D*. So, at the time of the failure, the logical IP network topology is reconfigured. As mentioned before, a special feature of the underlying optical network is needed for this: it must be able to provide a switched connection service to the client network quickly. Automatic Switched Optical Networks (ASONs) [89], or more generally Intelligent Optical Networks (IONs), have this particular feature.

9.6.4 Logical Topology Design

A challenge with dynamic multilayer recovery strategies involves the actual logical topologies to be realized. Typically the network scenario and traffic demand will favor a certain logical topology (and corresponding IP/MPLS routing) for the failure-free case. Network failures affecting part of this logical topology will require topology reconfigurations and rerouting to replace failing links and to circumvent the problem.

To illustrate the flexibility in logical topology design, two clearly distinct methods can be proposed. The method of global reconfiguration considers each failure scenario separately. The IP topology is recomputed from scratch for each failure scenario (after removal of the failing network elements). The method of local re-configuration follows a quite different approach. The topology design is now start-ing from the failure-free case. For a particular failure scenario, the affected IP links and/or routes are first removed, and only the affected traffic is rerouted, over the remaining topology (adding IP links where additional resources are needed). The idea behind local reconfiguration is that it lowers the amount of required re-configurations and rerouted traffic in case of failures, since the new logical topol-ogy is derived from the failure-free one.

In addition to these two approaches, the design of the logical topology requires a certain computational ability, and this can be performed online as well as offline. Online reconfiguration and rerouting is more suitable to adapt to a changing net-work scenario, and especially to changing traffic demands. Cross-layer traffic en-gineering techniques can be used to redistribute lower layer capacity to better cope with traffic, or to optimize bandwidth throughput regardless of traffic pattern. However, because of its online nature, computation time must remain limited for the logical topology update mechanism to retain its desired flexibility. The capa-bility to deal with network failures should arise naturally from the process' goal to solve changing situations in the network automatically.

In cases with a more static traffic demand, one may prefer offline logical topol-ogy design instead. In this case the traffic demand is used to calculate a failure-free logical topology and a set of failure topologies corresponding with possible network failures. Note that in this case one needs to decide in advance which fail-ures should be recoverable. Because the design is done offline, more time is avail-able so more extensive topology design algorithms can be utilized. This method is typically more optimal in terms of network cost, throughput, impact of failure on operation, etc.

9.7   Conclusions

Telecommunication infrastructure becomes more and more critical for the society, commerce, government, and education. An unbreakable connectivity is the expec-tation from users, and for some services is the actual requirement as well.

Since one cannot avoid occurrence of failures in the network, it is very impor-tant to provide survivability mechanisms that allow for rerouting the traffic around the failed facilities. Furthermore, the time between the occurrence of failure and the time at which the rerouting is completed should be minimized.

Efficiently dealing with security issues and failure management is of utmost importance in optical networks due to the tremendous amount of information they carry. The specific vulnerabilities of such networks, as well as the difficulties in monitoring incurred by transparency, make failure recovery a challenging task. Self-organizing concepts, in particular the common structural properties of self-organizing systems can potentially be applied to develop a more scalable and ro-bust failure management scheme. We have been investigating the possibilities of developing a small-world scale-free supervisory plane aimed at speeding-up communication between distant parts of the network for faster recovery.

We also we addressed the problem of efficient provisioning lightpaths with different protection requirements in a dynamic WDM network environment.

The integration of different network technologies into a multilayer network, as in Internet based networks carried by optical transport networks, creates new opportunities but also challenges with respect to network survivability. In the different network layers, recovery mechanisms are active that can be exploited jointly, to reach a more efficient or faster recovery from failures. This interworking is also indispensable to overcome the variety of failure scenarios that can occur in the multilayer network environment. A well-considered coordination between the different layers and their recovery mechanisms is crucial to attain high performance recovery.

References

1. Wosinska, L., Svensson, T.K.: Analysis of Connection Availability in an All-Optical Mesh Network. Fibre and Integrated Optics 26(2), 99–110 (2007)
2. Li, C.-S., Ramaswami, R.: Automatic fault detection, isolation, and recovery in transparent all-optical networks. J. Lightwave Technol. 15(10), 1784–1793 (1997)
3. *** Terms and Definitions, CCITT Blue Book, vol. I, Fascicle I.3, Geneva (1989)
4. Médard, M., Marquis, D., Barry, R., Finn, S.: Security issues in all-optical networks. IEEE Network 11(3), 42–48 (1997)
5. Mas, C., Tomkos, I., Tonguz, O.: Failure location algorithm for transparent optical networks. IEEE J. Select. Areas Commun. 23(8), 1508–1519 (2005)
6. Mas, C., Thiran, P.: An efficient algorithm for locating soft and hard failures in WDM networks. IEEE J. Select. Areas Commun. 18(10), 1900–1911 (2000)
7. Wu, T., Somani, A.: Cross-talk attack monitoring and localization in all-optical networks. IEEE/ACM Trans. Networking 13(6), 1390–1401 (2005)
8. Bergman, R., Médard, M., Chan, S.: Distributed algorithms for attack localization in all-optical networks. In: Network and Distributed System Security (NDSS'98) Symposium, San Diego, Cal., USA, Mar. 1998, session 3, paper 2 (1998)
9. Jurdana, I., Mikac, B.: An availability analysis of optical cables. In: Proceedings of the Workshop on All-Optical Networks, WAON '98, Zagreb, Croatia, pp. 153–160 (1998)
10. Lackovic, M., Mikac, B.: Analytical vs. simulation approach to availability calculation of circuit switched optical transmission network. In: Proceedings of ConTEL, Zagreb, Croatia, June 11-13, 2003, pp. 743–750 (2003)
11. Muchanga, A., Bagula, A.B., Wosinska, L.: On Using Fast Signalling to Improve Restoration in Multilayer Networks. In: Proc. IEEE/OSA Optical Fibre Communication/National Fibre Optic Engineers Conference, OFC/NFOEC'07, Anaheim, CA (March 2007)
12. Wang, X., Chen, G.: Complex networks: Small worlds, scale-free and beyond. IEEE Circuits Syst. Mag. 3(1), 6–20 (2003)
13. Watts, D., Strogatz, S.: Collective dynamics of 'small-world' networks. Nature 393, 440–442 (1998)

14. Barabasi, A.-L., Bonaneau, E.: Scale-free networks. Scientific American 288, 50–59 (2003)
15. Skorin-Kapov, N., Tonguz, O., Puech, N.: Towards efficient failure management for reliable transparent optical networks. Submitted to IEEE Comm. Mag.
16. Inkret, R., Kuchar, A., Mikac, B.: Advanced Infrastructure for Photonic Networks: Extended Final Report of COST Action 266. Faculty of Electrical Engineering and Computing, University of Zagreb, Zagreb, pp. 19–21 (2003)
17. Skorin-Kapov, N., Puech, N.: A self-organizing control plane for failure management in transparent optical networks. In: Hutchison, D., Katz, R.H. (eds.) IWSOS 2007. LNCS, vol. 4725, pp. 131–145. Springer, Heidelberg (2007)
18. Li, J., Yeung, K.L.: A Novel Two-Step Approach to Restorable Dynamic QoS Routing. J. Lightw. Tech. 23, 3663–3670 (2005)
19. Pandi, Z., Tacca, M., Fumagalli, A., Wosinska, L.: Dynamic Provisioning of Availability-Constrained Optical Circuits in the Presence of Optical Node Failures. IEEE/OSA Journal of Lightwave Technology 24(9), 3268–3279 (2006)
20. Markidis, G., Tzanakaki, A.: Network Performance Improvement through Differentiated Survivability Services in WDM Networks. To appear in Journal of Optical Networking (2008)
21. Tzanakaki, A., Markidis, G., Katrinis, K.: Supporting differentiated survivability services in WDM optical networks. In: ICTON 2008, WAOR, invited paper, Athens, Greece (2008)
22. Batchelor, P., et al.: Study on the Implementation of Optical Transparent Transport Networks
23. Mas, C., Tomkos, I., Tonguz, O.: Optical networks security: A failure management framework. In: Proc. of ITCom, Orlando, FL, USA, Sep. 2003, pp. 230–241 (2003)
24. Médard, M., Chinn, S., Saengudomlerti, P.: Node wrappers for QoS monitoring in transparent optical nodes. Journal of High Speed Networks 10(4), 247–268 (2001)
25. Skorin-Kapov, N., Tonguz, O., Puech, N.: Self-organization in transparent optical networks: A new approach to security. In: Proc. of the 9th International Conference on Telecommunications (Contel 2007), Zagreb, Croatia, invited paper, pp. 7–14 (2007)
26. Vasseur, J.-P., Pickavet, M., Demeester, P.: Network Recovery. Morgan Kaufmann, San Francisco (2004)
27. Papadimitriou, D., et al.: Shared risk link groups encoding and processing. Internet draft, work in progress (June 2002), http://www.ietf.org
28. Wauters, N., Ocakoglu, G., Struyve, K., Falcao Fonseca, P.: Survivability in a New Pan-European Carriers' Carrier Network Based on WDM and SDH Technology: Current Implementation and Future Requirements. IEEE Comm. Mag. 6(8), 63–69 (1999)
29. Guillemot, C.: VTHD French NGI Initiative: IP and WDM Interworking with WDM Channel Protection. Presented at the, IP over DWDM conf. (2000)
30. Demeester, P., et al.: Resilience in multi-layer networks. IEEE Communications Magazine 37(8), 70–76 (1998)
31. ITU-T Recommendation G.807/Y.1302, Requirements for automatic switched transport networks (ASTN). ITU-T Standardization Organization (July 2001), http://www.itu.int

10 Optical Storage Area Networks

T.E.H. El-Gorashi (chapter editor) and J.M.H. Elmirghani‾

Abstract Storage area networks (SANs) are a promising technology to efficiently manage the ever-increasing amount of business data. Extending SANs over large distances becomes essential to facilitate data protection and sharing storage resources over large geographic distances. A WDM metropolitan ring network is examined as a suitable extension for SANs where it is shown that sectioning the ring can help deal with traffic asymmetry and hot node (SAN node on ring) scenarios. Several network architectures are studied: One of the architectures accommodates a single SAN and its mirror connected through a sectioning link. Another architecture accommodates two pairs of SANs and their mirrors with sectioning links connecting each pair. Issues investigated include impact of the number of co-existing IP (non SAN) nodes; traffic models: Poisson and self-similar; slotted regime: fixed-size (FS), variable-size (VS) and super-size (SS) slot schemes; MAC protocol design; handling traffic asymmetry and performance measures.

10.1 Introduction

The days when storage systems were expected only to store and retrieve randomly accessible data are long gone. Today storage systems are expected to play an integral role in supporting high levels of flexibility, scalability and data availability. Storage area networks (SANs) [1,2] are emerging as the storage management structure to meet these requirements. SANs were initially designed to work within distance limited environments such as a campus. As the effect of natural disasters such as earthquakes, fires and floods, power outage, and terrorist attacks can be severely destructive in a limited distance environment; the need for extending SANs over large distances has become essential to protect data against loss or damage and to share storage resources among a larger number of users over large geographic areas. Most of the existing literature covering SAN extension is mainly concerned with long-haul overlay. The proposed solutions include optical-based extension solutions and IP-based extension solutions. The optical-based extension solutions include extending SAN over the synchronous optical network (SONET) and over wavelength division multiplexing (WDM). IP based extension solutions encapsulate data units of SAN traffic into standard IP frames to be transported over core networks. In [3], models were developed to compare the reliability of SONET-based extensions with IP-based extensions. It was found that SONET-based solutions are more able to satisfy customers while IP-based solutions have

I. Tomkos et al. (Eds.): COST 291 – Towards Digital Optical Networks, LNCS 5412, pp. 285–302, 2009.
© COST 2009

service interruptions due to hardware/software failure. In this work, a WDM-based SAN extension is considered in a metropolitan sectioned ring scenario.

In earlier work [4,5,6], a WDM slotted ring architecture with a single SAN node was proposed and evaluated. In this work a sectioning link is added to the ring to help deal with traffic asymmetry and hot node (SAN node on ring) scenario. Also a novel technique is introduced to mirror the SAN node to another node (having identical capacity) considered as a secondary SAN. The presence of two SANs and their associated mirrors is also examined in an architecture with an additional sectioning link.

An assumption was made in [7,8,9] that the packet size is fixed. However, in reality the packet size in data communication traffic is variable. In this work, in addition to the fixed-size (FS) slot scheme, two schemes accommodating variable size packets are evaluated—variable-size (VS) slot and super-size (SS) slot schemes [10].

This chapter is organized as follows: In Section 10.1.1 SANs and their protocols are briefly reviewed. Data mirroring techniques are presented in Section 10.1.2. The proposed architectures are presented in Section 10.2. Section 10.3 introduces the proposed mirroring technique. In Section 10.4 simulation results relating to both architectures are presented and analyzed. The conclusions are given in Section 10.5.

10.1.1 Storage Area Networks (SANS)

SANs are emerging as alternatives to traditional direct-attached storage. SANs take storage devices away from servers and connect them directly to the network, simplifying the management of large and complex storage systems. Sharing storage resources over SANs offers a number of benefits. SANs provide storage integration by allowing users to share storage resources across a wide geographic area. SANs also provide high availability and fast recovery from catastrophic disasters via backup techniques. SANs represent the ideal solution for large organizations with growing storage demands as they provide high scalability by eliminating the physical limitations of I/O buses.

A transport network should be designed to provide high throughput, reliability, scalability in terms of distance and number of nodes, and full accessibility to the secondary locations. The fibre channel protocol (FCP) [11] has been considered for years as the premier SAN protocol to transport the SCSI commands used to deliver block storage. FCP provides a reliable, fast, low latency, and high throughput transport mechanism. However, FC was designed to work within environments limited to a few hundred meters however natural disasters, power outage, and terrorist attacks can be severely destructive. In addition, installing FC networks requires separate physical infrastructure and new network management skills. To remedy the distance limitation two extensions of the FCP were developed—FC over TCP/IP (FCIP) and Internet FCP (iFCP) [3]. FCIP is a tunneled solution that interconnects FC SAN islands by encapsulating FC block data and subsequently transporting it over a TCP socket. On the other hand, iFCP is a routed gateway-to-gateway protocol that enables attachment of existing FC devices to an IP network by transporting FC frames over TCP/IP switching and routing elements.

However, FC extensions are still associated with high cost as they assume that the user has already invested in FC components. The IETF developed a new SAN protocol, internet SCSI (iSCSI) [12], to overcome the drawbacks of FC protocols. iSCSI transports SCSI data using already existing networks by encapsulating it in TCP/IP packets which makes iSCSI versatile and affordable even for small businesses.

10.1.2 Data Mirroring Techniques

The concept of back up is vital to improve reliability of data storage systems. Backup methods are usually based on data mirroring [13, 14] where exact replicas of the original data are created to be sent to secondary storage systems in far locations. In the event of lost or corrupted data in the primary storage site, data is retrieved from the secondary storage sites. Traditionally, this process was performed manually by writing data to tapes at the primary site, and then using a vehicle to transport the tapes to remote sites. Manual backup is a time-consuming, disruptive process that requires dedicated personnel resources and equipment. SANs extensions facilitate automatic performance of backups and disaster recovery functions across the MAN or WAN. Data mirroring is usually implemented by one of two strategies—synchronous and asynchronous. In synchronous mirroring data is transmitted from the transmitting node to the two SANs simultaneously; i.e. the state of the SANs is synchronized. However, in addition to the high bandwidth requirements, synchronization can introduce significant delays for large distances. On the other hand, under asynchronous mirroring data is initially transmitted to the primary SAN, and then the primary SAN replicates it to the secondary SAN. Usually asynchronous mirroring is scheduled to run after peak hours when the network bandwidth is idle to save peak hour bandwidth. Therefore it is efficient and cost effective. However it unsuitable for critical applications as the state of the storage locations is not synchronized.

10.2 Network Architectures

The two network architectures considered are illustrated in Fig. 10.1. Both architectures are metropolitan WDM ring networks with a unidirectional (clock wise) multi-channel slotted fibre. The networks connect a number of access nodes within a circumference of 138 kilometers. There are two types of nodes: *access* and *SAN* nodes operating at 1 Gb/s and 5 Gb/s, respectively.

For the first architecture, shown in Fig. 10.1(a), the ring is sectioned using a 44 km point-to-point link passing through the center of the ring. The link consists of a pair of fibres with opposite propagation directions and directly connects two nodes, a primary SAN and its secondary SAN. The sectioning link provides a shorter path for some source-destination pairs to communicate through instead of going through the entire ring and helps deal with traffic asymmetry and the hot-node scenario created by the SANs. The SANs are located at the sectioning points of the link to make use of the links and to ensure that they are separated by the maximum distance to survive disasters in a limited distance scenario.

For the second architecture, shown in Fig. 10.1(b), the ring contains two section-ing links and two pairs of SANs and their mirrors. In [6, 7, 8], an assumption was made that the packet size is fixed and equal to the slot size (Ethernet maximum transfer unit (MTU), defined as 1500 bytes). However, in reality the packet size in data communication traffic is not fixed. According to measurements on the Sprint IP backbone [15], there are mainly five major sizes of packets in data traffic, i.e. 40, 211, 572, 820 and 1500 bytes. The 40 bytes size is for TCP ACKs. The 572 bytes and 1500 bytes are the most common default MTUs. The 211 bytes packets correspond to a content distribution network (CDN) proprietary user datagram protocol (UDP) application that uses an unregistered port and carries a single 211 bytes packet. The packets of around 820 bytes are generated by media streaming applications. Obviously, the original architecture is not suitable for this situation, in which a huge percentage of slot space will be wasted. In this paper, in addition to the FS slot scheme, the performance of the sectioned ring is evaluated under two different schemes that accommodate variable size packet traffic—variable-size (VS) slot and super-size (SS) slot schemes. For the VS slot scheme, slots of five different sizes circulate around the ring. The slots sizes correspond to the five different sizes of data traffic packets from the access links, which are 40, 211, 572, 820 and 1500 bytes, with probabilities of 0.1, 0.2, 0.1, 0.2 and 0.4, respectively [15]. The total number of slots on the ring is 240. The number of slots for each size (related to the probability distribution of packet size on the access links) is 24, 48, 24, 48 and 96, respectively. The average size for a slot is 870.44 bytes. In the simu-

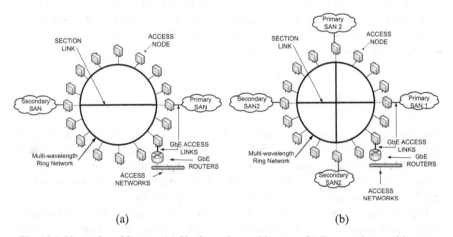

(a)   (b)

Fig. 10.1. Network architecture (a) Single section architecture (b) Two sections architecture

lation, the different size slots are generated with a fixed sequence. The 240 slots are divided into 24 groups, in each group there are 10 slots with one slot of 40 bytes, two slots of 211 byes, one slot of 572 bytes, two slots of 820 bytes and four slots of 1500 bytes. In order to simplify the calculation and rotation, the length of the ring is changed to 133 kilometres. In the SS slot scheme, the ring is divided into several super-size slot (24 super slots), which are much larger than the packets in the traffic (9000 bytes). Obviously, this approach is more realistic and suitable for a general situation than the VS slot scheme as it is not based on the packet size distribution. In this scheme, the length of the ring does not change.

Fig. 10.2 illustrates the logical topology of the two architectures. The two architectures can be considered as a number of logical rings: three for the single section architecture and five for the two sections architecture. The logical ring for each source-destination pair over the time slotted ring is chosen according to a shortest path algorithm. To demonstrate the effects of network loading, asymmetric traffic, hot-node scenario and the impact of sectioning, the number of wavelength is limited to 4 for the single section ring which is the optimum number needed given this architecture. The minimum number of wavelength can be decided by considering the three logical rings. Considering a 24 node network, there are 11 nodes common to rings A and B each operating at 1 Gb/s. If the wavelength rate is 2.5 Gb/s, then 4 wavelengths are needed, ignoring the statistical multiplexing gain achieved by the ring, i.e. 2 wavelengths per ring. Rings B and C can use the same two wavelengths as the sectioning link is made up of two counter propagating fibres. Therefore a total (minimum) of 4 wavelengths is needed in this architecture offering each of the three rings 2 wavelengths. A minimum number of 6 wavelengths can be calculated in the same way for the two sections architecture.

Under the single section architecture, each access node is equipped with two fixed transmitters to connect it to the two logical rings available to it. SAN nodes

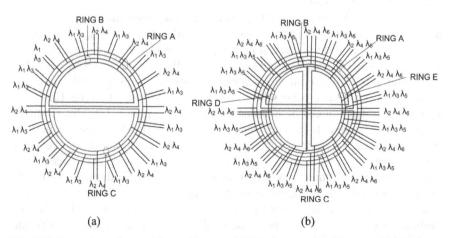

(a)   (b)

Fig. 10.2. Logical topology of the architecture (a) Single section architecture (b) Two sections architecture

placed at the sectioning points require an additional transmitter as each has access to the three logical rings. As the number of nodes is greater than the number of wavelengths, each wavelength is shared by a number of nodes for transmission. Four fixed receivers are used to allow nodes to share wavelengths for reception which results in higher scalability compared to the single fixed receiver architecture where the number of wavelengths is equal to the number of nodes. The use of a tunable receiver is possible but results in receiver collisions.

Under the two sections architecture, each node is equipped with three fixed transmitters and six fixed receivers. The SAN nodes also require an additional transmitter as they have access to four logical rings.

10.3 Proposed Mirroring Technique

A simple technique is used to mirror each primary SAN node to its corresponding secondary SAN node. Under this mirroring technique, the secondary SAN nodes do not send any traffic to the primary SAN nodes and ordinary nodes do not send any traffic to the secondary SAN nodes. However, the secondary SAN nodes ultimately receive all the traffic addressed to the primary SAN nodes as the primary SAN node remove a packet from a slot upon reception only if its corresponding secondary SAN node has already received this packet. Otherwise it will let the packet remain in the ring to go to the secondary SAN node. Therefore those packets passing by the primary SAN node first will travel further in the network to be mirrored in the corresponding secondary SAN node which means extra bandwidth is used. However, on average this proposed mirroring scheme saves bandwidth and introduces efficiency in that separate transmissions are not needed to synchronise the SAN and its mirror. The Two remain synchronised at all time subject to the ring propagation delay.

According to the original MAC protocol, for the single section architecture, nodes in the upper part of the network can use either ring B or ring A to send to the primary SAN node. Although both rings result in the same distance to the primary SAN node, ring A results in a longer distance to the secondary SAN node. This extra distance increases mirroring time and leads to inefficient bandwidth usage. To reduce bandwidth usage and the mirroring time for the upper nodes, a modification can be introduced to the MAC protocol. In this modified version of the protocol, the upper nodes have to use ring B to send to the primary SAN node. To overcome the extra load introduced to ring B, a wavelength can be taken from ring A and assigned to rings B and C. Ring A can accommodate its traffic in a single wavelength as less traffic travels through it (40% to 60% is assumed to be destined to the SAN node in the asymmetric scenario). The mirroring technique with the modified MAC protocol is applied to the two sections ring. Fig. 10.3 illustrates the modified mirroring technique with the modified MAC protocol for both architectures.

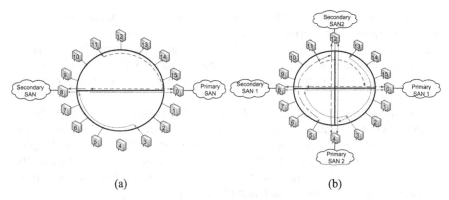

                    (a)   (b)

Fig. 10.3. Mirroring Technique (a) Single section architecture (b) Two sections architecture

10.4 Performance Evaluation

Simulation is carried out to evaluate the node throughput and queuing delay of the two architectures under the proposed mirroring technique. Two different traffic models are used—Poisson and self-similar. It has been shown in [16] that LAN and WAN traffic is better modelled using statistically self-similar processes. However, Poisson models are still used because they are analytically tractable and can be modelled easily compared to self-similar processes; see e.g. [17,18].

Networks of 16 and 24 nodes are simulated. The performance of the networks is evaluated under varying levels of traffic loads. The presence of the SANs creates traffic asymmetry and hot-node scenario i.e. the access nodes send to the primary SANs with a relatively higher probability while they send to each other with equal probabilities. For the single section ring, the primary SAN receives 40% or 60% of the total traffic. For the two sections ring this amount is divided equally between the two primary SANs.

10.4.1 Single Section Ring Architecture

Two networks of 16 and 24 nodes are simulated for the single section ring. To reflect the effect of traffic asymmetry and mirroring on the performance of the networks, average results relating to access nodes in the upper part of the network are shown separately from those of access nodes in the lower part of the network. Also results of the primary SAN are shown separately from those of the secondary SAN.

10.4.1.1 Results of a 16 Node Network

In the case of a 16 node network, an aggregated data rate of 20 Gb/s (5 Gb/s from the SAN node and 15 Gb/s from other ordinary nodes) is generated by nodes without the mirroring technique and 24 Gb/s (10 Gb/s from the SAN nodes and 14 Gb/s from other ordinary nodes) with the mirroring technique. This maximum traffic represents a normalized load of 1, L=1. Note that several choices exist and the two extreme cases (mirror does not transmit data, and mirror transmits to all nodes at the same rate as the primary SAN) are considered. Intermediate scenarios exist where for example nodes choose to retrieve data from the SAN or mirror according to proximity. In this case the SAN and the mirror each transmit only to a subset of the total nodes.

It worth mentioning that the total bandwidth capacity of the WDM ring is around 15 Gb/s (2.5 Gb/s × 6, where for rough guidance only we assume that the introduction of the section is similar to the introduction of two extra wavelengths) and therefore a normalized load of 1 will create more traffic on the ring than the total carrying capacity of an unslotted WDM ring network, however the slotted regime introduces a further spatial multiplexing gain.

The 16 node network results emphasize the performance difference introduced by the mirroring technique under the FS slot scheme. The performance is shown under the mirroring technique with the original and the modified MAC protocols. All these cases are evaluated under Poisson traffic. For comparison, the performance of the network with mirroring under the modified MAC protocol is also evaluated under self-similar traffic. All results are shown under both 40% and 60% asymmetric traffic.

Fig. 10.4(a) shows the node throughput results of the upper nodes. They manage to achieve the maximum throughput with and without mirroring for both 40% and 60% asymmetric traffic. However, this is expected as under both the original and modified MAC protocol, the upper nodes are not affected by mirroring as the bandwidth available to them will not be used by any of the mirrored packets. Mirrored packets from the lower nodes will always go through ring C. The figure also shows that the use of self-similar traffic slightly reduces the achievable throughput, under 40% asymmetric traffic when compared with the case where Poisson traffic is used. This is expected due to the burstiness of the traffic produced by self-similar sources.

Without the mirroring strategy, the lower nodes (Fig. 10.4(b)) manage to achieve the maximum throughput for 40% asymmetric traffic while for 60% asymmetric traffic the throughput is slightly less than the maximum. This is understood as higher proportion of traffic sent to the primary SAN node means more load on ring C. It can also be seen from the figure that while under Poisson traffic the modified MAC protocol achieves the maximum throughput, the throughput very slightly decreases under self-similar traffic as with the upper nodes.

For the primary SAN node (Fig. 10.4(c)), it can be seen that mirroring with the original MAC protocol decreases the throughput achieved as half of the packets (those from the upper nodes), which used to be emptied and possibly reused by the primary SAN node, continue their way to the secondary SAN node to be emptied and possibly reused. Therefore the bandwidth available to the primary SAN node decreases and the bandwidth available to the secondary SAN node increases. It is noticed from the figure that without mirroring the original MAC protocol performs better under 60% asymmetric traffic compared to 40% asymmetric traffic as increasing the proportion of traffic going to the primary SAN node means more slots will be emptied and possibly reused by it. It can be seen from the figure that the performance of the primary SAN node has improved under the modified MAC protocol as the extra bandwidth introduced by the added wavelength is greater than the extra traffic added to ring B. However, this extra traffic is more in the case of 60% asymmetric traffic which makes performance under 60% asymmetric traffic worse than 40% asymmetric traffic although in the first case more packets are emptied and possibly reused by the primary SAN node. The figure also shows that the throughput of the modified MAC protocol decreases under self-similar traffic. It is noticed from the figure that under the mirroring cases the network becomes heavily loaded at load of 0.7 and therefore the throughput is almost constant for higher loads.

For the secondary SAN node (Fig. 10.4(d)), without the mirroring technique the throughput achieved under 40% asymmetric traffic is higher than 60% asymmetric traffic as under 60% asymmetric traffic ring C is more loaded. The throughput reaches its maximum at a load of 0.9 and then decreases as the network becomes heavily loaded. As mentioned before, under the mirroring technique the maximum transmission rate of the secondary SAN node increases to 5 Gb/s. Although under mirroring more packets will be emptied and possibly reused by the secondary

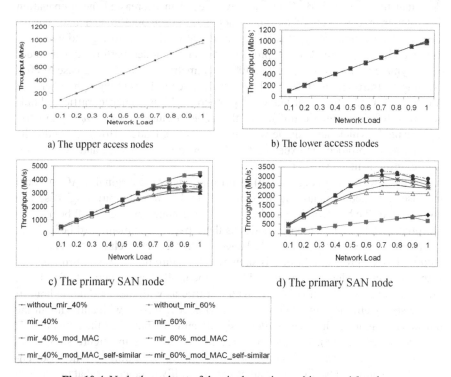

a) The upper access nodes

b) The lower access nodes

c) The primary SAN node

d) The primary SAN node

→ without_mir_40%	→ without_mir_60%
mir_40%	mir_60%
→ mir_40%_mod_MAC	→ mir_60%_mod_MAC
→ mir_40%_mod_MAC_self-similar	→ mir_60%_mod_MAC_self-similar

Fig. 10.4. Node throughput of the single section architecture-16 nodes.

SAN under 60% asymmetric traffic, still the high load on ring C makes the performance of the original MAC protocol better under 40% asymmetric traffic. It can be seen from the figure that the secondary SAN node under the original MAC protocol achieved its maximum throughput at a load of 0.6 then the through appears to be almost constant and then decreases as the network gets heavily loaded. It can be seen from the figure that while the modified MAC protocol manages to increase the throughput under 60% asymmetric traffic, the throughput decreases under 40% asymmetric traffic. This is understood as under 40% asymmetric traffic more traffic is going through ring A compared to 60% asymmetric traffic. Therefore taking a wavelength from ring A and giving it to ring C would result in decreasing the throughput. Again under self-similar traffic the throughput is reduced.

Fig. 10.5(a) shows the upper nodes average queuing delay. The increase in queuing delay introduced by mirroring is due to the increase in the transmission rate of the secondary SAN node which reduces the bandwidth available to the upper nodes. It can be seen from the figure that under mirroring the original MAC protocol outperforms the modified protocol. This is understood if we remember that under the original MAC protocol, traffic to the primary SAN node can go through either ring A or B, each having 2 wavelengths while under the modified MAC protocol only ring B is available with 3 wavelengths. However, it should be clear that the modified MAC protocol will result in decreasing the propagation time of mirrored packets of the upper nodes by an amount greater than the increase in the queuing delay. It is also noticed that without mirroring and with mirroring under the original MAC protocol, the upper nodes perform better under 60% asymmetric traffic compared to 40% asymmetric traffic as in the case of 60% asymmetric traffic more traffic will have the chance to be sent through either ring A or ring B. Under the modified MAC protocol, 60% asymmetric traffic still outperforms 40% asymmetric traffic but this time because under 40% asymmetric traffic ring A, which has a single wavelength, is more loaded. However, under the original MAC protocol as expected the difference between 40% and 60% asymmetric traffic is much smaller.

For the lower nodes (Fig. 10.5(b)), mirroring under the original MAC protocol increases the queuing delay slightly compared to the "without mirroring" case as a proportion of the bandwidth available to the lower nodes in ring A will be used by mirrored packets from the upper nodes as these packets, as mentioned before, can go through either ring B or ring A. However, this effect is eliminated under the modified MAC protocol. Also, it is shown in the figure that without mirroring and with mirroring under the original MAC protocol, 40% asymmetric traffic has a lower queuing delay than 60% asymmetric traffic as with the node throughput. It can be seen from the figure that for 60% asymmetric traffic, with and without the mirroring technique, the queuing delay significantly increases as the network gets heavily loaded (L>0.8). For the modified MAC protocol, the figure shows that while the queuing delay is reduced under 60% asymmetric traffic, it increases under 40% asymmetric traffic. This can be explained if we remember that under 40% asymmetric traffic more traffic goes through ring A compared to 60% asymmetric traffic. Therefore taking a wavelength from ring A and giving it to ring C results in deteriorating the performance.

Similar trends to those of node throughput are observed for the queuing delay of the primary SAN node (Fig. 10.5(c)) under Poisson traffic. While the network gets heavily loaded and the queuing delay significantly increases at loads higher than 0.7 without the mirroring technique, it gets heavily loaded and the queuing delay significantly increases at loads higher than 0.6 with the mirroring technique. It can be seen from the figure that without the mirroring technique under 60% asymmetric traffic and with the mirroring technique under the original MAC protocol that after a certain load the queuing delay appears to be almost constant as the buffers become full. The figure also shows that the use of self-similar traffic gives higher queuing delay than with Poisson traffic when the load is below the 0.8. For higher loads, the performance under self-similar traffic is better than under Poisson traffic. This is understood from the nature of Poisson traffic whose transmission rate, at high loads, becomes more constant as the packet interarrival duration decreases. Under a load of 1, the maximum transmission capacity of the network is reached; the congestion state becomes severe due to the almost non-variable, constant arrival of packets where the nodes transmission buffers are constantly filled with packets. Therefore, the queuing delay also increases rapidly when the congestion state is reached. On the other hand, under self-similar traffic, packet bursts tend to frequently increase the buffer loads and the average queuing delay value is seen to be higher than with Poisson traffic when $L<0.8$. However, as the maximum transmission capacity of the network is reached, the buffers are filled in an intermittent way and the congested state does not induce an abrupt decline in networking performance. Therefore, the queuing delay appears to be almost constant for higher loads.

Fig. 10.5(d) shows the queuing delay of the secondary SAN node. Due to the higher transmission rate, the queuing delay increases with mirroring at loads less than 0.9. For higher loads the queuing delay for the secondary SAN node without the mirroring technique is higher as ring C becomes highly loaded and fewer packets are emptied by the node compared to the mirroring technique. Applying self-similar traffic results in reducing the queuing delay for loads higher than 0.8 as with the primary SAN node. Also the queuing delay after a certain load appears to be almost constant as with the primary SAN node. Other trends in the figure are similar to those of the node throughput.

10.4.1.2 Results of a 24 Node Network

In the case of a network of 24 nodes, a maximum aggregated rate of 28 Gb/s (5 Gb/s from the SAN and 23 Gb/s from the access nodes) is generated by nodes without the mirroring technique, and 32 Gb/s (10 Gb/s from the SANs and 22 Gb/s from the access nodes) with the mirroring technique. Therefore the network is more loaded compared to the 16 node network case.

Results of the 24 node network illustrate the performance difference introduced by the modified mirroring technique under the FS slot scheme. Also they compare the performance of the VS and SS slot schemes. Performance of the mirroring technique with the FS slot scheme is also evaluated under self-similar traffic. All results are shown under both 40% and 60% asymmetric traffic.

Fig. 10.6(a) shows the node throughput of the upper access nodes. The maximum throughput is achieved without the mirroring technique under 40% asymmetric traffic. Under 60% asymmetric traffic the throughput is slightly less than the maximum under L= 1 as more traffic is destined to the primary SAN which heavily loads ring B. It can be seen that applying the mirroring technique reduces the throughput. This is due to the increase in the transmission rate of the secondary SAN which reduces the bandwidth available to the upper access nodes. The reduction in throughput is significant under 40% compared to 60% asymmetric traffic as under 40% asymmetric traffic ring A, which has a single wavelength, is more loaded. All these effects are not clear for the 16 node network as the network is less loaded. As with the 16 node network, Fig. 10.6(a) also shows that the use of self-similar traffic reduces the throughput significantly. The significant improvement obtained by applying the VS slot scheme is also clear in Fig. 10.6(a). This improvement is due to the increase in slot space utilization, which is defined as the ratio of used slot space to the total slot space.

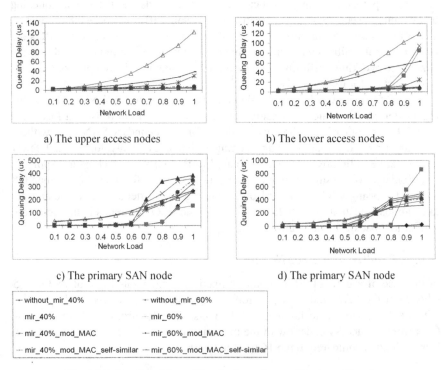

a) The upper access nodes   b) The lower access nodes

c) The primary SAN node   d) The primary SAN node

Fig. 10.5. Queuing delay of the single section architecture-16 nodes

As mentioned before, in the FS slot scheme, the size of the slot is 1500 bytes. However, in data communication traffic, the average size of the packets is equal to 867.4 bytes, which is about 58% of 1500 bytes. This means that only 58% of the slot space is used on average and the rest is wasted. In the VS slot scheme, the slot space is fully used. Therefore the maximum throughput of the FS slot scheme is predicted to be around 58% of that of the VS slot scheme. However the through-puts obtained in this case do not match these predictions as the maximum throughput is achieved with the VS slot scheme. However results in the two sections ring will match the predictions. The difference between VS and SS slot schemes will be clear in the queuing delay results.

Without the mirroring strategy, the lower access nodes (Fig. 10.6(b)) achieve lower throughput than that achieved by the upper access nodes. This is understood as packets from upper access nodes destined to the primary SAN can be transmitted through either ring A or B. However this is not the case for lower access nodes where these packets have to be transmitted through ring C. Lower access nodes perform better under 40% asymmetric traffic compared to 60% asymmetric traffic as higher proportion of traffic sent to the primary SAN means more load on ring C. Applying the mirroring technique results in reducing the achieved throughput under 40% asymmetric traffic. However, under 60% asymmetric traffic the throughput increases. This is a result of removing a wavelength from ring A and adding it to ring C as under 60% asymmetric traffic, traffic going through ring A can be accommodated in one wavelength i.e. giving more bandwidth to ring C traffic without affecting traffic on ring A. Again the difference introduced by mir-roring and different traffic asymmetry is clearer in the case of the 24 node network compared to the 16 node network. It can also be seen that self-similar traffic re-duces the throughput as with the 16 node network. The same trends of the VS and SS slot schemes for upper access nodes can be noticed for the lower access nodes.

As noticed for the 16 node network, it can be seen that the primary SAN (Fig. 10.6(c)) without mirroring performs better under 60% compared to 40% asymmet-ric traffic. Similar to the "without mirroring" case, 60% asymmetric traffic outper-forms 40% asymmetric traffic under the mirroring technique. The figure also shows that with the mirroring technique the throughput decreases under self-similar traffic. It can also be noticed that the SS slot scheme has resulted in further improvement in the throughput under loads less than or equal to 0.9. This can be understood if we remember that in the VS slot scheme, the different size slots are only allowed to carry packets with the corresponding size. However, if an empty slot arrives and the buffer, which contains its corresponding packets, is empty, the slot will be released to the downstream nodes which will affect the slot utilization probability (the ratio of the number of empty slots which have been used to the number of total empty slots). On the other hand, the SS slot scheme achieves a better slot utilization probability as the super slot can accommodate all packet sizes. However, it can be seen from the figures that under extremely high loads (L> 0.9) the VS slot scheme outperforms the SS slot scheme. It is noticed from the figure that for the mirroring cases the network becomes heavily loaded at loads higher than 0.5 where the throughput is almost constant up to a load of 0.8. For higher loads, the throughput starts to decrease.

For the secondary SAN (Fig. 10.6(d)), as with the 16 node network, without the mirroring technique the throughput under 40% asymmetric traffic is higher than 60% asymmetric traffic as ring C is more loaded under 60% asymmetric traffic. As mentioned before, under the mirroring technique the maximum transmission rate of the secondary SAN increases to 5 Gb/s. Other trends in Fig. 10.6(d) are similar to those of the primary SAN.

Fig. 10.7(a) shows the upper access nodes average queuing delay. It can be seen that with the mirroring technique under 40% asymmetric traffic after a certain load the queuing delay appears to be almost constant as the buffers become full. All the trends noticed for the throughput of FS slot scheme can be noticed for the queuing delay. The difference between the VS and SS slot schemes is clear. For the lower access nodes (Fig. 10.7(b)) similar trends are noticed.

Similar trends to those of node throughput are observed for the queuing delay of the primary SAN (Fig. 10.7(c)) under Poisson traffic. It can be seen from the figure that without the mirroring technique under 60% asymmetric traffic that after a certain load the queuing delay appears to be almost constant as the buffers become full. The figure also shows that as with the 16 node network the use of self-similar traffic gives higher queuing delay than with Poisson traffic when load is below 0.8.

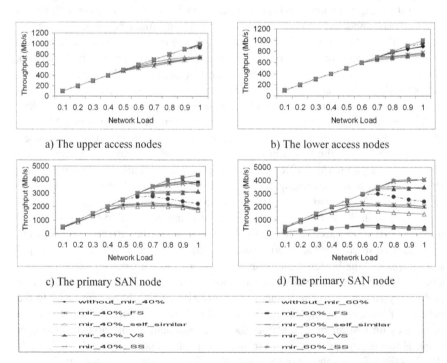

Fig. 10.6. Node throughput of the single section architecture-16 nodes

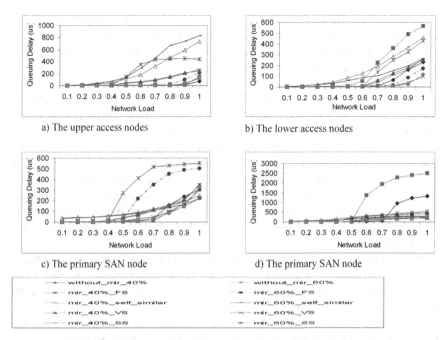

a) The upper access nodes

b) The lower access nodes

c) The primary SAN node

d) The primary SAN node

Fig. 10.7. Queuing delay of the single section architecture-16 nodes

Fig. 10.7(d) shows the queuing delay of the secondary SAN. As with the 16 node network, the queuing delay increases with mirroring at loads less than 0.9. For higher load the queuing delay for the secondary SAN without the mirroring technique is higher. Also applying self-similar traffic results in reducing the queuing delay for loads higher than 0.6 and the queuing delay after a certain load appears to be almost constant. The trends of VS and SS slot schemes are similar to those of the node throughput.

10.4.2 Two Sections Ring Architecture

The two sections architecture is evaluated for a network of 24 nodes. Simulation results of the two sections ring compare the performance of the different slot schemes under the mirroring technique. All the results are also shown under both 40% and 60% asymmetric Poison traffic. For comparison reasons result of the FS slot scheme are shown also under self-similar traffic.

For the two sections ring architecture, the aggregated data rate is 40 Gb/s (20 Gb/s from the SANs and 20 Gb/s from the access nodes). As mentioned before 6 is the minimum number of wavelengths for this architecture giving a total bandwidth around 25 Gb/s (2.5 Gb/s × 10, where we assume that there are 5 rings with 2 wavelengths per ring).

To reflect the effect of traffic asymmetry and mirroring on the performance of the network, average results of the access nodes are shown separately from those of the SANs (introducing the two sections create symmetry in the performance of different parts of the network).

Fig. 10.8(a) shows the average throughput of the access nodes. As with the single section architecture the significant improvement obtained by applying the VS slot scheme is clear. Also it can be seen that while the maximum average throughput achieved under the VS slot scheme reaches 980 and 930 Mb/s for 40% and 60% asymmetric traffic, respectively, it was limited under the FS slot scheme to 590 and 540 Mb/s for 40% and 60% asymmetric traffic, respectively, which is around the theoretical predictions mentioned previously. It is also noticed from the figure that under the three different slot schemes, the throughput achieved under 40% asymmetric traffic is higher than that under 60% asymmetric traffic. This is understood as higher proportion of traffic sent to the SANs unbalances traffic between logical rings.

Fig. 10.8(b) presents the average node throughput for the SANs. The significant increase in the throughput achieved by the VS slot scheme compared to the FS slot scheme is noticed. The average node throughput under L=1 increased from 2390 to 4000 Mb/s under 40% asymmetric traffic, and from 2550 to 4330 Mb/s under 60% asymmetric traffic. These values are not far from the predictions. Unlike the performance of the access nodes, it is noticed that 60% asymmetric traffic outperforms 40% asymmetric traffic as higher proportion of traffic sent to the SANs means more packets emptied and possibly reused by them. The difference in performance between the VS and SS scheme is not clear for the node throughput. It is noticeable for the queuing delay as discussed below.

The average queuing delay for the access nodes and the SANs are shown in Fig. 10.9(a) and Fig. 10.9(b), respectively. In both cases, it is clear that applying the VS slot scheme has significantly reduced the queuing delay. As with the single section ring, the SS slot scheme has resulted in further reduction in the queuing

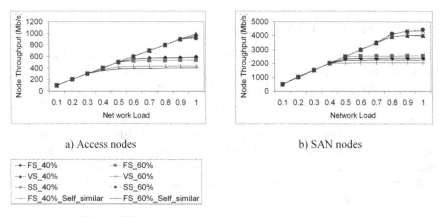

a) Access nodes   b) SAN nodes

—•— FS_40%	—■— FS_60%
—▲— VS_40%	—×— VS_60%
—×— SS_40%	—•— SS_60%
---- FS_40%_Self_similar	---- FS_60%_Self_similar

Fig. 10.8. Node throughput of the two sections architecture

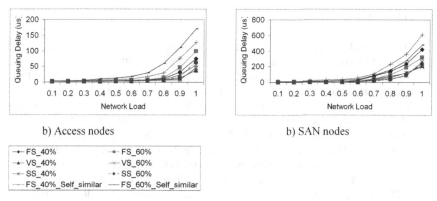

b) Access nodes   b) SAN nodes

Fig. 10.9. Queuing delay of the two sections architecture

delay under loads less or equal to 0.9. Other trends in Fig. 10.9 are similar to the throughput.

From Fig. 10.8 and Fig. 10.9, it is clear that applying self-similar traffic results in a worse performance as with the single section ring.

10.5 Conclusions

In this chapter, optical storage area networks and the performance of a number of novel mirroring techniques were evaluated. Simulation was carried out for two metropolitan WDM ring architectures under both Poisson and self-similar traffic. In addition to the FS slot scheme, performance was evaluated under two different slot schemes accommodating variable size packets — VS slot and SS slot schemes. Results of node throughput and queuing delay were presented and analyzed. For the single section architecture, two networks of 16 and 24 nodes were evaluated. The results showed that applying the proposed mirroring technique has different effects on different parts of the ring. For the upper access nodes mirroring degraded the performance. The deterioration was significant under 40% compared to 60% asymmetric traffic. For the lower access nodes, while applying the mirroring technique resulted in worse performance under 40% asymmetric traffic, under 60% asymmetric traffic the performance improved. The primary SAN performance was impact under the mirroring technique however the maximum transmission rate of the secondary SAN increased to 5 Gb/s.

For the two sections ring, a network of 24 nodes was evaluated. The results showed that the access nodes achieved good performance under the FS slot scheme. Because of their high transmission rate, the SAN nodes suffered from more performance degradation compared to the access nodes.

Significant improvements in the performance of both architectures were obtained under VS and SS slot schemes.

References

1. Qiu, X., et al.: Reliability and Availability Assessment of Storage Area Network Extension Solutions. IEEE Commun. Mag., pp. 80–85 (Mar. 2005)
2. Ehrhardt, A.: Extension of Storage Area Networks and Integration on Different Platforms of an Optical Transport Network. In: Proc. ICTON 2004, pp. 229–232 (2004)
3. Telikepalli, R., Drwiega, T., Yan, J.: Storage Area Network Extension Solutions and Their Performance Assessment. IEEE Commun. Mag., 56–63 (Apr. 2004)
4. Pranggono, B., Elmirghani, J.M.H.: Performance evaluation of INSTANT a metro WDM SAN under balanced and unbalanced traffic conditions. In: 7th International Conference on Transparent Optical Networks (ICTON'05), Barcelona (July 2005)
5. Pranggono, B., Elmirghani, J.M.H.: A Novel Optical SAN Implementation in a Metro WDM Setting. In: Proc. IEEE High Performance Switching and Routing (HPSR), Hong Kong (May 2005)
6. Pranggono, B., Elmirghani, J.M.H.: Design of storage area network based on metro WDM networking. In: Proceedings of IEEE London Communications Symposium, London, September 2005, vol. 1 (2005)
7. El-Gorashi, T.E.H., Prangonno, B., Elmirghani, J.M.H.: Multi-wavelength Metro WDM Sectioned Ring for SAN Extension under Hot Node Scenario and Variable Traffic Profiles. In: Proc. 8th International Conference on Transparent Optical Networks (ICTON'06), UK, 18-22 June (2006)
8. El-Gorashi, T., Pranggono, B., Elmirghani, J.M.H.: A Data Mirroring Technique for SANs in a Metro WDM Sectioned Ring. In: Proc. 12th Optical Network Design and Modelling (ONDM'08), Catalonia, Spain, March 12-14 (2008)
9. El-Gorashi, T.E.H., Prangonno, B., Elmirghani, J.M.H.: A Mirroring Strategy for SANs in a Metro WDM Sectioned Ring Architecture under Different Traffic Scenarios. Journal of Optical Communications 29 (2008)
10. Chen, B., Pranggono, B., Elmirghani, J.M.H.: A Metro WDM Multi-ring Network with Variable Packet Size. In: IEEE International Conference on Communications, ICC'06, Istanbul, 11-15 June (2006)
11. Yang, H.: Fibre Channel and IP SAN Integration. In: 12th NASA Goddard/21st IEEE Conf. Mass Storage Sys. And Tech. (Apr. 2004)
12. Lu, Y., Du, D.H.C.: Performance Study of iSCSI-Based Storage Subsystems. IEEE Commun. Mag. (Aug 2003)
13. Ji, M., Veitch, A., Wilkes, J.: Seneca: remote mirroring done write. In: Proceedings of USENIX Technical Conference, San Antonio,TX, June 2003, USENIX, Berkeley (2003)
14. Yan, R., Shu, J.-w., Wen, D.-c.: An implementation of semi-synchronous remote mirroring system for sANs. In: Jin, H., Pan, Y., Xiao, N., Sun, J. (eds.) GCC 2004. LNCS, vol. 3252, pp. 229–237. Springer, Heidelberg (2004)
15. Fraleigh, C., Moon, S., Lyles, B., Cotton, C., Khan, M., Moll, D., Rockell, R., Seely, T., Diot, S.C.: Packet-level traffic measurements from the Sprint IP backbone. IEEE Network 17, 6–16 (2003)
16. Baldi, M., Ofek, Y.: End-to-end delay analysis of videoconferencing over packet-switched networks. IEEE/ACM Trans. Networking 8, 479–492 (2000)
17. Paxson, V., Floyd, S.: Wide area traffic: The failure of Poisson modeling. IEEE/ACM Trans. Network. 3(3) (1995)
18. Marsan, M.A., Bianco, A., Leonardi, E., et al.: All-optical WDM multi-rings with differentiated QoS. IEEE Communications Magazine 37(2), 58–66 (1999)

Future Outlook (Part II)

M. Köhn (part editor)

In this part of the book, an abridgement has been given of the technical work performed in Working Group 2 and 3 of COST action 291. The chapters summarize the most important results and give insight into the investigated aspects of optical transport networks, ranging from network planning to operation, from traffic engineering to resilience aspects, from fundamental network concepts to application scenarios.

The design targets of all these architectures as well as the metrics of their evaluation focus on the overall performance—the quality of the services with all their aspects and the amount of traffic, that is carried by the network—or vice versa the amount of resources required to carry a given traffic and to provide a certain grade of service. Furthermore, the studies only focus on selected parts of the networks, i.e. core or metro networks. This was driven by the operators' requirement to reduce the capital expenditures and operational cost as well as the network operation strategies. Nevertheless, with the change of the requirements and the provisioned services new topics are coming up and gain higher impact.

One of the most prominent these days is the power consumption. Buzzwords like Green IT are of universal concern. While for a long time the processing capacity of network nodes was limited by the power supply and thermal power loss, in the future, the network design must additionally consider the overall power consumption due to global warming as well as lack of energy. This will also gain impact on new metrics like the energy consumed per bit in *Joule per Bit*.

A second prominent topic is the convergence of optical networks. While backbone and core networks mainly rely on optical networks and new optical solutions are deployed in access networks, these networks are separated. To deliver end-to-end services with performance guarantees, the convergence of the distinct network solutions is necessary to accomplish an integrated optical platform. This leads to a unified control and management framework which allows convergence on different levels.

Part III

Introduction (Part III)

B. Mikac (part editor), C. Gaumier, M. Karasek, and S. Rumley

Almost all researchers and designers in the field of the optical communication systems and networks use commercial, open-source or in house developed software tools for modelling, analysis, design, dimensioning and optimisation, and performance evaluation. The two chapters in Part III present the activities of the partners in COST Action 291 on the development and application of the tools in both optical physical and network layers. This part is the collection of not only individual contributions of the partners based on variety of the languages, platforms and software systems, but also the experiences and proposals to improve the interaction between different network tools aimed at achieving synergy in tool applications between different project teams. The need for "tool networking" is especially emphasised in the research at the optical network level.

Modelling of a physical layer in the optical communication (Chapter 11) can be roughly divided into simulation and design of the optical devices for the communication systems and numerical modelling of the systems or sub-systems. Development of the software tools for the physical layer modelling started in the 1980s with the rapid advance in the optical communication systems. Numerical analysis of the optical components proved very useful during development of the optical fibres, starting from a multimode to a single mode, and special fibres such as dispersion compensating and rare-earth doped fibres. Similar evolution underwent the semiconductor lasers, semiconductor amplifiers, waveguide amplifiers, fibre Bragg gratings and many other optical components. This strong interest in numerical simulations and modelling is attributed to the prohibitive prototype fabrication cost of the majority of optical components for the communication systems.

Modelling of the optical communication modules provides explanation and interpretation of the effect of several characteristics (e.g. chromatic and polarisation dispersion, Kerr effect) on the transmission of various signals (differing in bit rate, modulation format, input signal power, number of channels and channel separation in case of the WDM regime).

Several software packages for simulation of the optical components and communication systems are commercially available and widely used by the R&D community. The Working Group I was focused on the development of own software tools for applications uncovered by the commercially available packages. Chapter 11 summarises the results obtained in simulation of the semiconductor lasers and laser amplifiers, interactive simulation of the optical communication systems, numerical analysis of transient effects in the Raman fibre amplifiers, and efficiency and accuracy improvement of the split-step Fourier analysis suitable for solving the non-linear Schrödinger equation.

Chapter 12 describes utilisation of the software tools at the network layer. Due to the expensive nature of the optical networks, their design, dimensioning and optimisation require extreme care. Therefore, network planning is the important component of the software tools. On the other hand, optical communications networks being distributed in many locations, stretching over thousands of kilometres, and encompassing multiple different technologies and schemes, are extremely complex systems. Purely analytical models struggle to encompass this complexity, while the field experiments, similarly to the physical layer, have prohibitive costs. The remaining part of software tools, thus, regroups numerical models and simulators. They provide better understanding of the existing systems, identification of their flaws and allow for rapid prototyping of the novel systems.

In the first two sections of Chapter 12, several principles of modelling are described, and a classification of the modelling methods and of the software-based simulation techniques is presented. The software tools being subject to multiple problems related to their deployment, execution or integration with a third party software, a software engineering-oriented insight is also given. A tool integration is seen as a potential solution to the complexity of the present and future tools.

The next two sections present simulation models based on the commercial (OPNET) or open-source (OMNET++) software. A case study of Medium Access Control protocol development for Gigabit Passive Optical Network shows that the existing libraries of the network elements available in OPNET considerably facilitate modelling. Section 12.5 shows how performance of the automatically switched optical networks carrying IP/GMPLS are assessed using the OMNET++ simulator.

In the last sections of this Chapter, the proprietary tools developed and used by the partners in this action are presented. MatPlanWDM, an open source academic tool implementing several routing heuristics and algorithms, wavelength assignment and virtual topology design is presented. With its user-friendly graphical interface it does not require prior knowledge of the internal function which makes it very suitable for training purposes. On the other hand, being easily configurable and extendable it can be applied in research. Finally, the software-based experiments, e.g. simulations or network planning method assessments, can be conducted by programming everything from a scratch, but development is facilitated with the re-use of common libraries shared among the projects. Therefore, the last two sections present two of these libraries, the Javanco environment and the IKR Simlib, and show structuring of such packages.

This Part is concluded by the overview of the work realised in software development and with the proposed directions for future research.

11 Software Tools and Methods for Modelling Physical Layer Issues

M. Karasek (chapter editor), S. Aleksic, M. Jaworski, and J. Leibrich

Abstract. We concentrated on the development of our own software for simulating semiconductor lasers and laser amplifiers, interactive simulation of optical communication systems, numerical analysis of transient effects in Raman fibre amplifiers and improving the efficiency and accuracy of split-step Fourier analysis suitable for solution of non-linear Schrödinger equation.

11.1 Modelling of Optoelectronic Components (Lasers and Semiconductor Optical Amplifiers)

11.1.1 Introduction

The choice of an appropriate model for monolithic semiconductor-based optical components is often affected by the degree of accuracy required for a particular application. Analytical models describe the main physical effects very well. They are able to predict the main parameters under steady-state conditions. However, some applications require a prediction of transients. The prediction of dynamics and operation under both periodic and aperiodic conditions requires often the use of a numerical model. Here, computation time and complexity of the algorithm are important parameters for the model selection.

The numerical models can be classified into two groups, namely *time-domain* and *frequency domain* models [1]. The frequency-domain modelling techniques are most suitable for a general description and understanding of the physics of mode-locking. On the other hand, the large-signal distributed time-domain modelling techniques are likely to be most suitable for a practical laser design because they take into account dynamics and spatial inhomogeneities of the active layer. They can also be used to model other active optoelectronic components such as different kind of lasers, optical amplifiers and modulators.

11.1.2 Frequency-Domain Approaches

Frequency-domain techniques base on a description of the optical field evolution in the laser cavity in the form:

$$E(r,t) = \psi(x,y) \sum_j E_j(t) u_j(z) e^{i[\omega_j t + \varphi_j(t)]} \tag{11.1}$$

I. Tomkos et al. (Eds.): COST 291 – Towards Digital Optical Networks, LNCS 5412, pp. 309–330, 2009.
© COST 2009

with $\Psi x, y)$ being the transverse waveguide mode profile and $E_j(t)$, ω_j, and $\varphi(t)$ being the amplitude, frequency, and phase of the mode j, respectively. $u_j(z)$ are defined as the functions that satisfy the resonant condition for the laser cavity. These equations are nothing else but the wave equations with reflection boundary conditions at the laser facets. Although in a realistic cavity there is a weakly time dependence of the functions, this dependence is usually neglected in the frequency-domain approach.

The dynamics of the gain and saturable absorption is considered to be slow, i.e. with time constants longer that the round-trip time of the cavity, so that the carrier density in the gain and saturable absorber sections are described by the standard rate equations with no z dependence

$$\frac{dN_e(t)}{dt} = \frac{J(t)}{ed} - \frac{N_e(t)}{\tau_g} - v_g g_g S(t),$$

$$\frac{dN_e(t)}{dt} = -\frac{N_e(t)}{\tau_a} - v_g g_a S(t),$$

(11.2)

where N_e denotes the carrier density and J is the pump current density. $\tau_{g,a}$ and $g_{g,a}$ are, respectively, the spontaneous recombination lifetimes and the gain coefficients for the gain (index g) and absorber (index a) sections. e is the electron charge and d is the thickness of the active layer. Note that the gain coefficient within SA sections, g_a, has a negative sign and could be replaced by $-a_s$, where a_s denotes the absorption coefficient. $v_g = c/n_e$ is the group velocity of light in the laser waveguide and $S(t)$ is the photon flux density given by

$$S(t) = \sum_j \left| E_j(t) \right|^2$$

(11.3)

11.1.3 Time-Domain Models

A model in the time domain describes mode-locking as the propagation of an optical pulse through a waveguide. The optical field in the laser cavity is usually represented using two components propagating in opposite directions, say in the right and the left longitudinal (i.e., z) direction. Thus, the evolution of the optical field in the laser cavity is given by

$$E(r,t) = \psi(x,y)\left[E_R \exp(-ik_0 z) + E_L \exp(ik_0 z)\right]\exp(i\omega_c t)$$

(11.4)

where, respectively, ω_c and $k_0 = n(\omega_c)\omega_c/c$ are the optical central frequency and corresponding wave vector. The starting point of almost all time-domain models is a set of reduced equations, which includes the following expression for slowly varying amplitudes $E_{L,R}$

$$\pm\frac{1}{v_g}\frac{\partial E_{L,R}}{\partial z}+\frac{\partial E_{L,R}}{\partial t}=\Gamma(\hat{g}+i\Delta\hat{\beta}-\alpha_{\mathrm{int}})E_{L,R}+iK_{RL,LR}E_{R,L}. \tag{11.5}$$

In the above equation, both saturation nonlinearity and frequency dependence are included by introducing the operators $\hat{g}$ and $\Delta\hat{\beta}$, which describe the gain and the variable part of the propagation constant, respectively. The last term need to be included only for the sections containing a grating like in DBR and DFB sections. The model is completed by standard rate equations for carrier densities (within the gain and absorber sections) and photon densities.

$$\frac{\partial N_e(z,t)}{\partial t}=\frac{J(z,t)}{ed}-\frac{N_e(z,t)}{\tau_g}-v_g g_g S(z,t),$$

$$\frac{\partial N_e(z,t)}{\partial t}=-\frac{N_e(z,t)}{\tau_a}-v_g g_a S(z,t), \tag{11.6}$$

$$S(z,t)=\left|E_R(z,t)\right|^2+\left|E_L(z,t)\right|^2.$$

The equations above are the simplest and most common form of the rate equations. They can be upgraded in order to achieve more accurate and complete models. For example, by changing the first pumping term in the first equation and introducing a separate rate equation for the carrier density in the cladding layer, carrier transport effects can be taken into account [2]. The dependence of gain and saturable absorption on carrier density and wavelength can be taken into account by using either the quadratic or the cubic approximation. Finally, the most common way to include fast nonlinearities is to use the phenomenological nonlinear coefficients $\varepsilon_{g,a}$, which are also known as the gain compression coefficients. Thus, the following expression for gain can be obtained by using the quadratic formula and including the fast nonlinearities.

$$g_{g,a}=\frac{a_{a,g}(N_e-N_{et}^{g,a})-b_{g,a}(\lambda-\lambda_p^{g,a})^2}{1+\varepsilon_{g,a}S}. \tag{11.7}$$

11.1.4 Lumped-Element Models

In the specific case that pulses experience a small gain over one round-trip in the cavity and the pulse width is much smaller than the cavity length, the model can be simplified significantly. The amplification and dispersion may then be separately treated in two independent stages. Thereby, the distributed amplifier in the gain section can be approximately substituted by a lumped gain element. Analogous to the gain section, the saturable absorber can also be considered as a lumped element. These two elements perform the functions of amplification/absorption and self-phase modulation (SPM). Usually, they take into account only "slow" components of the gain/absorption saturation, although there are few approaches that also include "fast" effects. The dispersion of material gain and refractive index as well as other dispersive elements present in the cavity such as DBR or DFB gratings are combined in a lumped dispersive element. The main equation of the model can be solved analytically by making an additional assumption that the pulse energy is much smaller than the saturation energy of the amplifier. This assumption leads to a model, which can only be used for the small-signal regime. To overcome this limitation, the main equation can be solved numerically without making the small-signal approximation. The gain and saturable absorption operators are kept unchanged and realistic laser geometries are taken into account. Such numerical lumped-element models are large-signal models, which are easily extendable.

Frequency-domain and semi-analytical lumped time-domain models can be used for a general insight into the physics of mode-locking. They are suitable for easy and fast determining the main parameters. However, they are not suitable for modelling of dynamic regimes involving transients in the laser cavity, especially not for practical laser design. For this purpose, large-signal distributed time-domain models can be used.

11.1.5 Distributed Time-Domain Models

In fully distributed time-domain (DTD) models, the signal propagation equation and the rate equations are solved directly numerically. Using these models, a realistic picture of fast laser dynamics can be obtained. Unlike the lumped time-domain methods, the distributed models need no trial pulse at the beginning. They are able to produce optical pulses from spontaneous noise after the laser is turned on. Therefore, they can straightforwardly model the transients associated with the initial phase of the laser operation. To achieve this self-starting, the propagation equation is extended with a noise source that describes spontaneous emission. Usually, it is implemented numerically by adding complex random numbers to the propagating fields. Thereby, not only transients in the starting phase can be modelled, but also the noise/jitter properties can be accurately predicted. Several studies of laser dynamics using DTD models have pointed out some important results. First, the fast nonlinearities have less influence on simulations of long-cavities. However, it can be observed that their inclusion has an effect on pulse parameters. Secondly, SPM caused by carrier density variation and refractive index change in the active medium is the main reason for the chirp. The dispersive properties of the cavity play a less important role in the chirp formation. The results obtained by simulations show a good agreement with experiments when the laser parameters are precisely adjusted.

11.1.6 Modeling of Hybrid Mode-Locked Lasers

Hybrid mode-locked lasers allow generation of very short pulses with high repetition frequencies and a small chirp. An external cavity configuration enables large tunability in wavelength, pulse width, and repetition frequency by changing the resonator length mechanically [3]. However, the mechanical elements in the laser cavity can induce instabilities due to the environmental factors such as change in temperature and mechanical stress.

The structure of a monolithic hybrid MLL is shown in Fig. 11.1(a). It consists of three sections, namely saturable absorption (SA), gain and distributed Bragg reflector (DBR) section. The device is modelled using a distributed time-domain model based on a modified transmission line laser modelling method [4]. A model section is represented by a scattering matrix consisting of a photon model and a carrier model. In photon models, the instantaneous photon density is calculated from incident fields. The instantaneous photon density and injection current are used in the carrier density models to calculate the change in carrier density. The incident wave is then amplified by the frequency dependent gain, G, that is calculated in the carrier density models. The spontaneous emission noise is added to each section. The saturable absorption (SA) region is modelled using p sections represented by scattering matrices, which are of the same form as the matrices for the gain (active) region with a slight difference that the gain coefficient, g, is replaced by $-a_s$, a_s being the saturable absorption coefficient. The gain region is re-

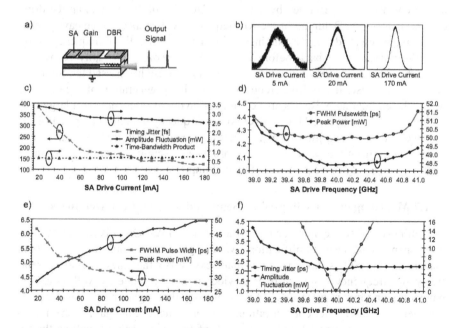

Fig. 11.1. Simulation results obtained by using a distributed time-domain model of a three-section hybrid mode–locked laser

presented by r active sections (active region matrices). The active DBR region sections consist of s active scattering matrices.

Some typical results of the model are shown in Fig. 11.1(b)-(f). In this particular case, the lengths of the DBR, saturable absorption, and active region were chosen to be 200 μm, 100 μm, and 500 μm, respectively. The laser is divided into 64 model sections. The generated pulse train that is obtained by feeding a 40 GHz radio frequency (RF) signal into the SA region. Pulses as short as 4.34 ps (FWHM) are produced by applying a current of 150 mA. Optical spectrum of the output signal is obtained by transforming the time-domain signal into the frequency domain using the fast Fourier transformation (FFT). Both timing jitter and fluctuation of the pulse amplitude can be minimized by applying high currents to the SA section. In Fig. 11.1(b), eye diagrams for three different currents are shown. A very large timing jitter and pulse broadening is observed for low currents (e.g., for 5 mA), while high-quality short pulses with a low timing jitter (125 fs) can be generated by applying a current of 170 mA.

The dependence of the main pulse parameters on the SA drive current is shown in Fig. 11.1(c) and (e). The pulses can be shortened from 6.15 ps to 4.23 ps by increasing the current from 20 mA to 180 mA. A larger peak power can be obtained for higher currents and shorter pulses. The amplitude fluctuation is reduced from 3.34 mW (by 20 mA) to 2.46 mW (by 180 mA), while the time-bandwidth product is slightly increased (from 0.58 to 0.67) by increasing the current.

The effects of the frequency detuning are illustrated in Fig. 11.1(d) and (f). The measurements were performed by applying a DC bias of 70 mA and an RF drive current of 150 mA. Detuning of the RF frequency with respect to the cavity resonance frequency produces a cyclic instability. That is, if the drive frequency is too low, the pulses arrive back before the saturable absorber has fully recovered. Thus, the growth of a new pulse after the returning pulse has passed the SA is favoured. Otherwise, if the drive frequency is too high, generation of a new pulse before the circulating pulse has arrived back is favoured. This induces a large timing jitter as shown in Fig. 11.1(f). Even lower frequency detunings (e.g., 50 MHz) produce a large timing jitter (2 ps). A further increase (or decrease) in drive frequency away from the cavity resonance produces instabilities. Thus, an accurate tuning of the RF drive frequency is of crucial importance.

11.1.7 Modelling of Travelling-Wave Semiconductor Optical Amplifiers

Travelling-wave semiconductor optical amplifier (TW-SOAs) is one of the key components for linear amplification, wavelength conversion, and fast optical switching. TW-SOAs have a nonresonant cavity, so that they provide a relatively flat frequency response over a wide frequency range as well as a high bandwidth and low noise. However, an input signal at higher power level causes a depletion of the carrier concentration, and consequently, the gain of the amplifier saturates. The spectral width and peak of the gain curve as well as the refractive index depend on the carrier concentration. The carrier dynamics has to be taken into account in order to model accurately the transient response of the SOA under modulation and fast intensity changes of the injected optical signal. Inhomogeneities of the active medium and variation of the carrier and photon densities along the cavity influence the gain saturation. The same distributed time-domain model as that used for modelling hybrid MLL, which takes into account all of the above mentioned effects, can be used to predict temporal and spectral responses of SOAs.

The bidirectional amplifier model consists of q sections placed between two anti-reflection coated facets. The sections are represented by scattering nodes, which are placed at regular intervals along the transmission line string. The length of a model section, ΔL, is related to the time step, $\Delta \tau_s$, and the group velocity inside the active region, v_g, by $\Delta L = \Delta \tau_s \cdot v_g = \Delta \tau_s \cdot c/n_e$, where c and n_e are the velocity of light in vacuum and the group effective index of the active waveguide. Note that this model supports only x-polarization mode, thus polarization dependence can not be taken into account.

In the simulations we performed to show the typical results of the model, the bulk SOA device is assumed with some arbitrariness to be 500 μm long. It is divided into 32 sections by setting the time step to 195.3 fs. The gain compression coefficient, ε, which represents the gain dynamics due to SHB and CH effects, is set to $1.0 \times 10^{23}\,\mathrm{m}^3$.

To investigate the response of carrier density in a SOA to a fast change in optical intensity at the input of the device, an optical sech^2 pulse as short as 2.4 ps FWHM with a central wavelength equal to the gain peak wavelength of the SOA (1.55 μm) was injected into the device. The change of the mean carrier density averaged through all model's sections was observed. The time dependence of the mean carrier density for three bias currents and three peak powers of the injected pulse can be seen in Fig. 11.2(b) and (c). The intensity of the current applied to the SOA influences significantly the carrier injection process. Therefore, the recovery of carrier density can be accelerated by applying a higher current. The carrier density is fully recovered in 250 ps when a current of 200 mA is applied, while only 30 % of the carriers recover in the same time period if 100 mA are applied. Here, the peak power of the injected pulse was set to 10 mW. The strength of the nonlinear effects including spectral hole burning, carrier heating, and ultrafast processes such as two-photon absorption and Kerr effect depends on the injected optical power. In Fig. 11.2(c), the influence of these processes on carrier density pulse response for different peak powers of the injected pulse can be seen. A high-power pulse induces a strong depletion of carriers driven by the fast processes. The carrier density recovers faster for lower powers of the injected pulses because the carrier depletion is weaker. For example, by injecting a pulse with a peak power of 2 mW, a reduction of the carrier density of $0.27 \times 10^{24}\,\mathrm{m}^{-3}$ can be observed within 23 ps. The carrier density is fully recovered within 200 ps. If a pulse with 40 mW peak power is injected, the carrier density is reduced from 3.3 to $2.72 \times 10^{24}\,\mathrm{m}^{-3}$ (a change of $0.58 \times 10^{24}\,\mathrm{m}^{-3}$) in 16 ps and fully recovered within 325 ps.

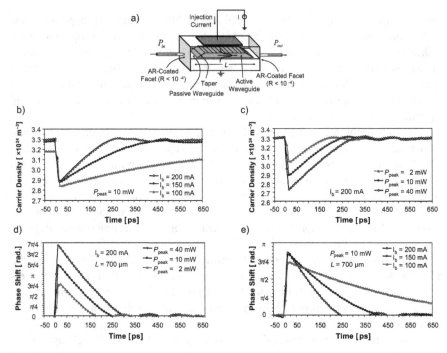

Fig. 11.2. Simulation results obtained by using a distributed time-domain model of a semi-conductor optical amplifier

Fig. 11.2(d) and (e) show the change in optical phase due to a carrier density induced refractive index variation. A phase shift of about $3\pi/4$ can be obtained when a pulse with 10 mW peak power is injected. Applying stronger bias current results in a slightly higher phase shift, but it decreases significantly the time in which the phase shift falls back to zero.

As shown in Fig. 11.2(d), the higher the pulse energy, the larger is the phase shift. With the SOA being 700 μm long, a phase shift of nearly $5\pi/4$ can be obtained by applying a bias current of 200 mA and by injecting an optical pulse with 10 mW peak power (in contrast to $3\pi/4$ phase shift obtained in a 500 μm long SOA - see Fig. 11.2(d). A larger phase shift in a SOA can be obtained by using longer devices and employing higher pulse powers. However, if a train of short pulses at a high repetition rate is injected, the carrier density in SOA cannot recover completely in the time period between two successive pulses and the change in phase is lower than the maximal achievable phase shift in the case of a single pulse. Therefore the energy of control pulses needed to achieve a π phase shift is usually higher in fast all-optical switches when they are used as OTDM demultiplexers than in arrangements where a single short pulse is occasionally injected into the SOA.

11.2 Simulation Tool MOVE-IT

The in-house software tool MOVE-IT (=**mo**dular **ve**rsatile **i**nteractive simulation **t**ool) [5] of University of Kiel (CAU) is a collection of MATLAB™ m-files, m-functions, and C-subroutines, providing a complete simulation environment for communication systems. Since MOVE-IT runs under MATLAB™, the large variety of its function libraries is available yielding platform-independent usability. Key features of MOVE-IT are given in the following:

- The simulation is completely MATLAB™-code-driven, i.e. the creation of the simulation results is fully comprehensible by the user who is able to examine the underlying source code. Only the time-consuming parts of the simulation are calculated by C-subroutines.
- The creation and modification of a simulation setup as well as the manipulation of the simulation parameters can be done in two ways:
  - o By means of the graphical user interfaces (GUI), the simulation parameters are accessible by schematic editing.
  - o Alternatively, all simulation parameters can be manipulated by changing MATLAB™ m-file code directly.

Both methods are equivalent, i.e. a modification using the GUI changes the m-file code, and vice versa. Fig. 11.3 shows the structure of the GUI.

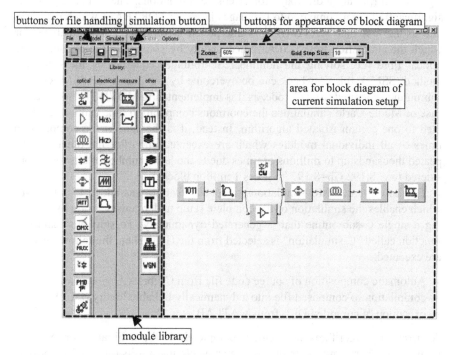

Fig. 11.3. Graphical user interface of MOVE-IT.

The topology of the communication systems under investigation may have any transversal structure, since MOVE-IT has a modular architecture, i.e. in MOVE-IT for each block an individual function is provided. The only task the simulation environment has to do is to call these functions in correct order and handle the data transition between modules according to the simulation setup. Due to this simple structure, MOVE-IT can be extended very easily by the user through adding new module functions to the existing ones. This extension does not have any impact on simulation setups that exist already.

Additional features of MOVE-IT are itemized as follows:

- Any number of global parameters to be used in several modules can be defined.
- For the values of parameters, any valid MATLAB^{TM} expression is allowed including functions and operators.
- Definition and execution of so-called user-defined loops allow for parameter optimization by carrying out the same setup several times while varying one or more simulation parameters (e.g. fibre length and/or filtering bandwidth).
- For each individual module as well as for the whole simulation environment, online help and online documentation as well as a tutorial are available.
- The simulation results are available as MATLAB^{TM}-variable for any further post processing.

Although the fact that MOVE-IT is based on MATLAB^{TM} shows a lot of advantages, it incorporates a drawback for extensive simulations: The simulation time is significantly longer compared to languages like C or FORTRAN, which mainly is due to the fact that MATLAB^{TM} is interpreted at runtime. For simulations for which the simulation time is determined mainly by computing the impact of the optical fibre [i.e. solving the nonlinear Schrödinger equation by the split step method (SSM)], this drawback can be overcome by computing the SSM within a subroutine generated from C-code, as it is implemented in MOVE-IT. However, in case of Monte-Carlo simulations the enormous computation time cannot be attributed to one certain isolated algorithm. Instead, it is the sum of the computation times of all individual modules which are executed one after the other and repeated thousands up to millions of times due to the large number of bits to be considered (e.g. 8.192 Gb=8.192 kb times 1 million blocks).

To allow for fast MC-simulations, a special feature was added to MOVE-IT which enables the simulation of the complete setup under consideration by executing a single C-subroutine that is generated dynamically. To start this feature, a function called "C-simulation" is selected from the GUI. Then, the following steps are executed:

- Automatic composition of source code file from C-library,
- compilation of composed file into a dynamically linkable library (DLL) and
- execution of compiled DLL from MATLAB^{TM}.

All necessary algorithms are included completely in the C-library. For example, each module performing a frequency-domain filtering has its own FFT algorithm.

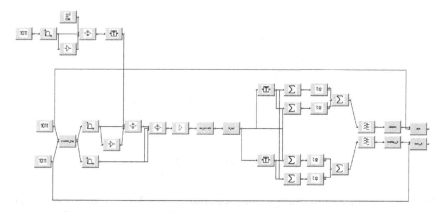

Fig. 11.4. Block diagram for MC simulation to determine receiver sensitivity for RZ-DQPSK. The following elements are observable: In the upper left corner, the RZ-pulse carver can be seen, which supplies the subsequent two MZM with a train of RZ-pulses. Two data streams (lower left corner) are pre-coded and modulated onto the pulse train by means of serial DQPSK modulation. After noise loading (center of figure), balanced detection and error counting is performed (right hand side of block diagram).

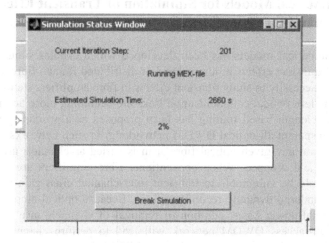

Fig. 11.5. MOVE-ITs simulation status window.

Although it results in lengthening of the source code, this way execution time is optimized. As a result, a complete source code file is generated which can be compiled by a C-compiler. Speed improvement by a factor of 5-10 is achieved compared to pure MATLAB^{TM} -code. The simulation of more than 1 Gb per day for simple setups is possible, which is the same as a simulation speed of more than 10 kb/s. This allows for investigation of those systems using full MC-simulation down to a BER of 10^{-9} which to the best of our knowledge has not been shown so far by any commercial numerical simulator for optical communication systems.

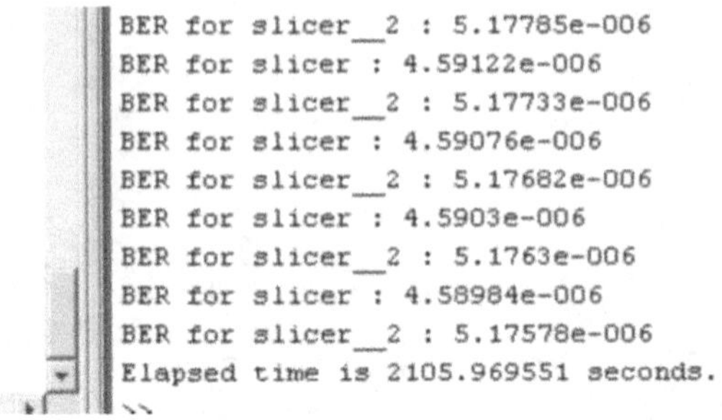

```
BER for slicer__2 : 5.17785e-006
BER for slicer : 4.59122e-006
BER for slicer__2 : 5.17733e-006
BER for slicer : 4.59076e-006
BER for slicer__2 : 5.17682e-006
BER for slicer : 4.5903e-006
BER for slicer__2 : 5.1763e-006
BER for slicer : 4.58984e-006
BER for slicer__2 : 5.17578e-006
Elapsed time is 2105.969551 seconds.
>>
```

Fig. 11.6. Display of MC simulation result.

In the following, a short example for application of fast MC-simulation in MOVE-IT is given. The block diagram of the back-to-back system with RZ-DQPSK modulation is given in Fig. 11.4. This model was used for generating results published in [6]. The proceeding of simulation can be followed in Fig. 11.5 while results are given in Fig. 11.6.

11.3 Numerical Models for Simulation of Transient Effect in Raman Fibre Amplifiers

A set of numerical models has been developed which enables simulation and analysis of transient effects in lumped and/or distributed Raman fibre amplifiers (RFA). The necessity to study transient effects in fibre amplifiers stems from the fact that modern optical communication links are becoming more and more dynamic. Wavelength based routing has been proposed as a promising approach towards transparent all-optical DWDM networking. In such networks, the channel load within any given optical fibre span is varied as channels are dynamically added and dropped. When conventional fibre amplifiers are used, such networks could be vulnerable to transient inter-channel cross-gain modulation when they undergo dynamic reconfigurations. Because optical amplifiers saturate on a total-power basis, addition or removal of channels in a wavelength routing multi-access DWDM network will tend to perturb channels at other wavelengths that share all or part of the route. When the network is reconfigured and wavelength channels are added or dropped, cross-gain modulation in fibre amplifiers may induce power transients in the surviving channels that can cause serious service impairment.

Our general RFA model is based on the solution of partial differential equations describing propagation of pumps, signals and both the downstream and upstream propagating amplified spontaneous emission powers (ASE) in RFA. The set of coupled partial differential equations for forward and backward propagating pumps, signals, and spectral components of ASE powers, P^+ (z,t,v), P^- (z,t,v), describing their evolution in space and time acquire the form of eq.(11.1), [7]

$$\frac{\partial P^\pm(z,t,v)}{\partial z} \mp \frac{1}{V_g(v)}\frac{\partial P^\pm(z,t,v)}{\partial t} = \mp\alpha(v)P^\pm(z,t,v) \pm \gamma(v)P^\mp(z,t,v) \pm$$

$$\pm P^\pm(z,t,v) \bullet \sum_{\xi>v}\frac{g_R(v-\xi)}{K_{eff}A_{eff}}[P^\pm(z,t,v)+P^\mp(z,t,v)] \pm$$

$$\pm hv\sum_{\xi>v}\frac{g_R(v-\xi)}{A_{eff}}[P^\pm(z,t,v)+P^\mp(z,t,v)]\bullet[1+\frac{1}{e^{h(\xi-v)kT}-1}]\Delta v \mp \qquad (11.8)$$

$$\mp P^\pm(z,t,v) \bullet \sum_{\xi<v}\frac{v}{\xi}\frac{g_R(v-\xi)}{K_{eff}A_{eff}}[P^\pm(z,t,v)+P^\mp(z,t,v)] \mp$$

$$\mp 2hv P^\pm(z,t,v) \bullet \sum_{\xi<v}\frac{g_R(v-\xi)}{A_{eff}}[1+\frac{1}{e^{h(v-\xi)kT}-1}]\Delta v$$

In order to eliminate the transfer of random intensity noise from the pump to the signals, RFA are mainly counter-directionally pumped. Therefore, $P^\pm(z,t,v)$ represents downstream signals, $P_s^+(z,t,v_s)$, upstream pumps, $P_p^-(z,t,v_p)$, and downstream and upstream spectral components of ASE power, $P_{ase}^\pm(z,t,\xi)$, contained in frequency slot Δv. $V_g(v)$ is the frequency dependent group velocity, $\alpha(v)$ is the fibre background loss, $\gamma(v)$ is the Rayleigh back scattering coefficient, $g_R(v-\xi)$ is the Raman gain coefficient between waves with frequency v and ξ, K_{eff} is the factor taking into account the polarization relation between pumps and Stokes signals, A_{eff} is the effective interaction area of the fibre, h is the Planck's constant, k is the Boltzman's constant and T the absolute temperature of the fibre.

$$
\begin{array}{ll}
P_{s0}^+(v_s,t) \dotfill & for\ i=1 \\
P_{s,i}^+(z=0,v_s,t) = \beta \bullet P_{s,i-1}^+(z=L_{i-1},v_s,t) \dotfill & for\ i=2,3
\end{array}
$$

$$P_{p,i}^-(z=L_i)=P_{p0,i}^- \quad \dotfill \quad for\ i=1,2,3$$

$$(11.9)$$

$$
\begin{array}{ll}
0 \dotfill & for\ i=1 \\
P_{ase,i}^+(z=0,v,t) = \beta \bullet P_{ase,i-1}^+(z=L_{i-1},v,t) \dotfill & for\ i=2,3
\end{array}
$$

$$
\begin{array}{ll}
P_{ase,i}^-(z=L_i,v,t) = \beta \bullet P_{ase,i+1}^-(z=0,v,t) \dotfill & for\ i=1,2 \\
0 \dotfill & for\ i=3
\end{array}
$$

Boundary conditions for optical powers must reflect the specific simulated case: single RFA, cascade of several RFAs, all-optical gain clamped discrete RFA [8]. If, e.g. 3 spans of communication fibre are counter-directionally pumped, signals and ASE spectral components are only slightly attenuated at span interfaces by the insertion loss β of the WDM pump couplers ($\beta = 1.6$ dB). Pump powers are introduced at the end points of individual spans, at $z=L_i$, and the boundary conditions take the form of eq.(11.2), [10].

When second-order pumping or all-optical gain clamping in the form of ring resonator is to be analyzed, boundary conditions must be modified accordingly [9].

For the steady-state solution of eq.(11.1) an iterative procedure based either on fourth-order Runge-Kutta routine or on Average Power Analysis approach has been used [9]. Once the steady-state distribution of forward- and backward-propagating optical powers is calculated, direct integration according to eq.(11.3) is used to obtain time evolution of individual optical powers $P^{+/-}(z,v,t)$ along the optical fibres in response to channel removal/addition. The time derivative $\partial P^{+/-}(z,v,t)/\partial t$ is separated from eq.(11.1).

$$P^{\pm}(z,v,t+\delta t) = P^{\pm}(z,v,t) + \frac{\partial P^{\pm}(z,v,t)}{\partial t} \cdot \Delta t \qquad (11.10)$$

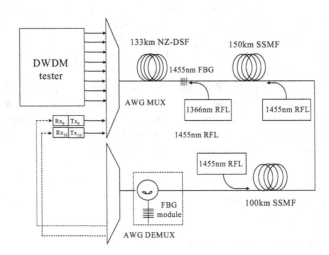

Fig. 11.7. Schematic diagram of the experimental setup for simulation of channel addition / removal.

We will now present the results of numerical simulations of an experimental link shown in Fig. 11.7. The first span of the link consists of 130 km of G.655 fibre and was counter-directionally second-order pumped by Raman fibre laser (RFL) at 1366 nm. High reflectivity (95 %) fibre Bragg grating (FBG) at 1455 nm and distributed Rayleigh back scattering in the G.655 fibre are responsible for the development of the final 1455 nm pump power which then amplifies signals in the 1550 nm wavelength range. The second and the third span consist of 150 and 100 km of SSMF, respectively. Both spans are counter-directionally pumped by 1455 nm RFLs. Pump powers of individual RFA's were 1260 mW at 1366 nm for span #1, 930 mW, and 943 mW at 1455 nm for spans #2, and #3, respectively as used in the experiments.

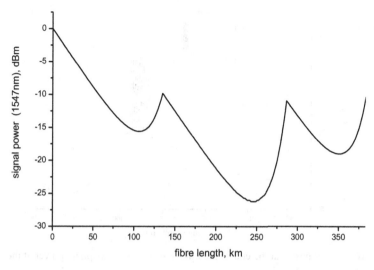

Fig. 11.8. Power distribution of channel at 1555 nm along the link of 383 km.

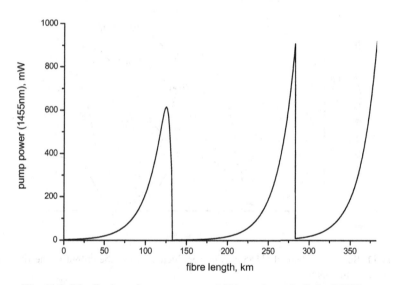

Fig. 11.9. Distribution of pump power at 1455 nm along the link of 383 km.

Ten channels starting at 1546 nm with channel spacing of 1 nm, were transmitted through the link. First 8 channels were 100 % square-wave modulated at 500 Hz to simulate channel addition/removal. Fig. 11.8 plots signal distribution at 1547 nm along the link. It can be seen that at the end of each span the signal power was ≈ -10dBm. Pump distribution along the link is shown in Fig 11.9. The first span is pumped at 1366 nm, the 1455 nm is generated through distributed Rayleigh back-scattering and reflection from the 1455 nm FBG.

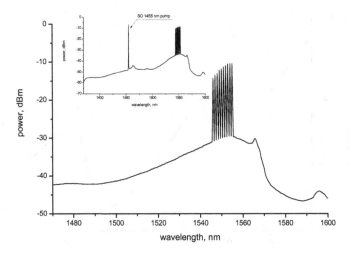

Fig. 11.10. Optical power at the end of the link. The inset shows optical power at the end of span#1.

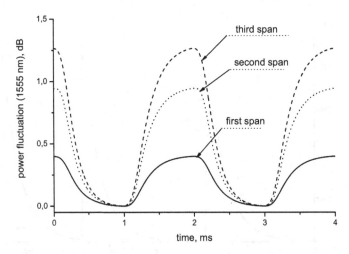

Fig. 11.11. Surviving channel (1555 nm) power fluctuations at the output of the first, second, and the third span.

Part of the optical spectrum calculated at the end of the link is displayed in Fig. 11.10. The inset to Fig. 11.10 shows the spectrum at the end of span#1 to demonstrate generation of the 1455 nm pump. Power fluctuations of one of the surviving channels (1555 nm) at the output of individual spans are plotted in Fig. 11.11. Reasonable qualitative agreement with the experimental results has been reached with the exception of span #2, where the calculated power overshot was ≈ 1.4 times higher than the experimental value.

11.4   Split-Step-Fourier-Method in Modeling of WDM Links

Modern WDM systems contain large number of channels and occupy very wide bandwidth, which cause difficulties in simulations due to spurious FWM and walk-off effect. Two class of methods are distinguished: single-band [11]-[17], [22] – in which full-bandwidth of WDM transmission is simulated, and multi-band [18]-[21] – in which separate channels are simulated, taking into consideration an influence of adjacent channels (Fig. 11.12). Single-band methods give an exact solution of the nonlinear Schrödinger equation (NLSE), i.e. include the impact of nonlinear phenomena, like: SPM, XPM, FWM, but on the other hand are used mainly in narrow bandwidth cases due to its high simulation time. Multi-band methods are faster, but give only limited information of nonlinear phenomena (SPM, XPM but not FWM) derived from other channels and are more flexible.

Split-step-Fourier-method (SSFM) is commonly used for simulating of light propagation in an optical fibre, described by the nonlinear Schrödinger equation (NLSE) [11], due to its high numerical efficiency. In many publications optimisation of the simulation time and accuracy is considered [12-22]. Higher order numerical methods (i.e. explicit Adams–Bashforth and implicit Adams–Moulton, etc.) or predictor-corrector methods [12] are used. Comparing to conventional symmetrical SSFM, the numerical effectiveness of higher order methods increases with higher required accuracy. These methods are especially useful for simulations of soliton propagation, where linear (L) and nonlinear (N) operators in SSFM are self-balanced.

Typically, there are higher dispersion and lower nonlinearity in WDM transmission, comparing to soliton transmission. As a consequence, special tailored methods should be applied to simulation of signal propagation in WDM links. Additionally, due to relatively low required accuracy (of the order of $10^{-2} - 10^{-3}$), the symmetrized SSFM (S-SSFM) of order $O(h^2)$ is preferred for WDM signal simulations. Besides common used S-SSFM, another methods are used in special cases, e.g. split-step wavelet collocation is faster then S-SSFM in very wideband simulations [13], but is applicable only for zero dispersion slope ($\beta_3 = 010$).

Local-Error-Method (LEM) is especially useful in single-band simulations, because it automatically adjust simulation step for required accuracy [13]. In this method step size is selected by calculating the relative local error of each single step, taking into account the error estimation and linear extrapolation. It provides higher accuracy than above-mentioned methods, since it is method of third order. Simulations are conducted simultaneously with coarse ($2h$) and fine (h) steps.

For large number of WDM channels all single-band methods, including LEM, show prohibitively long simulation time [21]. In this case multi-band methods are used [18]-[21]. Different multi-band methods have been evaluated in [21] and application of LEM method to cross-phase modulation (XPM) simulation in place of fixed step was proposed, which improves simulation accuracy and speed up to 30%.

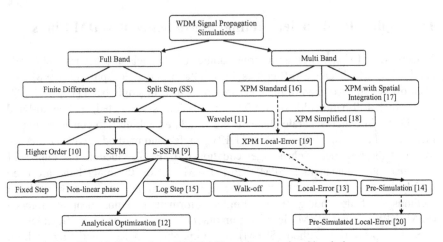

Fig. 11.12. Review of WDM Signal Propagation Simulations.

Optimal step size in S-SSFM is of uttermost importance to improve numerical efficiency. Lately, methods known in quantum mechanics were used to step size calculation [14]. The optimal step size $h_{optimal}$ can be estimated analytically for required global error δ_G. This procedure is fast in the case of lossless fibre. In more realistic case with lossy fibre, the optimal step size can be estimated as well, but with additional computational effort [14].

In pre-simulation method the step size is selected by calculating the global error δ_G in a series of fixed-step S SSFM pre-simulations with signal spectrum averaging [16]:

$$\left|U_n^{red}\right| = \sqrt{\sum_{i=n \cdot N_{red}}^{n \cdot N_{red}+N_{red}-1} \left|U_i\right|^2}, \quad \arg\left(U_n^{red}\right) = \arg\left[\sum_{i=n \cdot N_{red}}^{n \cdot N_{red}+N_{red}-1} \left(\left|U_i\right| \cdot U_i\right)\right] \quad (11.11)$$

where $U = \Im(u)$ is the Fourier transform of the signal. For reduced number of samples N_{red}, split-step pre-simulation on the test signal can be much faster ($> N_{red}$) than the corresponding simulation on the full signal. Several pre-simulations must be carried-out iteratively to calculate optimal step size $h_{optimal}$, required to achieve desired global accuracy. Pre-simulations typically take 30% of full spectrum simulation time [16]

11.4.1 Pre-simulated Local Errors S-SSMF

We proposed novel simulation method which comprises two stages: step optimization $h_{optimal}(z)$ is carried out in the initial stage, combining local-error and pre-simulation methods and in the second stage conventional S-SSFM is used, applying optimal steps obtained in the initial stage. Overall time savings up to 50% are realistic, depending of simulated system scenario. We called this novel procedure Pre-simulated Local Error S-SSFM (PsLE S-SSFM).

In PsLE S-SSFM LEM algorithm [15] is used with averaged signal spectrum (1) [16]. In [15] method of order $O(h^3)$ is utilized by taking fraction of coarse u_c and fine solutions to calculate the next step. In PsLE S-SSFM only fine solution u_f is used, which gives better stability and does not degrade accuracy in the case of WDM simulations, where the global error is low, of the order of 10^{-3}. The initial stage duration is only small percentage (2%) of the second stage, in which full-band simulation is carried on using fixed-step method.

11.4.2 Results

We have explored the applicability of PsLE method to WDM systems with different number of channels. The method was used for simulation of WDM link with various number of channels and the following parameters: bit rate of 40 Gb/s, channel spacing of 100 GHz, channel power of 1 mW, simulated bandwidth of 320 GHz/channel and bit sequence length of 2^9. Transmission line comprises 100 km of Standard Single Mode Fibres (SSMF), with parameters given in Table 11.1.

Results shown in Fig. 11.13 indicate that PsLE S-SSFM is up to 50% faster than walk-off method in all simulated cases in important global error range of 10^{-2}-10^{-3}. Relation between method parameter and global error was considered for fixed-step and PsLE methods (Fig. 11.14).

Table 11.1. Fiber parameters used in the simulation.

Parameter	Unit	SSMF
Attenuation	dB/km	0.22
Dispersion	ps/(nm·km)	16.00
Dispersion slope	ps/(nm·km)^2	0.08
Nonlinear coefficient	1/(W·km)	1.32

The method parameter is the parameter in a split-step method that should be varied to obtain required accuracy. For required global error $\delta_G = 10^{-3}$ the local error (*i.e.* the parameter of PsLE method) varies from $2\cdot10^{-5}$ to $3\cdot10^{-4}$ for different number of simulated channels, in the same conditions the step size (*i.e.* the parameter of fixed-step method) varies in wider range – from 8 m to 5000 m. It is clear that local error in PsLE method is better criterion to assess global error than step size in fixed-step method. The same is true for walk-off method, which in fact, is fixed.

PsLE method has two basic advantages: shorter simulation time of up to 50% in comparison with walk-off method, which is known as the most efficient in WDM simulations [13] and offers simply accuracy criterion *i.e.* local error, which is a good indicator of the global accuracy.

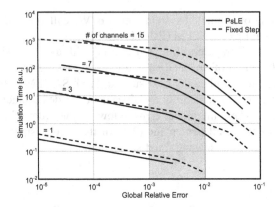

Fig. 11.13. Simulation time vs. global relative error for fixed-step (dashed line) and PsLE (solid line) method.

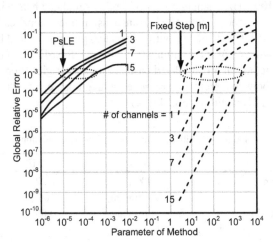

Fig. 11.14. Global relative error vs. method parameter: local error for PsLE and step size for fixed-step.

11.4.3 Conclusions

Pre-simulated local-error S-SSFM halves simulation time of conventional S-SSFM. Moreover, local-error used in pre-simulation seems to be a good indicator of the global accuracy. To the best of our knowledge PsLE S-SSFM is the fastest method for simulations of light propagation in WDM links.

References

1. Avrutin, E.A., Marsh, J.H., Portnoi, B.L.: Monolithic and Multi-GigaHertz Mode-Locked Semiconductor Lasers: Constructions, Experiments, Models and Applications. IEE Proceedings-Optoelectronics 147, 251–278 (2000)
2. Vassilovski, D., Georges, J.B., Lau, K.Y.: Carrier Transport Effects in Active and Passive Modelocking of Monolithic Quantum-Well Lasers at Millimeter-Wave Frequencies. IEEE Photonics Technology Letters 8, 1603–1605 (1996)
3. Schmidt, C., Dietrich, E., Diez, S., Ehrke, H.J., Feiste, U., Küller, L., Ludwig, R., Weber, H.G.: Mode-Locked Semiconductor Lasers and Their Applications for Optical Signal Processing. In: Conference on Lasers and Electro-Optics (CLEO 1999), Baltimore, Maryland, USA, pp. 348–349 (1999)
4. Lowery, A.J.: New Dynamic Multimode Model for External Cavity Semiconductor Lasers. IEE Proceedings - Photonics Journal 136, 229–237 (1989)
5. Leibrich, J., et al.: MOVE-IT: modular versatile evaluating and interactive simulation tool. In: ECOC 2000, workshop 1, Munich (2000)
6. Leibrich, J., et al.: Receiver sensitivity of advanced modulation formats for 40 Gb/s DWDM transmission with and without FEC. In: ECOC 2006, Cannes, paper We3P.108 (2006)
7. Karasek, M., Kanka, J., Radil, J., Vojtech, J.: Large signal model of TDM-pumped Raman fibre amplifier. IEEE Phot. Technol. Lett. 17, 1848–1850 (2005)
8. Karasek, M., Kanka, J., Khan, G.R., Radil, J.: Design of all-optical gain-clamped lumped Raman fibre amplifier for optimal dynamic performance. IEE Proc. Optoelectronics 152, 223–229 (2005)
9. Karasek, M., Kanka, J., Bohac, L., Krcmarik, D., Radil, J., Vojtech, J.: Surviving channel power transients in second-order pumped lumped Raman fibre amplifier: Experimentation and modeling. J. Lightwave Technol. 25, 664–672 (2007)
10. Karasek, M., Vojtech, J., Radil, J.: Channel addition-removal response in a cascade of three distributed Raman fibre amplifiers transmitting 10×10 GE channels: experimentation and modeling. J. Optical. Networking~7, 15--24 (2008)
11. Agrawal, G.P.: Nonlinear Fiber Optics, 3rd edn. Academic, San Diego (2001)
12. Liu, X., Lee, B.: A fast method for nonlinear Schrödinger equation. IEEE Photon. Technol. Lett. 15, 1551–1552 (2003)
13. Kremp, T.: Split-step wavelet collocation methods for linear and nonlinear optical wave propagation. Ph.D. dissertation, High-Frequency and Quantum Electronics Laboratory, University of Karlsruhe, Cuvillier Verlag, Göttingen (2002)
14. Rieznik, A.A., Tolisano, T., Callegari, F.A., Grosz, D.F., Fragnito, H.L.: Uncertainty relation for the optimization of optical-fiber transmission systems simulations. Optics Express 13, 3834–3840 (2005)
15. Sinkin, O.V., Holzlöhner, R., Zweck, J., Menyuk, C.R.: Optimization of the split-step Fourier method in modeling optical-fiber communications systems. J. of Lightwave Technology 21, 83–88 (2003)
16. Rasmussen, C.J.: Simple and fast method for step size determination in computations of signal propagation through nonlinear fibres. In: Proc. of OFC 2001, WDD29-1 (2001)
17. Bosco, G., Carena, A., Curri, V., Gaudino, R., Poggiolini, P., Benedetto, S.: Suppression of spurious tones induced by the split-step method in fiber systems simulation. IEEE Photon. Technol. Lett. 12, 489–491 (2000)

18. Yu, T., Reimer, W.M., Grigoryan, V.S., Menyuk, C.R.: A mean field approach for simulating wavelength-division multiplexed systems. IEEE Photon. Technol. Lett 12, 443–445 (2000)
19. Leibrich, J., Rosenkranz, W.: Efficient numerical simulation of multichannel WDM transmission systems limited by XPM. IEEE Photon. Technol. Lett. 15, 395–397 (2003)
20. Pendock, G.J., Shieh, W.: Fast simulation of WDM transmission in fiber. IEEE Photon. Technol. Lett. 18, 1639–1641 (2006)
21. Jaworski, M., Chochol, M.: Split-step-Fourier-method in modeling wavelength-division-multiplexed links. In: Proc. of ICTON 2007, Rome, Italy, paper Mo.P.13, 4, pp. 47–50 (2007)
22. Jaworski, M., Marciniak, M.: Pre-simulated Local-Error-Method for modelling of light propagation in Wavelength-Division-Multiplexed Links. In: Proc. of ICTON-MW 2007, Sousse, Tunisia, paper Fr4B.4, pp. 1–4 (2007)

12 Software Tools and Methods for Research and Education in Optical Networks

S. Rumley and C. Gaumier (chapter editors), R. Aparicio-Pardo,
C.-H. Chang, W. Colitti, B. Garcia-Manrubia, P. Kourtessis,
J.A. Martínez-León, A. Nowé, P. Pavón-Mariño, J. Scharf, and
K. Steenhaut

Abstract. Recent advances in photonic communication networks require planning, modelling and simulation tools of ever increasing scope and complexity. Based on valid and credible models, simulators are used heavily to investigate and assess new solutions before implementing testbeds and field trials. On the other hand, tools relying on heuristics algorithms or analytical models are widely used for network planning and dimensioning. This chapter reviews some recent trends in conception and utilisation of tools for modelling and planning, and reports several developments performed with commercial or academic tools and frameworks within the COST action 291.

12.1 Models and Simulations

One of the key roles of engineers is to understand the behaviour of a specific system and also to provide quantitative results for it. An example for this is performance evaluation of a communication network. Measurements, analysis and simulation are three different methods, which are suitable to achieve such quantitative results. While the prerequisite for measurements is a real system, the two other approaches are based on a model of the investigated system.

All three methods have different characteristics. One method is often better suited for a specific problem than another one. In the following several aspects are highlighted, which help to determine which approach suits best.

In general, measurements in a real system are superior to analysis and simulation with respect to accuracy. This is due to the fact, that the model building the foundation for analysis and simulation is in most cases simplified and thus does not reflect all properties of the real system in a perfect way.

The major advantage of simulations and analysis is that they can easily deal with virtual situations. Therefore, it is not necessary to provoke situations, which for example might endanger human lives or nature. This aspect is also important for predictions, as bringing about the scenario to be investigated might be hard or even impossible.

Beside this fundamental aspect, there are further criteria, which might argue against measurements:

I. Tomkos et al. (Eds.): COST 291 – Towards Digital Optical Networks, LNCS 5412, pp. 331–364, 2009.

- *Costs:* Prerequisite for a measurement is a working real system with all necessary components as well as measurement equipment. However the required costs might be prohibitive high for evaluation of the addressed issues. Furthermore, some components might be not available, e.g., when dealing with a future technology.
- *Complexity and time consumption:* Configuration of real systems is always a time consuming task and prone to failures. Furthermore it is not possible to study multiple configurations at the same time on one single real system. In contrast to this it is very simple to modify the parameterization of a simulation. Also, run multiple simulations in parallel is in general no problem at all.
- *Measurability:* Some effects occur only seldom in reality. Thus it is hardly possible or at least very time consuming to make meaningful measurements of such an effect. Besides the possibility to have an increased simulated time, there are also simulation techniques dealing with this problem in general (so called rare event simulation).
- *Configurability:* A real system has distinct properties, e.g. a buffer has a certain length. Although it might be interesting to evaluate effects when changing these properties, this can become arbitrarily costly. Furthermore changing input variables for some real systems is more or less impossible. This also holds for physical constants. These limitations do not exist for a simulation.
- *Reproducibility:* In general measurements come along with a random measurement error and reproducing the exact conditions of a previous measurement is often not possible. This makes for example troubleshooting difficult. An analytic approach should always come to the same results. This should also be the case for simulations, when no real random processes influence the result (i.e., there are at most pseudo random processes).

Besides these criteria, the choice of the appropriate method also depends on further side constraints. One example is that the functionality has to be demonstrated in reality, e.g. with a prototype system. In such a case it is self-evident that measurements are the adequate method.

In summary there are good reasons for preference of measurements if none of the above listed reasons makes it impracticable. Due to their flexibility, simulations allow achieving results in much more cases than it is possible with measurements. In principle this would also hold for the analytic approach, but such an analysis is in most cases only practical for models with low complexity.

The accuracy of results achieved with analysis or simulations depends strongly on the underlying model. If the model is inappropriate the results might differ from real values.

12.1.1 Modelling

A model describes a real system with respect to certain aspects. It also describes procedures within the system and the reaction to input variables or events.

As it would be too complex – and also unnecessary – to model a system in all details, the model omits parts of the real system and simplifies the remaining components and dependencies [1]. In doing so, all relevant effects and relationships for the object of investigation should be considered, while those having only a minor impact can be neglected. The model is thus an abstraction from reality and the high art of modelling is to keep it as simple as possible and simultaneous as complex as necessary. Thereby the experience of the modeller plays an important role.

Even when dealing with the same system the model might vary for different problem definitions. A communication network with traffic serves as example to illustrate this. Considering a packet-oriented network, possible problems exist on different time scales, i.e., flow level, packet level, bit level down to sub-bit level. While on sub-bit or bit level physical effects play the dominant role, account for these effects for a simulation on flow level will be overkill. For some questions it is necessary to model the whole network. For others a single node might be sufficient. Components might fail if reliability and resilience issues are taken into account, otherwise failures can be neglected. Last but not least such a communication network has not only a data plane but also a control plane, which raises further interesting questions. This listing does not claim to be exhaustive but should nevertheless show the multitude of problem definitions and therefore required appropriate models.

12.1.2 Simulation Techniques

In general the used simulation technique depends on the actual model and which kind of problem is investigated (e.g., transient or stationary behaviour). In the following, the two main techniques are introduced, namely event driven simulation and time based simulation.

In an event driven simulation the modelled system changes its state only at distinct points in time (so called events) [2]. A calendar records all events and executes them consecutively. Thus, it is not necessary to simulate the time between events. Typically simulations on flow and packet level in a communication network are such event driven simulations.

In contrast to the modelled system of an event driven simulation, a system may change its state continuously. This leads to a time based simulation. As a simulation with infinitesimal time advances will run eternally, it is necessary to discretise time for the model which results in finite time steps. The appropriate size of the time steps depends on the variability of the system. This technique can be used for simulations on the (sub-) bit level for physical effects.

Besides these two techniques there is a further distinction based on the occurrence of random processes within the simulation. In deterministic simulations there are no such random processes and repetitions of the simulations will come to identical results. In contrast to this, random processes occur in stochastic simulation and might also impact the results.

The Monte-Carlo simulation is probably the most prominent stochastic simulation method. Here an experiment is repeated again and again for identical initial system states but different randomness. After many repetitions it is possible to make quantitative conclusions about the experiment.

12.1.3 Simulation and Model Verification

The outcome of such simulations is an improved knowledge of the system as well as quantitative results in form of numbers. However several effects might impair the significance of the outcome and especially the numbers. These effects can be divided into such caused by the method of simulation itself and such caused by errors.

The actual simulation time is always limited. However there are many simulations which could – at least in theory – run forever. The actual results might differ from the results, which would be achieved for such an infinite run time and ideal conditions (e.g., ideal random number generators). Especially in case of random processes within the simulation it is therefore only possible to give a probability for the results being in a certain interval (in doing so further assumptions are necessary). Increasing simulation time eases this effect.

The next reason for a limited significance is that the implementation of the model might contain errors. Discovering such errors can be a very tough task. Being able to parameterise the simulation such that the outcome is known in advance gives the best chances for revealing errors. Simulations with other implementations of the model, analytic evaluation and measurements can build the base for the reference values. Additionally sanity checks may help besides methods from software design (e.g. code review).

Finally, the underlying model itself can be incorrect or contain unrealistic assumptions, which is the worst case. Overlooking of such an error is easily possible as the model itself works fine. This is especially true when no reference values are available. Consequently, in these cases, all results might be wrong or at least falsified.

12.1.4 Summary on Modelling

When dealing with complex systems, simulations based on models are often the only possibility to achieve quantitative results. The design of the underlying model is a very demanding task and needs a lot of experience as well as knowledge. In order to achieve significant results, a verification of the model and its implementation, and a validation of its results are mandatory. However, simulations are not the universal remedy. There are good reasons for using analytic approaches and especially measurements if this is possible. Having a real working system is a benefit which should not be underestimated.

12.2 Tool Integration Perspectives

The development of any new software tool is a costly activity, principally in terms of time, and these costs are increasing together with the software complexity. In the context of optical networks, both network planning functions and network model simulators are facing an increase of complexity. An approach to reduce this complexity is thus presented in this section.

The proposed approach is inline with the component oriented architecture (COA), and is based on two key principles. First, a complex tool should be clearly divided in a collection of independent and modular building blocks. Second, new tools should be created by integrating existing components together, rather than starting the implementation from scratch. Consequently, the generic term of tool integration is used.

The advantages provided by this approach are multiple. By dividing the project in several components, workload can be easily distributed among different persons and the implementation tasks can be assigned according to individual skills. In addition, resulting components are smaller and thus easier to develop, to verify, and to maintain. Furthermore, integration of existing and already tested components shortens the total development time, and the reusability of components lessens the part of work which is repeated. The component reusability is particularly important given that many models in optical networks and more generally in telecommunication networks share inherent similarities [3].

The separation in components eases the verification and validation process. Furthermore, several components addressing the same problem can be juxtaposed, permitting a benchmarking which might help to detect invalid results. This eventually increases the credibility of each component. In a similar way, if several components are individually available in different versions, it becomes possible to test various combinations.

Another incentive to perform component integration is the increasing interest for studies covering multiple layers. This emergence is partly due to:

- *the network convergence trend* which fades separations between the layers and makes the requirements for each one difficult to establish
- *the latest emerging internet services* (YouTube, Voice-over-IP, Video on demand) which are claiming for end-to-end Quality-of-Service (QoS) in addition to high bandwidth and coerce to consider the QoS problem from a multilayer point of view.

This interest for multilayer modelling and performance analysis [4] is pushing towards integration of different models, and implicitly of the various tools associated to these models.

Though it is very interesting on the paper, the concept of software integration is hindered in practice by incompatibilities between the components. Hence, the main challenge of software integration consists in finding various ways to ease the flow of data between them. In the rest of this section, the concept of integration itself is technically defined, as well as the different degree of achievement it can reach. The multiple obstacles that can be encountered when integrating two components are also listed and several potential solutions are cited.

12.2.1 Integration: Definitions

An integration of two components is achieved when data processed by a module considered as independent is transferred to another module, for further processing. The degree at which different tools or components are integrated is variable. Three degrees of integration are distinguished here: manual, virtual, and full.

- *Manual integration (lowest):* interactions between components require human operations. Data exchanges between components are made by means of files. In the worst case, the output of one tool must be manually recomposed or adapted before being passed to the next component. In the best case, the user is only requested to launch each tool and to load the file(s) separately. Common example of manual integration: a simulator first outputs simulation results into a text file; this data is then loaded in a reporting software to setup charts.
- *Virtual integration:* components, while offering no direct support for their integration with other ones, are orchestrated in an automatic manner by third party software (middleware). The use of UNIX pipes, which permit to transmit the results of one component directly as input for the next one, or of shell scripts triggering successive execution of various stand-alone executable, are examples of virtual integration. Although the file system stays the common way to pass information between components, other techniques of transmission are possible (again UNIX pipes, Sockets). Within sophisticated middleware systems, data can also be shared between components using a database system.
- *Full integration (highest):* components make explicit calls to other components within a single process. This higher level of integration is often difficult to reach, as it generally implies the utilisation of the same programming language or at least the same execution environment.

Besides the integration level, one can distinguish local integration from distributed integration. While a local integration only connects components available on a single computer, a distributed integration connects components available of different ones.

A manual distributed integration is possible and even very common (e.g. results obtained with one tool are emailed to somebody, who will use these results as input). To realise a virtual distributed integration, middleware must be extended with distributed programming capabilities (CORBA, RMI, SOAP / Web Services) [5].

A full distributed integration is excluded, as processes are bind to physical machines. Some middleware is mandatory to synchronise local process with remote ones, and therefore, only virtual distributed integration can be achieved.

12.2.2 Obstacles to Integration and Possible Diversions

Software integration is in general hindered by many factors, acting at different levels. "Data plane" factors are the first and the most limiting ones, as they have to be addressed to permit any kind of integration. The main factor of this group is the I/O compatibility and file format compatibility between components. Components performing similar tasks (e.g. routing) on the same input variables should share a common definition of the problem and dispose of a common file format. Perfect compatibility supposes the use of highly standardised file format, what does not exist in the domain of network simulation. However, perfect compatibility is not required, as an "adapted" compatibility is often sufficient. If a particular file format is sufficiently documented and/or self-explanatory, filters or interfaces toward another format can be written. The use of the XML format, which permits to include some meta-data and documentation inside the document, can be of great interest in these situations.

The problem of the file format for network based experiment is addressed in [6], where a flexible format called Multilayer Network Description (MND) is also presented. Besides I/O compatibility, size and/or amount of documents transiting between components can be a limiting factor as well, especially in case of manual integration. In [7], authors propose to delegate the management of intermediate files to specific software, in order to release the limitations.

An obvious prerequisite for joint utilisation of several tools is the ability to execute them, ideally on the same workstation (to later involve virtual integration). "Execution plane" factors are those hindering the utilisation of software on a particular environment. This can be due to the absence of required libraries or wrong configuration (missing environment variables), or simply due to incompatibilities, e.g. at the operating system level. Further limitations can appear if the software must be compiled and built locally. Execution plane factors can rarely be completely avoided: most of the time, drawbacks must be balanced to reach the required trade-off. For instance, tools available as binaries or executables do not require any build process and are thus easier to deploy, but are more prone to platform incompatibilities. On the contrary, open source available projects require a build process which may fail, but can be compiled according to platform specification (CPU, input/output) and thus offer more flexibility. C/C++ based projects generally offer good performance but are less transportable than those based on interpreted languages as Java or Python. Within a distributed integration, the problem of software deployment is partly solved. A unique installation of one tool is installed and running on one dedicated machine, while the application is available for request as distributed service. Thus, no further installation processes are required, except of the client software, but which can be a simple web browser. On the server machine, once the running environment has been setup, it must only be maintained. The potential of such an approach has been evaluated and presented in [8].

Several other factors are going against component integration, regrouped as "software management plane" factors:

- *Intellectual property:* In house developed software form assets that authors may not want to vilify. In a smaller extend, an author may grant the access to binaries versions of a component only, keeping source code hidden. These situations exclude de facto a full integration of the released component within a high order system, and strongly limit virtual integration. To grant the usage of a tool without releasing it, distributed integration may also be applied [8].

- *Project management:* Component oriented software design, while simplifying the development process of the atomic components, transfers a part of the responsibilities and the complexity on the project managers. A large integration of many components within a unique system is hardly conceivable without some centralised project management. This can be a hardly limiting factor, especially in the academic sector, as the collaborations between groups are often made informally. Pushed by the recent fervour for collaborative open-source projects, software engineering concepts as continuous integration [9] are pledging for decentralised component oriented development, and can offer some support in the present case.

- *Performances:* The weight of the middleware in the total computing resources consumption can be very high. In particular, the transmission of information between components may induce costly serialisation operations. In other cases, same operation will be conduced twice or more by independent components. In those situations, efforts should be consented to increase the level of integration between components.

The list of the factors is long and to eliminate all of them is not possible. The component oriented approach should therefore not be considered as a perfect solution. It should be rather taken as a set of implementation practices, whose benefits are expected over the long term.

12.2.3 Conclusions and Future Outlook

According to the global tendency in computer science which advocates for more homogenous and formalised information systems, and in order to address the giant complexity of future software requirements, problem of software integration has to be addressed, also in the context of scientific research.

Recently, tools developed to assist researcher for analytical studies and simulation in optical networks evolved from the simple ad-hoc network planner to complex and integrated frameworks. In parallel, there are more and more situations in which results obtained within one experiment can be reused as input for another study.

In this context, network planners, simulators and other common tools should not anymore be considered as separated entities running in close-circuit. On the contrary, they should be developed as constituting elements of a higher order information system, easing in this way a hypothetic future integration which third party components.

12.3 Modelling with OPNET: A Practical Example

OPNET Modeller has been extensively used at the University of Hertfordshire to develop novel protocols for the implementation of very high capacity access networks and their enhancement to support long-reach, wide-splitting capabilities [10]-[13]. Over the past four years, three researchers at the Optical Networks research group have been concentrating in Gigabit Passive Optical Network (GPON) modelling to simulate dynamic algorithms for control optimisation and management of bandwidth, throughput, mean packet delay and loss rate in terms of diverse network capacity, queuing status, service differentiation and service level agreement (SLA) [10].

The protocols developed in this time frame, whose performance is shown in Fig. 12.1, reduce significantly transmission GPON frame idle periods associated with alternative dynamic bandwidth assignment protocols. They also diminish the packet waiting time in optical network unit (ONU) buffers by integrating the ONU transmission order and packet round trip time to display diverse long-reach GPON throughputs and quality of Service (QoS) with SLA [10], [12].

Protocol development in OPNET has been recently extended to study multi-wavelength GPON topologies for the support of Dynamic Wavelength Allocation (DWA) in addition to Dynamic Bandwidth Allocation (DBA) to demonstrate increased ONU scalability and bandwidth-on-demand provision over deployed, power-splitting standard GPON architectures.

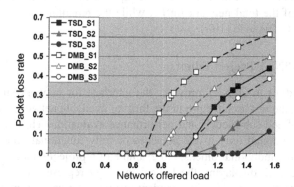

Fig. 12.1. Packet loss rate for Dynamic Minimum Bandwidth (DMB) and (Two-State Dynamic) TSD algorithms [10], [12].

12.3.1 OPNET Domains

OPNET employs a hierarchical domain structure comprising the project editor in the network domain to provide network connectivity, the node domain where the network node models are defined and the process domain where individual node models are programmed by means of equivalent process models to conduct specific operations. To achieve network integration, system designers normally need to define the protocol packet formats in the individual node models, the transition mechanism in each process model and the transceiver modules used in each node.

At top level complete networks are modelled in the network domain using the OPNET project editor offering a congregation of sub-networks, nodes and links. The functionalities of each node and link are specified in the OPNET node and link editors respectively. To integrate various operations in a node, each node model comprises several modules. Some of them are programmed in the process domain and edited in the OPNET process editor by using C programming or added directly from the OPNET library. Due to their domain architecture and programmable capacity, the network, node, and process modelling environments provide users with the flexibility to easily modify network behaviour by only reprogramming or redesigning the element of significance across layers.

To that extent, a GPON model has been simulated in OPNET modeller, based on a single OLT and 16 ONUs with varying weights, $1*W_3+5*W_2+10*W_1$, to represent accustomed service level loading, transporting 1 Gbit/s packets with simple headers up to 25 km link lengths. Network traffic is generated based on a Pareto, self-similar model available in OPNET with typical Hurst parameter of 0.8, maximum ONU channel capacity of 100 Mbit/s and basic bandwidth of 34 Mbit/s to effectively simulate bandwidth requirement of representative network services. To confirm dynamic bandwidth assignment for each ONU under various network traffic conditions, half of the network ONUs operate at fixed traffic loading, representing 33% of the total network traffic, while the remaining experience gradual increments in traffic load [10].

12.3.2 The OPNET Project Editor

The project editor in OPNET provides the interface for network simulations, by implementing the test architecture in the form of available network element libraries or user-defined sub-models and performance evaluation. These operations are successfully compiled only in the presence of fully functional node and process editors to define individual blocks. An example of a project editor environment is shown in Fig. 12.2 displaying the logical topology representation of a standard GPON tree architecture where independent bus lines are used to model the broadcasting downstream and time-sharing upstream data allocation processes. Also shown in Fig. 12.2 are the individual nodes comprising, the OLT and ONUs and also the downstream/upstream links.

Following configuration of the OLT and ONU node editors, the developed MAC algorithms are embedded accordingly to define the individual node processing models which define their and extensively the complete network's functionality.

To simulate a MAC protocol, the design of the OLT is mainly focused on a centralised upstream transmission control and statistics collection modules. Consequently the OLT node model comprises an upstream receiver, the OLT

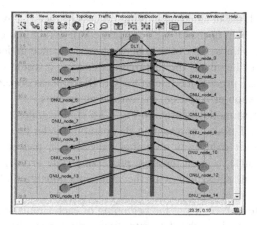

Fig. 12.2. Logical PON structure.

process module, the upstream packet sink and a downstream transmitter. As shown in Fig. 12.3, the bus receiver block, *Up_Rx*, receives and later applies the upstream packets into the OLT process module, *OLT_process*, programmed to define the node data processing principles. In that sense if a received packet represents upstream traffic, it would be used by the process module to draw instant performance characteristics, such as the channel throughput and packet delay, before it proceeds to the standard OPNET sink block, *Up_sink*, and the whole process terminated by discarding the received packet.

In contrast, if a packet represents a report message, used by the OLT to allocate transmission time slots to individual ONUs, the relevant information is extracted by the OLT process module to generate a grant message according to the incorporated MAC algorithm. Produced grant messages are then broadcasted using the standard OPNET bus transmitter, *Down_TX*, to the corresponding ONUs.

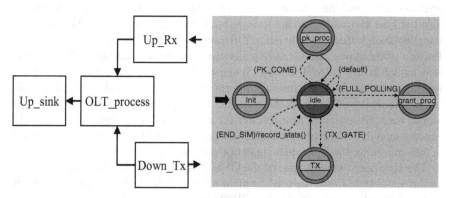

Fig. 12.3. OLT Node model.

Fig. 12.4. State transition diagram of the OLT process model.

In the node model, the detailed underlying process of the OLT process module is patterned with the help of the OPNET process editor taking into consideration the applied MAC algorithms. To evaluate network performance under different MAC protocols, which is the aim of the research in this area, it would require modification of just the OLT process module in the OPNET process domain to account for the resource allocation principles of the various algorithms.

The OLT process module shown in Fig. 12.4 demonstrates the state transition diagram of the OLT node, representing its operation under the control of various MAC algorithms. It comprises an initial state *Init*, a packet processing state *pk_proc*, an idle state, a grant message processing state *grant_proc* and a transmission state *TX* in agreement with the main functionalities of the OLT node model described above.

The initial state is executed at the beginning of each simulation to initialise the network variables, such as the data rate, frame size or number of ONUs in the PON. Subsequently the process model stays idle until an event has occurred. When for example an upstream packet arrives, the model will move from the idle state to the pack process state where the received packet is analysed and the information carried is processed depending on the MAC protocol principles. To deal with individual operations, triggered by the arrival of report packets, such as network resources allocation methodology and the generation of grant messages, the operation state will have to switch to the *grant_proc* state to arrange the upstream time slots. This whole process is controlled and defined by using C programming. Once the grant messages are produced, a self-interruption is triggered to transmit these grant messages to ONUs at the *TX* state.

Similar to the OLT node mode, the ONU node model also contains a transceiver module, the processor module and a sink module to receive, transmit and process packets. Furthermore, as shown in Fig. 12.5, with the intention of generating and buffering upstream data, the ONU node model displays two additional function modules known as the upstream traffic source, *Up_source*, to generate upstream data and the first-in-first-out queuing buffer, *Queue_buffer*, to buffer data until the assigned upstream time period is located, respectively. In relation to the simulation traffic characteristics, different traffic types, such as self-similar or Poisson distributions can be easily generated by using the standard traffic generation model in the OPNET library. When the ONU receives the assigned grant message from the OLT, the message information is extracted and arranged according to the deployed MAC protocol in the ONU process module. Once the upstream transmission cycle time reaches the assigned upstream time slot, the ONU process module will record the buffer queuing status into a report message which is then transmitted with the queuing data at the assigned time slot.

The state transition diagram of the ONU process model is shown in Fig. 12.6. In the initial state (*init*), the configuration of each ONU node and the registration of ONU specifications are obtained. While active, three main network processes take place according to the order of events. In the event of an upstream packet delivery, the process state moves from idle to the upstream data process state *fr_src*,

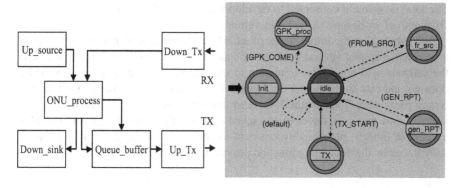

Fig. 12.5. The ONU node model.   **Fig. 12.6.** State transition diagram of the ONU process model.

in order to store or discard the upstream packet according to the ONU buffer queuing status, returning at the end of the process back to idle. If the packet arrived at the buffer is the assigned grant packet, the process state will move to the grant packet process state *GPK_proc*, to extract the upstream time slot window. Finally, as soon as the transmission cycle is in close proximity of the assigned upstream time slot, the process is self-interrupted to switch to the report packet generation state *gen_RPT* to store the ONU buffer queuing status information in a report message and consequently transmit it through the self-triggered transmission state *tx*, concurrently with the upstream data.

12.3.3 Developing Models with OPNET: Conclusion

The OPNET Modeler allows users to employ already existing libraries of commercialised network elements such as transceivers and extensive traffic models. This speeds up the modelling implementation process dramatically, while offers confidence and research output value since practical simulations close to real situation can be performed. Subsequently characteristics of channel throughput, packet delay or buffer size under heavy or low network loading conditions can be obtained with no limitations. Equally important: once a simulation platform for a specific technology has been constructed modifications of network conditions or evaluation of different algorithms becomes straightforward, since it requires simply the adaptation of the OLT and ONU process models to reflect the corresponding algorithms and the complete re-utilisation of the node and network models.

During initial stages of the research, development of the platform accounted for more than two third of the total time investment. However, compact and mature implementations representing only about 1k of line code resulted from this initial phase, allowing new studies to be done based on this platform. Nowadays, only a marginal part of the time invested is drained for its adaptation and maintenance.

The total investment consented to develop and test this framework represent about six month of work for a skilled programmer.

Nevertheless, in the initial stages of model development if all parameter values in reoccurrence among several domains or sub-networks do not correspond precisely, the software will not allow simulation to execute while the error prompt typically does not indicate to the actual error point. Consequently, users have to extract the wrong parameters from hundreds or even thousands lines of code which could potentially slow down the development speed significantly. Once the network model is executable though, functionality errors in the process models can be methodologically traced by the debugging function provided in the modeller.

12.4 Simulation of ASON/GMPLS Using OMNET++ Simulator

Automatically Switched Optical Networks (ASONs) are optical transport networks with dynamic connection capability. In an ASON the end-to-end provisioning process, traditionally addressed in the management plane, has become a process handled in a more intelligent control plane. This has been enabled by automatic control plane functions such as resource discovery and connection management (setting up, maintaining and tearing down) [14].

The ASON technology was standardised by the International Telecommunication Union, Telecommunications Standardization Sector (ITU-T). Rather than being a protocol or a collection of protocols, ASON is an architecture that defines the components of an optical control plane and the interactions between those components.

As the ASON specifications are protocol independent, the Internet Engineering Task Force (IETF) has defined the signalling and routing protocols under the standardisation of Generalized Multi Protocol Label Switching (GMPLS). GMPLS is the generalised and extended version of Multi Protocol Label Switching (MPLS) to cover both packet and circuit oriented switching technologies such as Time Division Multiplexing (TDM) and DWDM [15].

GMPLS proposes extensions to IP based routing and signalling protocols. Resource Reservation Protocol with Traffic Engineering (RSVP-TE) and Constraint-based Routing - Label Distribution Protocol (CR-LDP) are used for signalling purpose. Open Shortest Path First - Traffic Engineering (OSPF-TE) protocol and Intermediate System to Intermediate System - Traffic Engineering are the extended routing protocols. Due to the separation between the transport and the data plane, GMPLS also proposes extensions for the Link Management Protocol (LMP) for link management and discovery functions.

The ASON/GMPLS control technology has undoubtedly improved the interaction between the IP and optical layer. End-to-end optical connections can be setup and torn down in the order of milliseconds, which practically means that capacity can be dynamically created in case of high traffic load and removed in case of low traffic load. This technique, called Multilayer Traffic Engineering MTE, replaces

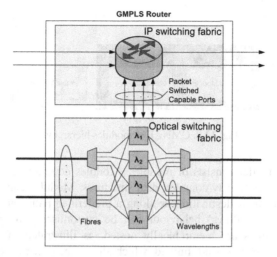

Fig. 12.7. IP/MPLS over ASON/GMPLS node model.

the traditional load balancing solely done by the IP layer [16]. MTE is further enhanced when the system is equipped with the IP/MPLS over ASON/GMPLS node model depicted in Fig. 12.7.

The model in Fig. 12.7 consists of a hybrid multi-granularity architecture encompassing an IP/MPLS packet switching fabric and a fibre/wavelength switching fabric controlled by an ASON/GMPLS control plane. The hybrid architecture allows a node to have either or both these two functions: source/destination point of an OLSP or switching point of an OLSP.

To the best of our knowledge, a simulation tool able to model and simulate IP/MPLS over ASON/GMPLS networks is not available neither among commercial products nor among open source ones. Since this model is believed to be widely adopted in future telecommunications networks, a tool that can model and simulate such features is highly desirable for research purpose. We have designed and implemented a simulator for IP/MPLS over ASON/GMPLS using the OMNET++ simulation environment and the INET framework [17], [18]. In Paragraph 2.4.1 we provide basic information about the OMNET++ simulation platform and INET package and in Paragraph 2.4.2 we give a detailed description of the building blocks of our simulator. In Paragraph 2.4.3 some conclusions are drawn.

12.4.1 The OMNET Simulator and the INET Framework

OMNeT++ is an object-oriented modular discrete event network simulator. The simulator can be used for modelling various components of a telecommunication network, such as traffic, protocols, queuing networks, multiprocessors and other distributed hardware systems.

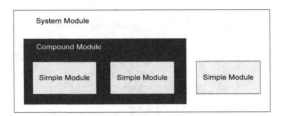

Fig. 12.8. OMNET++ modules hierarchy.

An OMNeT++ model consists of modules. Modules communicate through messages containing arbitrarily complex data structures. Modules can send messages either directly to their destination or along a predefined path, through gates and connections. The modules' behaviour can be customised through parameters. OMNET++ modules are hierarchically nested, as illustrated in Fig. 12.8. The depth of module nesting is not limited, which allows the user to accurately reflect the logical structure of the actual system.

Modules containing submodules are termed *compound modules*, as opposed to *simple modules* which are at the lowest level of the module hierarchy. Simple modules contain the algorithms in the model. The user implements the simple modules in C++, using the OMNeT++ simulation class library.

Each module has a structure that is described in its OMNeT++'s Network Element Descriptor (NED) file [17]. Fig. 12.9 illustrates a scheme reporting the process scheme executed by the system when building and running simulations under a LINUX distribution.

INET Framework is a collection of OMNET++ simulation modules implementing different layers of the Open Systems Interconnection (OSI) model. The INET Framework contains IPv4, IPv6, TCP, UDP protocol implementations, and several application models. The framework also includes an MPLS model with RSVP-TE and LDP signalling. Data link models are PPP, Ethernet and 802.11x.

The design and implementation of our ASON/GMPLS simulator in INET involved the following issues:

- Extending existing RSVP-TE and link state protocols to their generalised version (GMPLS).
- Extending RSVP-TE and link state to deal with the new optical environment in the case of ASON control plane.
- Designing and implementing the Optical Cross Connect (OXC) switching module.
- Including in the INET Framework network control algorithms for Multi-layer Traffic Engineering such as integrated routing policies.
- Eliminating the need of external static routing files for all the network's nodes, this is because the topology is not a priori defined but continuously changing during the simulation.

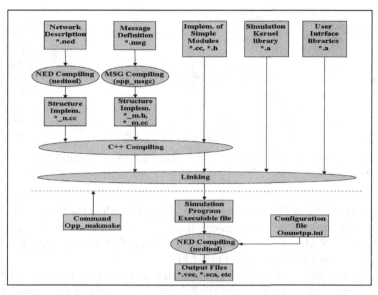

Fig. 12.9. OMNET++ simulation build and run under Linux.

- Modifying the interface modules to introduce Packet Switched Capable (PSC) ports that are the interfaces between the optical and electronic domains.
- Eliminating the need of external XML files to configure the MPLS modules of an ASON/GMPLS node.
- Integrating two independent RSVP routers in one node and synchronizing them to provide a virtual and optical layer.

12.4.2 IP/MPLS over ASON/GMPLS Simulator

An example of the simulator's building block is illustrated in Fig. 12.10, which shows a basic topology called *Test* consisting of 5 nodes and 7 bidirectional optical fibres. The test topology has been implemented only for testing purpose; the final experiments are always executed on more extended topologies [16].

The *statistics* module in Fig. 12.10 is a simple module storing in a file all the statistical information about the network's behaviour. The results can then be analyzed using different tools (e.g. Microsoft Excel, Plove, Matlab, Gnuplot).

The *generator* module is responsible for the traffic generation. It can generate traffic at Label Switched Path (LSP) level or at packet level. In the former case, the connections are generated according to a Poisson process with an average rate λ and connection holding time exponentially distributed with mean $1/\mu$. In the latter case, the packet generation follows a self-similar distribution. Also, the generator can generate connections of different priority, capacity and with randomly chosen source and destination.

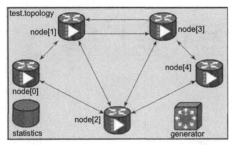
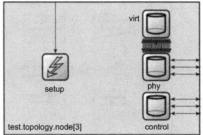

Fig. 12.10. IP/MPLS over ASON/GMPLS   **Fig. 12.11.** Node module's building blocks.
building blocks.

Amongst the building blocks of Fig. 12.10, the *node* module has been the key and most challenging module to implement. It embraces all the main functionalities of the integrated IP/MPLS over ASON/GMPLS router. The basic building blocks of the node module are illustrated in Fig. 12.11.

The *setup* module in Fig. 12.11 is responsible of the synchronisation and cooperation between the electronic IP/MPLS layer and the ASON controlled optical layer. It receives the connection requests from the generator and tries to accommodate them after choosing the most suited routing policy.

The *phy* module implements a wavelength conversion enabled OXC. This module is responsible for the transport of the traffic over a WDM network. The *phy* module consists of several submodules called *wdm*. A *wdm* module is attached to each optical fibre connected to the node.

The *virt* module implements the IP/MPLS electronic layer of the node model. This module has been implemented by assembling and suitably adapting modules already existing in the INET framework. A snapshot of the *virt* module is reported in Fig. 12.12.

A brief explanation of the building blocks of the *virt* module is reported in the following points:

- *linkStateRouting* module: it is the link state routing protocol agent for the IP/MPLS layer. It reads the network topology graph in the routing table and executes the shortest path algorithm when accommodating a connection. After a successful LSP set up the *linkstateRouting* module informs its peers about the change of the network state. The communication among link state agents is executed by flooding Link State Advertisements (LSAs).
- *rsvp module:* models the RSVP-TE protocol agent. It reserves resources to set up LSPs.
- *ted module:* models the Traffic Engineering Database (TED). The TED module stores information coming from the link state routing agent and it is queried whenever an LSP has to be set up.
- *Lib Table module:* models the Label Information Base table, in which the RSVP protocol stores the label operations (swap, pop and push) for all the LSPs that start from it, end in it or that pass through it.

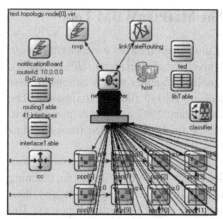 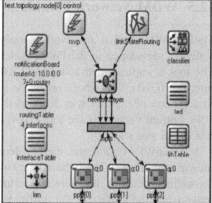

Fig. 12.12. Vrt module's building blocks.   **Fig. 12.13.** Control module's building blocks.

- *classifier module:* used by MPLS to sort incoming and outgoing packets as MPLS or normal IP ones.
- *network layer module:* the IPv4 layer of the IP/MPLS layer.
- *routing table module:* the IPv4 routing table used by the IP layer to forward packets.
- *ppp modules:* these modules model the interfaces of the IP/MPLS virtual layer router. Each interface represents a port of the optical/electronic interface between the physical and virtual layer.
- *cc module:* models the Connection Controller (CC) of an ASON network, which is responsible for the management of connection's setup, release and modification using Link Resource Manager.

The *control* block in Fig. 12.11 implements the ASON control agent that allows the router to communicate with the other ASON agents forming the ASON control plane. As a consequence of the GMPLS control plane, the ASON control module's architecture has been inspired by the architecture of the virt module in Fig. 12.12, as illustrated in Fig. 12.13. The blocks in Fig. 12.13 have the same functionalities as the *virt* module's blocks, suitably adapted to an optical domain. Therefore, a detailed explanation of the control module's building block is not provided.

12.4.3 Conclusions

The OMNET++ simulation tool is highly flexible and therefore suitable to modelling and simulating new generation intelligent networks. Being based on a modular architecture, OMNET++ can be easily extended with new modules. The modules are implemented with the C++ programming language. This increases the flexibility in adding whatever needed functionality. Moreover, OMNET++ is an open source tool and therefore the availability of the source code helps during the learning phase. By knowing the mechanism of already existing modules, it is easy to extend the tool with a variety of functionalities.

12.5 WDM Network Planning: The MatPlanWDM Tool

MatPlanWDM tool is a network planning tool for multilayer wavelength-routing networks, developed in the Polytechnic University of Cartagena (UPCT, Spain). The tool is implemented as a MATLAB toolbox together with a Graphical User Interface (GUI). The whole project is made up of 140 files approximately. The first version of the tool was distributed in January 2007 and it was presented in [19]. Current version of the tool (v0.4) can be freely downloaded from the MatlabCentral web site [20].

MatPlanWDM addresses the problem of static, dynamic, and multihour planning of multilayer lightpath based networks. In any of the three cases, the planning algorithm calculates the network virtual topology (set of lightpaths and their routes), and the routing of the traffic demand on top of the virtual topology. The differences among the three planning philosophies are described in the next sections.

12.5.1 Distinctions Between Planning Problems

Static planning: This mode addresses the static multilayer planning problem. Fig. 12.14 displays the GUI for this option. The input parameters for the planning process are specified in three files: 1) a text file describing the physical topology (*.phys* file), establishing the information per node (number of E/O transmitters, O/E receivers, wavelength converters - WC), the number of wavelengths per link, and the lightpath capacity (in Gbps). 2) a text file describing the traffic matrix (*.traff* file), indicating the average traffic offered between each node pair (measured in Gbps) 3) a Matlab file which implements the planning algorithm.

The solution calculated by the planning algorithm defines: i) the existing lightpaths and their routes on top of the physical topology, ii) the carried flows and their routes on top of the virtual topology.

The solution is returned as a Matlab structure. The tool kernel automates the calculation of the performance merits of the given solution, which is written into a report. The user can save the report in a *.results* text file for later use. In addition, the solution found can be examined in two panels. The *Physical Topology* panel plots the physical topology of the design. The user can select a particular lightpath, so that the traversing fibre links of the lightpath are highlighted in the physical topology, together with the wavelength in each hop. The *Virtual Topology* panel plots the virtual topology graph. The user can select a particular high level traffic flow, so that the traversing lightpaths are highlighted in the virtual topology.

The tool is released with a set of heuristic algorithms suitable for testing purposes, and an integrated MILP formulation based algorithm, which calculates the optimum solution for small problems. This can be very useful for comparison. Using this algorithm requires the installation of a TOMLAB/CPLEX license [21].

Dynamic Planning: This mode automates the performance evaluation of dynamic optimisation algorithms, which react to a temporal sequence of higher layer flow demands. This mode is explained in detail in [22]. The input parameters for the simulation are specified in three files: 1) a *.phys* file describing the physical topology. 2) a Matlab file which implements the traffic generator module. 3) a Matlab file which implements the planning module.

The kernel implements an event driven simulation which invokes the generator module for scheduling the flow arrivals and terminations, and invokes the planning modules to decide on the changes to the existing network state due to those flow arrivals and terminations. An example generator and a planning module are included in the distribution. The user can implement new ones as Matlab functions with a given signature.

MatPlanWDM automates the calculation of a large set of statistics. It reports the minimum, maximum and temporal average value of the following parameters:

- *Associated to the cost of the network:* number of used O/E, E/O, and WCs per node, number of used wavelengths per fibre, amount of traffic electronically processed in each node (Gbps).
- *Associated to the performance perceived by the carried traffic:* average number of physical hops, average number of virtual hops, single virtual hop traffic, average message propagation, network congestion (defined as the traffic utilisation of the lightpath with the highest traffic utilisation).
- *Associated to the blocking performance:* blocking probability, percentage of blocked traffic.

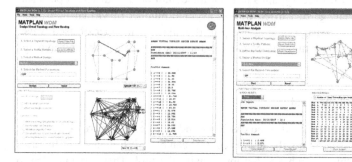

Fig. 12.14. Design Virtual Topology and Flow Routing Interface. **Fig. 12.15.** Multi-Hour Analysis Interface.

Multihour Planning: This mode allows the user to test planning algorithms which react to changes in the traffic demands following a multi-hour pattern. This type of traffic patterns appears typically in backbone networks that span over large geographical areas, where network nodes are situated in different time zones. In the input area (see Fig. 12.15), the user should define the input parameters for the simulation: 1) a text file describing the physical topology

(*.phys* file) 2) a text file describing the traffic matrix (*.traff* file) 3) the timezone each node belongs to 4) a Matlab code file (*.m* extension), implementing the planning algorithm to be applied.

The tool uses the given traffic matrix, and the timezone information for each node, to calculate 24 traffic matrixes, one for each hour of the day, as presented in [23]. The planning algorithm is responsible of calculating the 24 network states (virtual topology and flow routing) which minimise a given cost function. The algorithm can include reconfiguration costs.

After the simulation is completed, per-hour reports can be examined in the *Performance reports* panel. Also, per node information reports summarise the evolution of one selected metric across the nodes and along time. This provides fast access to relevant information like the time evolution in the number of used transmitters, receivers or converters.

12.5.2 Integrated Tool

MatPlanWDM offers two graphical interfaces to ease the creation or edition of .phys files and traffic matrixes files. Also, MatPlanWDM incorporates the "What If Analysis" functionality. This option allows the user to launch a series of simulations varying the input parameters in a given range (i.e. the number of transmitters, receivers or converters per node, number of wavelengths per fibre, traffic demand, etc.). This automates more exhaustive performance tests and comparisons. At the end of all the simulations, a summary report is provided. In addition, the user can select 7 different graphs to track the evolution of 7 different performance metrics: percentage of carried traffic, average number of virtual hops, network congestion, percentage of single hop traffic, average number of used wavelength channels, average number of used lightpaths per fibre link, and average propagation delay.

12.5.3 Extension of the Tool

MatPlanWDM tool is designed to be easily extended. The details are documented in the tool *Help*. Furthermore, the interested user can benefit from a set of libraries of classical algorithms which are distributed with the tool. They are intended to facilitate the implementation of user-made heuristic algorithms for the static, dynamic or multihour case. In addition, in authors' opinion, Matlab provides a powerful mathematical framework that allows a faster implementation of complex algorithms. Finally, integration of the algorithms in MatPlanWDM simply requires saving them in a designated directory. Therefore, the tool is well-suited for being used at research, and also for teaching at postgraduate level courses.

Current version of the tool is v.0.4. Version 0.5 is expected for July 2008. New functionalities under preparation include the definition of an XML format, to standardise the input and output files to the tool.

The techno-economic network planning studies address a more profound evaluation of the networks, which takes into account a concrete time period, where the capital expenditures (Capex), operational expenditures (Opex), demand assessment, and price policies, are analyzed.

Techno-economic planning involves very heterogeneous processes that strongly depend on the particular network scenario. Also, the goals of the techno-economic evaluations can be very different (e.g. define a pricing policy, evaluate financial feasibility, take a long term decision on a network technology). The traffic and services demand model that feed the analysis process is itself a matter of study. In some occasions, the demand is an input parameter, assumed to be estimated by experts (i.e. using a Delphi method by which the opinions of experts are systematically canvassed). In other occasions, complex prediction models should be included in the tool. The results of the tools are usually economic indicators such us Net Present Value (NPV) or the Payback Period, that are used for the sensitivity analysis and as a financial criteria.

In the optical networks scenario, the main research efforts in techno-economic tool development have been promoted by the European Commission, through the RACE programs. It can be said that the seed to all works was the TITAN Project (1992-1995) [24]. The objectives of the TITAN Project were to design a tool for the techno-economic evaluation and comparison of access network technologies, integrating engineering inputs with cost and demand forecast data. The TITAN project produced an Excel spreadsheet tool that incorporates a database of costs for components, civil works, operations and maintenance. Learning curve models are used to estimate cost trends of these elements.

One evolution of the TITAN Project was the OPTIMUM project [25]. In the OPTIMUM Project (1996-1998) the TITAN tool was adapted to deal with the assessment of multimedia business cases. So, it was addressed to understand the techno-economic factors governing the development of multimedia networks and services, both in the residential and business environments. Several multimedia network architectures were compared using the First Installed Cost, Life Cycle Cost and overall financial budget Risk.

On the other hand, the TITAN / OPTIMUM tool also evolved to evaluate the economic viability of advanced communication services and networks obtained from the ACTS projects and field trials. This project was called TERA Project [26], its aim was to clarify the impact of key cost elements, revenues and the broadband upgrade economics. The guidelines and results are targeted for the ACTS community, network and service providers, equipment suppliers, public authorities and regulatory bodies.

The developers group in the MatPlanWDM project sees the inclusion of techno-economic indicators and analysis in the MatPlanWDM tool as a medium term objective. The final goal is to provide a suitable tool for the analysis of optical multilayer networks.

12.6 The Javanco Environment

Javanco is a software project which began in 2006, at the Telecommunication Laboratory (TCOM) of EPFL. Several developers and students have been involved in its development, and the project regroups nowadays more than one hundred of classes, representing more than 20k line of code. Javanco is programmed within the java 1.6.0 platform, using the popular java language.

The main purpose of Javanco is to provide a coherent object oriented structure to represent graph and network topologies. Over this structure, several packages offer various features like graphical visualisation, support for disk serialisation of topologies or execution of common graph algorithms (e.g. Shortest Path, k-Shortest Path, Spanning Tree). Based on these core packages, user can rapidly develop and test network planning procedures on various topologies, or setup network animations (mainly for educational purpose). Additionally, Javanco provides support for simulation, allowing the construction of simulation models.

Javanco has been conceived to be as versatile as possible. Any new function integrated into the project for one experiment is generally reusable in other situations. Functions are also designed to be replaceable by variants in case of need. Object oriented programming concepts of genericity, inheritance, and the use of reflective programming [27] helps to achieve this goal.

12.6.1 History and Predecessors

In the past, TCOM preferred proprietary tools (CANPC [28], COSMOS [29]) to commercial or open source planning and modelling frameworks like OPNET or NS-2. Motivations for this choice are:

- *Educational orientation of the laboratory* (i.e. possibility to entrust part of the development to students; development of proprietary software considered as an academic goal).
- *High direct and indirect costs induced by the use of commercial software* (e.g. licence, maintenance, reporting).
- *Mistrust against open-source tools like NS-2 [30]* (e.g. instability, missing documentation).

Furthermore, TCOM accepted the high cost in terms of human resources related to a new development, assuming that it spares much of the efforts required to know how to use an unknown existing tools, and that the developers of the software will later be experts users, speeding up the realisation of experiments. This investment is expected to be profitable on the long term (3-7 years).

Development of Javanco has been mainly launched to replace the former planning tool of the laboratory, CANPC. The latter was employing an old and no longer supported version of Python. The first version of Javanco has been built based on a former tool dedicated to electrical networks [31]. Successive versions have later been influenced by different inspiration sources (principally COSMOS tool [32]).

The purposes expected when development started were: graphical interface with graph editing function, usage scripts file rather than code files to manipulate the networks, provision of general purpose algorithms (e.g. shortest path, path enumeration).

12.6.2 General Architecture

A general picture of the Javanco architecture is given in Fig. 12.16. The gravity point of the project is the *NetworkHandler* class. The main role of the *Network-Handler* is to organise the references toward each object composing the graph, i.e. the links, the nodes and the layers. Layers objects are used to group several links or nodes together. Javanco network regroups one or more of these layers. Other role of *NetworkHandler* is to provide the access to several managers and engines (user interface manager, serialisation manager, script engine). In this way, the access to the other packages is facilitated for the user.

Connectivity between nodes and links is handled by a dedicated subcomponent of *NetworkHandler* called *IncidenceManager*. Two implementations of it are available. The first is matrix (2-dimensional arrays) based and provides fast retrieval of a specific link, but do not scales well in terms of memory usage for large networks. The second, based on a tree structure, shows a better scalability, at the price of a little time overhead for object retrieval. Other alternative implementation can be easily incorporated in the framework.

The *NetworkHandler*, via the *IncidenceManager*, does not point over links or nodes objects directly, but over containers. The container itself stores the reference toward its contained element, and reversely, the element contains the reference of its container. In this way, each element has access to its direct neighbourhood thru the container, and to the whole topology, as the container also contains a reference to the *NetworkHandler* (Fig. 12.17).

Besides its associated object, a list of attributes is attached to each container. An attribute is a pair of character strings, one used as key and the other used as

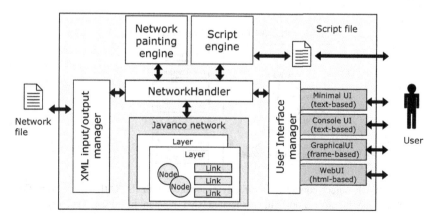

Fig. 12.16. General view of Javanco architecture.

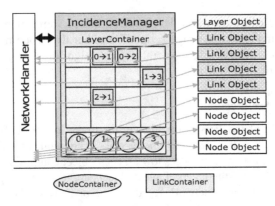

Fig. 12.17. Internal representation of a layer, using a matrix based structure. The Network-Handler uses the IncidenceManager to reference network object containers. Containers in turn contain network objects.

value. Attributes are mostly used to carry the information required for graphical representation, but may also store other values (e.g. link length or node population). As it is possible to replace the object contained in a container by another one, attributes may be used to temporary store parameters or results.

XML is deeply anchored inside Javanco. It not only permits to load and save network topologies using the XML format, but also connects the Document Object Model [33] structure with the topological one (Fig. 12.18). Each container is associated with an XML element, and all attributes owned by one container are directly attached to this XML element as XML attributes. Furthermore, any sub element contained into the associated XML element is accessible, thru the Container. In this way, user can load or store data from or to the XML tree is a transparent manner.

Javanco uses the Multilayer Network Description proposed in [6] as XML file format.

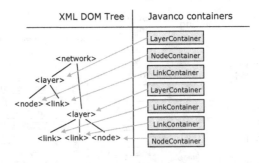

Fig. 12.18. Each object container is directly associated with its XML counterpart. Contained object can therefore modify the associated XML element to store or load data.

Each *NetworkHandler* can be connected to one or more User Interfaces. An interface can be graphical or text based, and can be unidirectional or bidirectional. In the unidirectional interfaces, user is only informed of what exists or of what it is happening, while in the bidirectional, user can interact with the framework. Furthermore, particular interfaces, as frame based ones, support the visualisation and handling of multiple networks in the same time.

Table 12.1 lists the available implementation of interfaces. The *GraphicalInterface* (Fig. 12.19) will be used to perform node and link placement, edit values of attributes, and get an insight of the topology. *MinimalInterface* is used when Javanco performs a resource consuming task and when limited feedback is required. *WebInterface* has been realised in order to use Javanco without prior installation, as a remote application.

Table 12.3. List of available interfaces in Javanco.

Interface name	Direction	Type
MinimalInterface	unidirectional	text based
ConsoleInterface	bidirectional	text based (can export graph pictures)
GraphicalInterface	bidirectional	frame based, graphical
WebInterface	bidirectional	graphical, available thru a WebBrowser

Functionalities providing a graphical representation of the networks have been regrouped inside a specific package dedicated for network painting. The painters are decoupled from the user interface, and can be easily interchanged. Alternate graphical representations of the network are therefore possible. The default 2D painter, whose result appears inside the Graphical Interface of Fig. 12.19, is generic and paints only the containers (i.e. independently from the contained link or node object). It is sensible to several attributes which define the appearance of each object (e.g. the positions, sizes, icons or colour of the nodes, the width and curve of the links). These attributes can be modified at any time, through the Graphical User Interface, or programmatically (for instance, the result of a shortest path calculation can be displayed, by changing the colour of each segment of the shortest path). Recently, a 3D painting engine has also been developed.

Javanco embeds a script engine which allows calls to any functionality of the environment. In this way, the user is dispensed to write complete java classes and is thus not required to install the full Java Development Kit (JDK) nor an Integrated Development Environment (IDE). Scripts are written in the Groovy language [34], a dynamic language for the Java Virtual Machine inspired by languages like Python, Ruby and Smalltalk. Groovy scripts can be edited using a simple text editor.

Scripts can be used in various situations: to construct an arbitrary topology from scratch; to compute simple parameter like links lengths (using flat geographical distances or the Haversine transform); to assign a specific colour to overloaded links; to test new functions; to configure and launch specific applications.

The Script engine supports step by step execution. If such a script is launched via the graphical interface, a control panel appears, allowing the control of the execution speed. This functionality has been implemented overall for educational purpose (e.g. to show the packet routing mechanisms and the collisions), but can also be used for debugging purposes. The frame based graphical interface permits to browse the local file system and select the script to be executed, using the network displayed in the active frame. It is also possible to give the network file name and the script file name as command line parameter when launching Javanco, to directly load a network and launch a script on it. In this particular case, the aforementioned *MinimalInterface* can be selected since no action is required from the user during the execution.

Various situations require the involvement of more than one package (e.g. the Graphical Interface triggers a XML serialisation operation; the script manager triggers the display of a message in the Graphical User Interface). To keep coherence between all components, the NetworkHandler acts as broker, centralizing all the calls, and offer methods permitting to: access and modify internal structure (create a new node, retrieve a link); use the XML I/O package to read and save actual network; display messages to user thru the User Interface, or to modify the appearance (zoom, size of nodes); launch a script. As certain calls to NetworkHandler may change the state of the network, the NetworkHandler notifies the changes to the other components using an event mechanism. An example is detailed in Fig. 12.20. The method *newNode()* of the NetworkHandler has been called (1). This last includes the new node inside the structure (2) and then fires an event which notifies the XML Manager (3) and the User Interfaces (4) that a new node has been created. If a graphical interface is used, the graphical information is updated (5) to paint the network coherently.

Functionalities dedicated to simulation have been integrated in the project. Links, nodes, or layers classes representing object with a dynamic behaviour should provide several methods and implement the *Simulable* interface. Using this distinction, the simulation engine can setup the list of object having a specific behaviour.

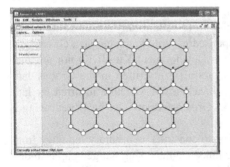
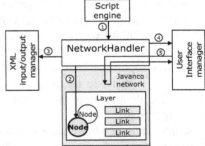

Fig. 12.19. Node and link placement with the Graphical Interface.

Fig. 12.20. Operation sequence resulting to a node creation.

Simulation is operated in a discrete manner. The state of each object involved in the simulation is changing only at a countable number of points in time [2]. Three type of discrete simulation are supported: discrete state (synchronous), discrete event (asynchronous), and hybrid. The discrete state simulation is complex but easy to parallelise, as the next state of each component is computed independently. The next event time progression is the most intuitive but is difficult to parallelise, as it required a common event list. The mixed simulation mode combines both modes. The synchronisation of the hybrid mode permits to take advantage of the parallelisation while the flexibility and the precision of the event are conserved. It however implies a few limitations. For instance, any event concerning another object than the one who triggered it should expire at least after the duration of a step. Therefore, the transmission time of all links of the topology should be higher than the duration of a step.

12.6.3 Utilisations

Javanco has been used in various situations so far: for dynamical networks simulation and random graph experiments; for development and test of routing and wavelength assignment (RWA) algorithms [8]; as a base for the JAVOBS simulator [35]; for the setup of graphical and animated demonstration of the Dijkstra algorithm, and of the packet routing mechanism.

Recently, Javanco is also being used by several students enrolled in EPFL's mathematics master program. These student are not confident with object oriented languages nor IDE, but use nevertheless the functionalities of Javanco, using the groovy language, which require low commitment.

12.6.4 Future Developments and Conclusion

After almost three years of development, Javanco is now yielding its benefits. Having available a flexible framework which allow a rapid prototyping, new ideas or concepts are easier to test, and invalid ideas are identified earlier. This eventually increases the productivity of the research activity, while providing many benefits in terms of software programming experience.

The authors wish to acknowledge the following students who contributed to the Javanco project: Loana Chatelain (XML functionalities), Christophe Trefois and Eric Kankwende Zazi (html-based interface), François Moulin (graphical rendering), Oscar Pedrola (simulation package) and Stefania Tanasescu (random topologies).

12.7 IKR Simulation Library

The IKR Simulation Library [36] is a tool which is mainly used for event-driven simulation of complex systems in the area of communications engineering. It is deployed as a C++ class library, which is publicly available under the GNU Lesser

General Public License (LGPL) and thus allows changes within the library itself as well as proprietary programs to use it.

The library consists of more than 400 classes and around 40000 lines of pure code. It should run under all Linux platforms without any problems (also e.g. CygWin under Windows). As the library has almost no platform dependent code, it should be a manageable effort to port it to other platforms. It was originally developed by Hartmut Kocher in 1993 during his dissertation at the IKR [37]. At this time it was an object-oriented alternative to the Pascal simulation library which had been previously in use at the IKR.

Continuous enhancements and improvements of the C++ library have been realised in order to improve the original design. Driver for this development is the wide usage at the institute as well as the involvement of many programmers. The library was used and is used for several public and private funded projects. Simulations based on this library are also performed in student projects and up to now more than one hundred of these student projects have been finished. Furthermore, industrial partners of the IKR use the library for complex simulations.

Since the launch the library showed its applicability for performance evaluation in a multitude of areas, e.g. for IP, photonic, mobile, signalling, and P2P networks.

12.7.1 Conceptual Structure

The IKR Simulation Library [38] consists of three main parts as shown in Fig. 12.21 on the left side. First of all there are basic concepts, which include mechanisms that support the simulation. One of these mechanisms is the simulation control, that handles the initialisation, i.e., when to stop the transient and begin with the actual performance evaluation phase and finally when to stop the simulation batches. The control also signals the according changes to all objects needing this information. As it is a library for an event-driven simulation, the basic concepts offer inherent support for event handling, e.g. by providing a calendar. One important aspect is also the distribution-oriented random number generation. The library implements many continuous and discrete distributions. Statistical evaluation is also supported by many different statistics. Besides simple statistics like a sample or counter statistic there are also more complex ones, e.g. an integral statistic. One distinguishing feature from many other simulation tools is the provisioning of metrics dealing with the statistical significance, i.e., a confidence interval is calculated based on the student's t-test. Finally the basic concepts contain mechanisms for reading parameters and printing results.

The next main part of the library provides modelling concepts. In general a model has a hierarchical structure with several components that communicate with each other. These components are called entities within the simulation library. The port concept realises the interface for message exchange between these entities. Filters and meters are connected to ports. Filters inspect and may change messages based on certain rules. In contrast to this meters primarily update statistics, with values derived from the messages, e.g., the length or time of arrival.

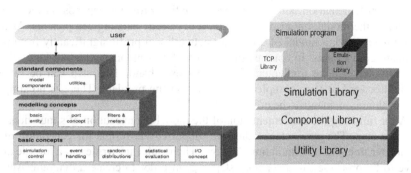

Fig. 12.21. Structure of the IKR Simulation Library (left) and relationship of simulation program and libraries (right). Both are taken from [39]).

Standard components are the third part. Ready-to-use model components like queues, servers and multiplexers together with further utilities allow a simple model generation, especially for queuing networks.

12.7.2 Libraries

The IKR Simulation Library utilises two other libraries, namely the IKR Component Library [40] and the IKR Utility Library [41] as shown in Fig. 12.21 on the right side. These two libraries provide amongst others simple to use strings, memory management, an argument parser as well as data structures like dynamic lists, arrays and matrices. With respect to these data structures, the usage of the C++ Standard Template Library STL would be an alternative. Two further libraries are built on the Simulation Library. This is on the one hand the IKR TCP Library [42]. This library offers a basic TCP implementation, which allows simulation of elastic applications and elastic traffic flows. On the other hand there is the IKR Emulation Library [43]. The emulation library can emulate a system that is specified as a simulation model, i.e., the same model can be used in simulation and emulation. For the emulation, messages in the simulator are sent as real packets, and vice versa. The simulation calendar is substituted by a real-time calendar. By this approach complex simulation models can be transformed to a network emulation model with little effort.

The simulation program itself uses all these libraries and possibly further external ones. The usage of the Emulation Library and TCP Library is optional.

12.7.3 Application of the Simulation Library

Writing a simulation program based on the simulation library requires a basic understanding of the library. The simulation library comes along with extensive documentation and comprehensible tutorials and examples. Both help to get fast an understanding of the library.

The implementation complexity the simulation program depends on the complexity of the model, but also on the extent of already existing components. As already mentioned, a model of a queuing network can profit significantly from the standard components. In contrast to this, complex components and all kind of algorithms have to be implemented by one's own hand. This step can be rather straightforward or very complex, depending on the problem. There is no simple rule of thumb to quantify the effort.

After finishing the implementation of the model the execution of simulations is the next step. The parameter studies are supported via scripts for generation of a parameter tree and for simulating this tree. The results are written to a log file and have to be processed in a separate step, e.g. in order to create a diagram. Again this step is well documented.

12.7.4 Summary

The IKR Simulation Library is well suited for event-driven simulations but can also be used for other kinds of simulations, e.g. Monte-Carlo simulation. It is publicly available and continuously improved. Its key advantages compared to other simulators are the clear design and the powerful statistical evaluation support.

The library showed its applicability and flexibility in many projects. Thereby it is not only usable for experts in the field of simulation but also for beginners due to the documentation and tutorials.

References

1. Law, A.M., Kelton, W.D.: Simulation Modeling and Analysis, 3rd edn. McGraw-Hill, New York (2000)
2. Phillips, C.: A Review of High Performance Simulation Tools and Modeling Concepts. In: Recent Advances in Modeling and Simulation Tools for Communication Networks and Services, pp. 29–48. Springer, Heidelberg (2007)
3. Lackovic, M., Bungarzeanu, C.: A Component Approach to Optical Transmission Network Design. In: Modelling and Simulation Tools for Emerging Telecommunications Networks, pp. 335–355. Springer, Heidelberg (2006)
4. Ince, A.N.: European Concerted Research Action COST 285 Modeling and Simulation Tools for Research in Emerging Multiservice Telecommunications. In: Modelling and Simulation Tools for Emerging Telecommunications Networks, pp. 1–18. Springer, Heidelberg (2006)
5. Schantz, R.E., Schmidt, D.C.: Middleware for Distributed Systems. Evolving the Common Structure for Network-centric Applications. In: Encyclopedia of Software Engineering, Wiley, Chichester (2001)
6. Rumley, S., Gaumier, C.: Multilayer Description of Large Scale Communication Networks. In: Recent Advances in Modeling and Simulation Tools for Communication Networks and Services, pp. 121–135. Springer, Heidelberg (2008)
7. Savić, D., Pustisek, M., Potorti, F.: A tool for packaging and exchanging simulation results. In: First International Conference on Performance Evaluation Methodologies and Tools Valuetools (October 2006)

8. Rumley, S., Gaumier, C.: Routing and Wavelength Assignment via Web-Services. In: Proc. of the conference on Optical Network Design and Modelling (March 2008)
9. Fowler, M., Foemmel, M.: Continuous integration, http://www.martinfowler.com/articles/continuousIntegration.html
10. Hung, C., Kourtessis, P., Senior, J.M.: GPON Service Level Agreement based Dynamic Bandwidth Assignment Protocol. IET Electronic Letters 42(20), 1173–1174 (2006)
11. Shachaf, Y., Chang, C.-H., Kourtessis, P., Senior, J.M.: Multi-PON access network using a coarse AWG for smooth migration from TDM to WDM PON. Journal of OPTICS EXPRESS 15, 7840–7844 (2007)
12. Hung, C., Kourtessis, P., Senior, J.M.: Dynamic Bandwidth assignment for Multiservice access in long-reach GPON. In: 33rd European Conference and Exhibition on Optical Communication (ECOC2007), Berlin, paper 8.4.3, pp. 277–278 (2007)
13. Shachaf, Y., Kourtessis, P., Senior, J.M.: A Full-duplex Access Network based on CWDM-routed PONs. In: Optical Fiber communication/National Fiber Optic Engineers Conference, OFC/NFOEC (2008)
14. ITU-T Rec. G.8080/Y130411: Architecture for the Automatic Switched Optical Networks (ASON) (2001)
15. Mannie, E. (ed.): Generalized multi-protocol label switching architecture, draft-ietf-ccamp-gmpls-architecture-07.txt (2003)
16. Colitti, W., Gurzì, P., Steenhaut, K.: Ann Nowé, Adaptive Multilayer Routing in the Next Generation GMPLS Internet. In: Proc. of Second Workshop on Intelligent Networks: Adaptation, Communication and Reconfiguration (2008)
17. Varga, A.: OMNeT++ User Manual,
    http://www.omnetpp.org/doc/manual/usman.html
18. INET Framework for OMNeT++/OMNEST – release 2006-10-12 –
    http://www.omnetpp.org/doc/INET/neddoc/index.html
19. Pavon-Mariño, P., Aparicio-Pardo, R., Moreno-Muñoz, G., Garcia-Haro, J., Veiga-Gontan, J.: MatPlanWDM: An educational tool for network planning in wavelength-routing networks. In: Tomkos, I., Neri, F., Solé Pareta, J., Masip Bruin, X., Sánchez Lopez, S. (eds.) ONDM 2007. LNCS, vol. 4534, pp. 58–67. Springer, Heidelberg (2007)
20. MATLAB Central: http://www.matlabcentral.com, (last access: 9th April, 2008)
21. http://tomopt.com/tomlab/ (last access: 9th April 2008)
22. Pavon-Mariño, P., Garcia-Manrubia, B., Aparicio-Pardo, R., Garcia-Haro, J., Moreno-Muñoz, G.: MatPlanWDM: An educational RWA network planning tool for dynamic flows. In: Proc. 7th Workshop in G/MPLS networks, co-located with 12th International ICST Conference on Optical Network Design and Modelling, Vilanova i la Geltrú, Spain (2008)
23. Pavon-Mariño, P., Aparicio-Pardo, R., Garcia-Manrubia, B., Garcia-Haro, J.: WDM networks planning under multi-hour traffic demand with the MatPlanWDM tool. In: Proc. Industry Track "Simulation Works" co-located with SIMUTools 2008, Marseille, France (2008)
24. Stordahl, K., Murphy, E.: Forecasting Long-Term Demand for Services in the Residential Market
25. http://www.telenor.no/fou/prosjekter/optimum/ (last access: 9th April 2008)
26. http://www.telenor.no/fou/prosjekter/tera/index.htm (last access: 9th April 2008)

27. Maes, P.: Concepts and experiments in computational reflection. SIGPLAN Not. 22(12), 147–155 (1987)
28. Lackovic, M., Bungarzeanu, C., et al.: Advanced Infrastructure for Photonic Networks – Tools. Extended Final Report of the COST Action 266, pp. 192–198 (2003)
29. Lackovic, M., Inkret, R.: Network Design, Optimization and Simulation Tool Cosmos. In: Proceedings of the 2nd International Workshop on All-Optical Networks, Zagreb, Croatia (2001)
30. Kubinidze, N., Ganchev, I., O'Droma, M.: Network Simulator NS2: Shortcomings, Potential Development and Enhancement Strategies. In: Modelling and Simulation Tools for Emerging Telecommunications Networks, pp. 263–277. Springer, Heidelberg (2006)
31. Rumley, S.: Application of Multi-Agent Techniques to the Optimal Configuration of Electric Distribution Systems Including Dispersed Generation (2005), http://www2.ing.puc.cl/power/paperspdf/rumley.pdf
32. Rumley, S.: Short-term Scientific Mission Report, http://www.ait.gr/cost291/STSM_reports_pdfs/20.pdf
33. http://www.w3.org/DOM/
34. http://groovy.codehaus.org/
35. Pedrola, O., Rumley, S., Klinkowski, M., Gaumier, C., Sole-Pareta, J.: Flexible Simulators for novel OBS experiments. In: Proceedings of the 10th ICTON conference (2008)
36. http://www.ikr.uni-stuttgart.de/Content/IKRSimLib
37. Kocher, H.: Entwurf und Implementierung einer Simulationsbibliothek unter Anwendungobjektorientierter Methoden. Dissertation, University of Stuttgart (1994)
38. Bodamer, S., Dolzer, K., Gauger, C., Barisch, M.: IKR Simulation Library 2.6 User Guide. IKR, University of Stuttgart (2006)
39. Bodamer, S., Dolzer, K., Gauger, C., Necker, M.: Object-Oriented Simulation – The IKR Simulation Library. IKR, University of Stuttgart (2005)
40. Bodamer, S., Dolzer, K., Gauger, C., Kutter, M., Steinert, T., Barisch, M.: IKR Component Library 2.6 User Guide. IKR, University of Stuttgart (2006)
41. Bodamer, S., Dolzer, K., Gauger, C., Kutter, M., Steinert, T., Barisch, M.: IKR Utility Library 2.6 User Guide. IKR, University of Stuttgart (2006)
42. Bodamer, S., Lorang, M., Barisch, M.: IKR TCP Library 1.2 User Guide. IKR, University of Stuttgart (2004)
43. Necker, M., Reiser, U.: IKR Emulation Library 1.0 User Guide. IKR, University of Stuttgart (2006)

Future Outlook (Part III)

B. Mikac (part editor) and S. Rumley

The computers and the associated software have acquired considerable importance in the last decades, due to their capacity to execute precisely the very exhausting and repetitive tasks in scientific research and in many other fields of human activity. Allowing delegation of abundant tasks, they amplify and facilitate human intellectual work and provide invaluable assistance.

In the two Chapters of this Part, a selection of the tool conceived and used in the research into the optical communication is presented, and some improvements in the interaction between different tools are proposed. These Chapters assist the reader in the demanding selection, design, configuration and utilisation of the tools. Indeed, the intellectual leverage effect is maximised if the appropriate software tools is applied when resolving specific tasks.

There are two fundamental issues to be addressed whenever starting up a tool-based experiment. The first one is whether this experiment should be based on the existing stand-alone software package (commercial or open-source, as OPNET or OMNET++), created from a scratch with all its dependencies, or designed with the proprietary or the third party libraries. There is no universal solution to this issue, because it encounters multiple parameters, such as research group size, former experience with a particular tool, or even a tool license price. Unlike the proprietary software packages, the commercial ones may provide more guarantees in terms of results validity, but they may lack flexibility. Nonetheless, each option must be balanced, and the selection made after a complete analysis. Nothing is worse than simple arbitrary selection of one.

The second issue is about the "the influence of the tool on the data". Every tool can be viewed as a "contributor", which takes some data out of a pool and unleashes its results back into this pool. Like in many other cases, it is particularly important to structure the data contained in the pool and how each tool will produce its results. As mentioned in Chapter 12, the usage of the XML format, which acts as a stand-alone and self-explanatory database, can provide such a structure. However, this is only the first step towards a fully structured information system, within which the data flow smoothly between different contributing tools. This sounds as a promising field of investigation that may provide convergence in the current very fragmented and heterogonous community studies.

Future Outlook

I. Tomkos, K. Kanonakis, and M. Spyropoulou

The COST 291 Action has investigated many topical issues in the field of optical communications. The advances in photonic technology up to the beginning of the current decade endowed the optical transmission plant with data rates in the scale of Terabits/s and hundreds of available wavelengths per optical fibre, while all-optical amplification allowed spanning hundreds to thousands of kilometres without the need for signal regeneration. During the last years, research focused on network elements and techniques that would make the exploitation of the available technology possible. This resulted in the advent of reconfigurable all-optical networks, making use of devices such as Reconfigurable Optical Add/Drop Multiplexers (ROADMs) and Wavelength Selective Switches (WSSs), while revolutionary concepts such as Optical Burst Switching (OBS) and Optical Packet Switching (OPS) promised to offer a greater degree of dynamicity to the optical core. In addition, optical technology has spread also in the access part of the network, with the most promising solutions so far being Passive Optical Networks (PONs) and Active Ethernet (or Point-to-Point Ethernet).

At the same time, emerging services like grid networking, High-Definition TV (HDTV), real-time video streaming, video conferencing, remote medicine and interactive gaming impose requirements higher than ever in terms of bandwidth, Quality of Service (QoS) performance and reliability. Hence, with most of the underlying technological advances already in place, the most fundamental challenge at this point of optical networking evolution is to achieve convergence, and this is needed to happen in multiple levels.

The concept of a converged and transparent core/metro/access infrastructure is of paramount importance for efficient end-to-end service delivery with performance guarantees supporting both single- and multi-operator scenarios. In addition, a convergence of the various existing optical switching techniques (optical circuit/flow/burst/packet switching) is mandatory in order to achieve optimal exploitation of the network resources. Convergence between multiple layers in the protocol stack - namely optical, network, transport and application - is also imperative in order to handle efficiently the huge available capacity. Thus, protocols belonging to multiple levels of the stack should be further optimized for their best possible interoperability. Finally, considerable enhancements should be made to the access part of the network in order to include the convergence of optical and wireless technology reaching a hybrid optical/wireless access infrastructure that will facilitate user mobility and support the vast number of devices and sensors that will need to connect to the internet from the user premises, while the simultaneous introduction of WDM technology will help to increase bandwidth and enhance network upgradeability.

I. Tomkos et al. (Eds.): COST 291 – Towards Digital Optical Networks, LNCS 5412, p. 367, 2009.